SOTHIS PRESS

Sothis Press aims to completely reimagine the public domain book with an emphasis on universal access to captivatingly designed print editions. We believe that technology will enable every single book ever published to be perpetually accessible in beautiful print form. Sothis is beauty in ubiquity.

www.sothis.press
1 Waterhouse Square, London, EC1N 2ST

A Rudimentary Treatise on Clocks and Watches and Bells

Edmund Beckett

SOTHIS PRESS

ISBN 978-1-0365-0696-4

A Rudimentary Treatise on Clocks and Watches and Bells is in the public domain.

A RUDIMENTARY TREATISE

ON

CLOCKS AND WATCHES AND BELLS

BY

SIR EDMUND BECKETT, BART.

LL.D., Q.C., F.R.A.S.

PRESIDENT OF THE BRITISH HOROLOGICAL INSTITUTE, AUTHOR OF 'ASTRONOMY WITHOUT MATHEMATICS,' ETC.

WITH NUMEROUS ILLUSTRATIONS

SEVENTH EDITION, REVISED AND ENLARGED

Capio Lumen

LONDON

CROSBY LOCKWOOD AND CO.

7, STATIONERS' HALL COURT, LUDGATE HILL

1883

PREFACE.

As this is not unlikely to be the last time that I shall revise this book, considering my age and the number of copies printed in each edition of late (3000), and as I have had more leisure than for many years, I have endeavoured to make it as complete as possible, and have introduced more new matter and alterations than in any edition since the fourth.

This was one of the first 'Rudimentary Treatises,' undertaken with great spirit by the late Mr. Weale, at the suggestion of my friend and connection, Colonel Sir W. Reid, K.C.B., the author of the 'Law of Storms,' and the chief manager of the Exhibition of 1851, at the request of the Prince Consort; and 7000 copies were printed of the first edition. The articles on Clocks and Watches, but not on Bells, in the eighth and ninth editions of the *Encyclopædia Britannica*, were abridgments of this book, and therefore make this edition practically the ninth written by me.

It should be understood that this professes to be a rudimentary treatise in the sense of teaching the principles of horology, and so much practical knowledge as may be useful both to clockmakers and to amateurs who wish to make, or direct the making of, their own clocks of superior character; and I have had abundant information that it has been useful in that way, besides vastly improving the general character of public clocks especially, in all the English-speaking world, and wherever large English clocks go.

Nobody can learn the details of watchmaking from a book, and at any rate in no such space as could be given to it in a volume like this. There have been, from time to time, useful letters on various details of the art in the *English Mechanic* and the *Horological Journal.* I have never heard of any amateur taking up watchmaking as many do the making or designing of clocks; and it depends much more on manual dexterity and practice. Therefore that part of the book deals more with principles than with working details.

I leave the chapter on Bells to speak for itself, as the only English treatise on their proper construction, shape, composition, and the best modes of hanging; and as the result of long experience and many special experiments, to revive an art which had sunk to a lower ebb thirty or forty years ago than it had ever reached in the thousand years or more since large bells were first made.

E. B.

BATCH WOOD, ST. ALBAN'S, AND
33, QUEEN ANNE STREET, W:
Jan. 1883.

CONTENTS.

		PAGE
MEASURES OF TIME		1
Sidereal day		1
Solar day and transit instrument		2
Mean time and year		3
Equation of time		4
Astronomical and civil day		5
Table of equation of time		6
Sidereal time		8
Table of local time		9
Sidereal and mean dials useless		10
Sun-dials		12
Meridian dial simplest	*fig.* 1	13
Dipleidoscope	*fig.* 2	14
Water and sand clocks		16
Clocks, History of		18
Early English ones		19
Common clock train	*figs.* 3 and 4	21
Fan-fly instead of pendulum		23
Conical pendulum, or governor		24
Relation of length to gravity		25
Barker's mill governor		26
Revolving pendulum	*fig.* 5	27
Balance-wheel, earliest	*fig.* 6	28
Pendulums, history of		29
Crown-wheel escapement	*fig.* 7	31
Cycloidal theory	*fig.* 8	32
Cheeks, a failure		33
Circular error		34
Mathematics of pendulum theory		35

PAGE
Lengths of various pendulums 36
Centre and radius of oscillation 37
Moment of inertia 38
Standards of length 39
French metre a bad measure 40
Egyptian and Jewish cubits 41
Short and slow pendulums 41
Shape and material of pendulums 42
Suspension of them . . . . *fig.* 9 44
Where they may be 46
Pendulum springs 47
Regulation of clocks 49

Compensated pendulums 52
Table of expansions and specific gravities 53
Zinc and steel pendulum 54
Weights of some 57
A mistake about compensation 58
Wood and lead compensation 59
Wood, zinc, and lead 61
Smeaton's glass pendulum 62
Compound bar compensation . . . . . . *fig.* 10 63
Homogeneous and Ellicott's . . . . . . *fig.* 11 64
Mercurial; Baily's mistake 65
A new kind of jar . . . . *fig.* 12 67
Cast-iron jars best; calculation for 68
Barometric error and compensation 70
Westminster pendulum fully compensated 72
Air-tight clock cases 72
Compensation by thermometer tube 73
Greenwich plan 75

Anchor pallets . . . . *fig.* 13 76
Harrison's recoil escapement 77
Clocks out of beat 78

Dead escapements . . . . *fig.* 14 80
General theory of 81
Importance of dead friction 82
Sir G. Airy's calculations 85
His conclusion erroneous 89
Proper construction 90
Half-dead escapement 91

PAGE

Loseby's isochronal spring fails 9
Large and small arcs 9
Value of firm fixing of pendulum 93
Materials for scapewheels 94
Weight of them 95
Pallets should be short 96
Rules for construction 97
Pin-wheel dead escapement . . . . *fig.* 15 99
Pin pallets. 100
Single-pin escapement. . . . . *fig.* 16 101
Sir E. Beckett's three-legged one . . . . . *fig.* 17 102
Another form of it . . . . *fig.* 18 104
Detached escapements, Airy's . . . . *fig.* 19 105
Beckett's . . . . *fig.* 20 106
Single-beat escapements waste no force 107

Gravity escapements, value of 108
Mudge's . . . . *fig.* 21 110
Sir G. Airy's calculations 111
Proper conclusion therefrom 114
Effect of a close case 114
Failure of early gravity escapements 115
Cumming's and Hardy's 116
Kater's, Gowland's, Gannery's 117
Bloxam's . . . . *fig.* 22 119
Correction of Airy's mistakes 120
Beckett's gravity escapements 121
Four-legged . . . . *fig.* 23 122
Banking pins, useful 126
Fly to be light and large 127
Double three-legged . . . . *fig.* 24 129
Most successful for large clocks 131

Going part of clocks, generally 132
Section of one . . . . *fig.* 25 134
Dial work 135
Regulator clocks 137
Winding keys should be long 138
Year clocks 138
Clock cases and Moon dials. 139
Day of the month clocks . . . . *fig.* 26 141
Chronoscopic dials without hands 144

PAGE
Equation of time clocks . . . . fig. 27 146
Maintaining powers 147
Endless chain . . . . fig. 28 148
Spring-going barrel . . . . fig. 29 149
Bolt and shutter 150
Beckett's improved one 150
Sun-and-planet power . . . . fig. 30 152
Spring clocks 152
American clocks and Austrian 153
Improved French clocks 154
Self-winding clocks 155
Water clocks 155
Electrical, Bain's 156
Shepherd's 157
Jones's controlled 158
Ritchie's . . . . fig. 31 159
His pendulums . . . . fig. 32 160
Lund & Blockley's 161
Timeballs and guns 162

Striking clocks, for **one** only . . . . . . fig. 33 163
Common striking part . . . . fig. 34 164
Repeating method 165
New French form of it 166
Strike and silent 167
Locking plate . . . . fig. 35 168
Half-hour striking one 170
With rack movement 171
Quarters on two bells 172
Quarter chimes 174
Best construction for them . . . . fig. 36 175
Tell-tale clocks, and musical 178

CHURCH OR TURRET CLOCKS 179
Pendulums long and heavy 180
Position of clock 180
Frame, old form 181
Some of the largest clocks 182
Small turret clock with gravity escapement . . fig. 37 183
Clock towers and architects 184
Wire ropes 187
Striking from great wheel 187

		PAGE
Two hammers		188
Stops for weights		189
Number of lines and pulleys		190
Bushes, various forms	*fig.* 38	192
Four-wheeled train clock	*fig.* 39	193
Size of fly important		194
Large clocks with quarters		195
Best pattern of one	*fig.* 40	196
Number of cams and teeth		197
Discharge of striking		199
Two-quarter hammer	*fig.* 41	201
Cambridge and Westminster quarters		202
Doncaster quarters		202
Worcester		203
Arrangement of chime barrel		204
Proper interval for chimes		205
Lifting off hammers		205
Quarters on two bells		206
Striking at exact time		206
Chime tunes		207
New method	*fig.* 42	209
Large and small bells together bad		210
Clock hammers	*fig.* 43	211
Fixed above bell	*fig.* 44	212
Cranks for them	*fig.* 45	214
Weight of hammers		215
Dials of turret clocks		216
List of some large ones		217
Materials for		219
Concave *v.* convex		220
Hands, construction of		221
Illumination of dials		222
Dial wheels		224
Universal joints and friction	*fig.* 46	226
Weathercocks and ventilation		227
Train remontoires		228
Bevelled-wheel remontoire	*fig.* 47	230
Continuous motions		231
Telescope drivers		232

PAGE
Royal Exchange remontoire . . . . . *fig.* 48 234
Simplest train remontoire . . . . . . *fig.* 49 235
Spring remontoire, Sir G. Airy's 236
Sir E. Beckett's . . . . *fig.* 50 237
Cast-iron wheels 239
Decay of brass in bad air 241
Jobbing in public clocks 242
Specifications for them 243
Degree of error to be allowed 247

The Westminster clock (see frontispiece) 248
Double-barrelled crab there . . . . . . *fig.* 51 249
Another form of it . . . . *fig.* 52 251
Mode of regulating 252
Pendulum and escapement 253
Maintaining power 254
The dials and hands 255
Quarters 257
Hour striking part 258
Mode of discharging exactly . . . . . . *fig.* 53 260
Winding of striking parts 261
Its rate 263
History of it 265
Sir C. Barry's hands for it 267
Mr. Cowper Temple about it 268
The bells and their history 269
Sir G. Airy's report on them 271
Mr. Cowper Temple again 272
Cost of clock and bells, and of the dials and bell-frame . . 273

Teeth of wheels 273
Helix teeth . . . . *fig.* 54 275
Epicycloidal . . . . *fig.* 55 276
Drivers and runners . . . . *fig.* 56 278
Lantern pinions . . . . *fig.* 57 280
Depthing tool 283
Internal wheels 283
Bevelled and skew-bevelled 284
Cams 285
Calculation for them 286
Simplest construction . . . . *fig.* 58 287
Oil for clocks 288

		PAGE
WATCHES AND CHRONOMETERS		290
History of early watches		290
Mainsprings		291
American spring barrel	*fig.* 59	292
Tipsy key		292
American watch	*fig.* 60	293
English	*fig.* 61	294
Winding stops		295
Up-and-down dial	*fig.* 62	296
Dial wheels		297
Balance spring		297
Balance free from gravity		298
Regulation of watches	*fig.* 63	300
Old plan	*fig.* 64	301
Glass balance springs		302
Timing for position		302
Gimbals	*fig.* 65	303
Ship timepieces		303
Compensated balances		304
Harrison's and Le Roy's		305
Common plan	*fig.* 66	306
Chronometrical thermometer		306
Secondary compensation		307
Mathematical theory of		308
Greenwich trials		309
Eiffe's compensation		309
Dent's	*fig.* 67	310
Loseby's	*fig.* 68	311
Kullberg's	*fig.* 69	313
Dent's prismatic	*fig.* 70	314
Escapements, vertical	*fig.* 71	315
Lever	*fig.* 72	317
Horizontal	*figs.* 73, 74, 319,	320
Duplex	*fig.* 75	322
Chronometer	*fig.* 76	324
Lever chronometer	*fig.* 77	325
Tourbillons		326
Remontoires useless		327
Repeaters		327
Keyless watches	*fig.* 78	329

PAGE
Pedometers and stop watches 330
Recording watch . . . . *fig.* 79 332
Dials and cases 333
Watch factories. 334

BELLS 335
History of bells 335
Founders, past and present 336
Foreign and English 339
Old bell metal 340
Notes and sizes 341
Notes and weights 343
Scales of thickness 344
Exeter, York, and Bow peals 347
St. Paul's and Worcester 348
Bradford and Manchester 349
Bad modern towers 350
Doncaster and Burton peals 351
Headingley and Gainford 352
Thickness of large and small bells 353
Maiden bells 354
Sizes for peals 354

Shape of bells, rules for 355
Section or sweep . . . . *fig.* 80 358

Composition, best 361
Atomic proportions 362
Cast steel, bad 363
Silver a vulgar error 363
Moulding, two modes of 365
Mending cracked bells impossible 367
New bell crown, Sir E. Beckett's . . . . . . *fig.* 81 368
Taylor's 369
Clapper bolt independent 369
Tucking up in stock 370
Tolling levers and gudgeons 371
Great Paul's hanging 371
Bell ropes 372
Ellacombe's chiming hammers 373
Stays and sliders 374
Gudgeons should be long 375

PAGE
Clappers and iron stocks 376
Bell frames 377
  Bottom should fit walls 377
  Arrangement of frames 378
  Folly of some constructions 378
Clappering cracks bells 379

Specifications for peals 380
Bell towers, proper size of 384
  Proper plan of bells in 385
  Architects destroy belfries 386
  Windows in 387
The great bells of Europe 388
  List of seventy-two, weights and sizes 389

ON THE MEASURES OF TIME.

Before considering the construction of horological machines it is desirable to understand what it is they have to measure. It is easy to reply, days and their subdivisions. But there are various kinds of days, all meaning one rotation of the earth, with reference to the stars, or the equinoctial point ♈, or the sun, or an imaginary sun going with the average speed of the real sun in its apparent annual motion round the earth.

A sidereal day is the first or the second of these, or is one absolute rotation of the earth, or the time between two successive passages of the same star over any given meridian; for the stars are so far off that the whole width of the earth's orbit does not make a difference of $1''$ in the apparent place of the nearest star: much less does the motion of the earth in one day make any sensible difference. And as this rotation is the only period that is measurable by any object which is both fixed and visible, all clock time is ultimately regulated by sidereal time, which is daily taken at observatories by observing when certain stars pass the meridian.

This is done by having a telescope fixed at right angles across a stiff axis with pivots at the ends, which is laid exactly east and west and horizontal on very firm supports, so as to be free from all vibration and disturbance. The telescope can then only move in the plane of meridian, and is called a *transit instrument*, and is the primary horological instrument for measuring whole days, while clocks measure their subdivisions. Some very fine wires are stretched vertically across the focus of the telescope, or the place where an image of the star is formed by the object-glass,* one of the wires being in the middle or meridian: the others are useful to measure the distance of the star as it approaches and leaves the meridian; and the magnifying power increases the apparent motion, and makes it so much easier to observe accurately.

But this is not strictly true: 0 or 24 o'clock by sidereal time is not when any particular star passes the standard meridian of the country (here of Greenwich), but when that imaginary point called ♈ passes, where the sun is at the vernal equinox, or the intersection of the ecliptic and the celestial equator. And that recedes 50″.1 a year among the stars; which is equivalent to 3⅓ sec. of time in the earth's rotation, but to 20⅓ min. in the annual revolution, which is the difference between the sidereal and the equinoctial or common year. Therefore a clock going rightly by the stars would be 3⅓ sec. slow at the end of the year by ♈, or by what is called sidereal but is really equinoctial time. The place of ♈ is always known, because the distance of every considerable star from it in sidereal time, or Right Ascension, is given in the Nautical almanac; and so clock time is actually measured by the stars at last, but with this correction, which you see is only the 110th of a second a day.

A solar day is the interval between two successive transits of the middle of the sun over the meridian; and before that can happen the earth must make more than one absolute rotation, because the sun is going (apparently or relatively) round it in the same direction as the earth rotates, —*i.e.*, from west to east; for the apparent motion of all the

* See my 'Astronomy without Mathematics,' p. 399, 7th ed.

heavenly bodies westward is only the consequence of the earth's rotation eastward. Therefore there is one more sidereal day in the year than there are solar days; or a sidereal day is a 366th, or nearly 3m. 56s. shorter than a solar day; or they are in the proportion of 1.00274 to 1, or 1 to .99727. A small wheel of 8 teeth revolving round a fixed large one of 2922 makes 366¼ absolute rotations, but 365¼ relatively to the large one: this is called a sun and planet motion accordingly.

The length of a solar day is continually varying a little from two causes. One is that the earth's orbit is an ellipse with the sun in one focus, and the earth moves slowest when it is farthest off—*i.e.* in July. The other, and much the larger cause is, that the sun moves in the ecliptic, and not in the equator; and if you bring each successive 10° (say) of the ecliptic on a globe up to the brass meridian, you will see that they seldom correspond to exactly 10° of the equator, which is the hour circle of the globe. The consequence is that the days are a minute longer in December, and half a minute longer in June than they are in April or September; and if a good clock is set with the sun-dial in November, it will be half an hour before the sun in February. Hence it is that the afternoons are so much darker in November than they are at the same number of days after Christmas. It was the fashion in France, as late as 1826, to make the public clocks show solar time, either by some complicated machinery or by altering them nearly every day.

Mean time—that is, a mean solar day—is the average length of all the solar days in the year; which is no exact number of days; for the sidereal year, or time of the sun's return to the same place among the stars, is 365.25634 mean days, and the equinoctial or common year, or mean time of the earth's return to ♈, is fixed by astronomers at 365.242216 mean solar days, or 365 days, 5 hours, 48m. 49½s. The difference between true and mean solar time for each day is called the *Equation of time*. It is usually given in almanacs in the form of 'clock before' or 'after sun.' But the following table, at p. 6, shows the time which a clock in the longitude of Greenwich, Boston, Louth, and Grimsby (for they are all in longitude 0°), ought to show when the sun is on their meridian. It is easily adapted to any place west of those, by adding to the time in the table 4m. for 1° of longitude, and so on; and for the few places in England east of Greenwich you must subtract at the same rate from the time in the table.

The equation is not quite the same every year, but moves on nearly a quarter of a day yearly, till leap-year comes and puts it very nearly back again. The reason why it does not do so quite is, that the year is not quite 365¼ days. That is the error which is attempted to be corrected by dropping the leap-day of every century except those divisible by 4. I have shown in my 'Astronomy' that a much more complete correction would be made by simply dropping every 32nd leap-day, or that of every 128th year; and that dropping the leap-day of every century except the fifth would be better than the fourth.

The table at p. 6 is made for the first year after leap-year; and it will practically serve for a long time to come (in case you have not an almanac with an equation table), if in the second year you subtract the difference between the equation for any given day and the next, when the time is increasing, and add when it is decreasing. Thus on 1 January 1881 the clock was 3m. 58s. before sun, and 4m. 26s. on 2 January—the difference being 28s. Therefore subtract 7s. for 1 January 1882, which makes the clock 3m. 51s. before sun. Again, for 1 January 1883, the third year after leap, subtract 14s. from the time in the table, and you get 3m. 44s.; for leap-year up to February 11, where the equation turns, subtract three quarters of the difference. From thence to the end of the month, as the time is decreasing, you must add the three quarters of difference, treating the 29th as March 1.

And generally, when the time is decreasing, you must add the proper fraction of the difference. Thus the time on 1 October 1881 is 11.49.38, and on 2 October is 11.49.19. For 1882 add a quarter of the difference (say 5s.) and you have the time on 2 October 11.49.24. For the third year after leap add 10s. for 2 October; but for leap-year itself back to 29 February, or the day before 1 March, subtract a quarter of the difference when the time is decreasing. Thus for 1 March of leap-year it is 0.12.30, and for 29 February it is 0.12.42; while for 28 February, beginning from the other end of the year, it became 0.12.54 by adding three quarters of the difference. I find by looking back at old almanacs that this always gives the equation as nearly right as you can observe the sun by any kind of dial: if you want to be more exact you must get an almanac with the equation for every day in the year.

In setting a mean time clock by sidereal observations you have nothing to do with the equation of time, which only relates to the sun. Some almanacs besides the Nautical give the mean time at which certain stars and planets pass the meridian of Greenwich, and then the operation of setting the clock is simple and obvious enough. Whitaker's almanac gives the sidereal clock time at each mean solar noon. But it should be observed that astronomers' mean time has no A.M. and P.M., but is reckoned up to 24 hours from the noon after the midnight at which the civil day is considered to begin. Thus 11h. A.M. 1 January 1880 of common life was 23h. 31 December 1879 with astronomers. Without some of the data which are only to be found in such an almanac it is of no use attempting to find the time from sidereal observations, and therefore it is not worth while to give any less simple modes of doing it from other data than those just mentioned.

TABLE OF THE

FOR THE FIRST

Showing Greenwich clock time when the Sun is on the

Day of Month	January.			February.			March.			April.			May.			June.		
	H.	M.	S.	H.	M.	S.	H.	M.	S.	H.	M.	S.	H.	M.	S.	H.	M.	S.
1	0	3	58	0	13	55	0	12	33	0	3	55	11	56	57	11	57	31
2		4	26		14	3		12	21		3	37		56	50		57	40
3		4	54		14	9		12	8		3	19		56	43		57	50
4		5	21		14	15		11	55		3	1		56	37		58	0
5		5	48		14	20		11	41		2	43		56	31		58	10
6		6	15		14	24		11	27		2	25		56	26		58	20
7		6	41		14	27		11	12		2	8		56	21		58	31
8		7	6		14	29		10	57		1	51		56	17		58	42
9		7	31		14	30		10	42		1	34		56	14		58	53
10		7	55		14	31		10	26		1	17		56	11		59	5
11		8	19		14	31		10	10		1	1		56	9		59	17
12		8	42		14	30		9	54		0	45		56	7		59	29
13		9	5		14	29		9	37		0	29		56	6		59	41
14		9	27		14	27		9	20		0	14		56	6		59	54
15		9	48		14	24		9	3	11	59	59		56	6	0	0	7
16		10	8		14	20		8	46		59	44		56	7		0	19
17		10	28		14	16		8	28		59	30		56	8		0	32
18		10	48		14	11		8	11		59	16		56	10		0	45
19		11	6		14	5		7	53		59	2		56	12		0	58
20		11	24		13	58		7	35		58	49		56	15		1	11
21		11	41		13	51		7	17		58	37		56	19		1	25
22		11	57		13	43		6	58		58	25		56	23		1	38
23		12	13		13	35		6	40		58	13		56	28		1	51
24		12	27		13	26		6	22		58	2		56	33		2	4
25		12	41		13	17		6	3		57	51		56	38		2	17
26		12	54		13	7		5	45		57	41		56	44		2	29
27		13	7		12	56		5	26		57	31		56	51		2	42
28		13	18		12	45		5	8		57	22		56	58		2	54
29		13	29					4	49		57	13		57	6		3	6
30		13	38					4	31		57	5		57	14		3	18
31		13	47					4	13					57	22			

EQUATION OF TIME,

YEAR AFTER LEAP YEAR.

meridian, after correcting for the longitude of the place.

Day of Month	July.			August.			September.			October			November.			December		
	H.	M.	S.	H.	M.	S.	H.	M.	S.	H.	M.	S.	H.	M.	S.	H.	M.	S.
1	0	3	30	0	6	2	11	59	50	11	49	38	11	43	43	11	49	19
2		3	41		5	58		59	31		49	19		43	42		49	42
3		3	52		5	53		59	11		49	0		43	42		50	6
4		4	3		5	48		58	52		48	42		43	43		50	30
5		4	13		5	43		58	32		48	24		43	45		50	55
6		4	24		5	36		58	12		48	6		43	47		51	20
7		4	33		5	29		57	52		47	49		43	50		51	46
8		4	43		5	22		57	31		47	32		43	55		52	12
9		4	52		5	14		57	11		47	16		44	0		52	39
10		5	0		5	5		56	50		47	0		44	6		53	6
11		5	9		4	56		56	29		46	45		44	12		53	34
12		5	16		4	46		56	9		46	30		44	20		54	2
13		5	24		4	36		55	48		46	16		44	28		54	31
14		5	30		4	25		55	27		46	2		44	38		55	0
15		5	37		4	14		55	6		45	49		44	48		55	29
16		5	43		4	2		54	44		45	36		44	59		55	58
17		5	48		3	49		54	23		45	24		45	11		56	28
18		5	53		3	36		54	2		45	13		45	24		56	57
19		5	58		3	23		53	41		45	2		45	37		57	27
20		6	2		3	9		53	20		44	52		45	51		57	57
21		6	5		2	55		52	59		44	42		46	7		58	27
22		6	8		2	40		52	39		44	33		46	23		58	57
23		6	10		2	25		52	18		44	25		46	39		59	27
24		6	11		2	9		51	57		44	17		46	57		59	57
25		6	12		1	53		51	37		44	11		47	15	0	0	27
26		6	13		1	37		51	16		44	5		47	34		0	57
27		6	13		1	20		50	56		43	59		47	54		1	27
28		6	12		1	3		50	36		43	54		48	14		1	56
29		6	10		0	45		50	17		43	50		48	35		2	25
30		6	8		0	27		49	57		43	47		48	57		2	54
31		6	5		0	8					43	45					3	23

But you may *regulate* the rate or going of a mean clock from sidereal observations without the aid of any tables or astronomical data, though you cannot *set* it to the actual time. The mean day is 3m. 56.5554s. sidereal longer than a sidereal day; and therefore sidereal hours, minutes, &c., may be turned into mean ones by multiplying them by .99727, or by subtracting 9.83 seconds from every hour, and 1 second from every 6m. 6s. And as the 3m. 56.55s. sidereal = 3m. 56s. mean, the mean clock ought to show so much less at the second transit of a star than the time it showed at the first transit.

This operation may also be performed with sufficient accuracy without a transit instrument. For if you make a small eye-hole in a thin plate, fixed looking south, and set up or find anywhere due south of it a perfectly vertical straight edge, the occultation or emergence of any given star against that straight edge will be seen through the eye-hole at exactly every 24 sidereal hours, or at every 23h. 56m. 4s. of mean time. And it is not necessary for merely regulating a clock, that the hole and the edge should be very exactly in the meridian, if you only use it for observing stars not far from the equator. But if you aim at using it for setting as well as regulating your clocks, with the help of astronomical tables, then you must take care to have the hole and the edge exactly in the meridian. Probably the best way of doing this is to fix the plate by carrying the time by two or three chronometers of known rate from the nearest observatory, using the almanac time of the 'southing' of one or more stars.

In all operations for setting clocks from any kind of celestial observations, you must remember the difference between your own longitude and that which is the standard of the country, which is here the Royal Observatory at Greenwich. The meridian you use must be the true one of the place, and not a false one adapted to Greenwich beforehand, unless the longitude is very small; in which case the error arising from the change in the sun's declination between summer and winter will be insignificant. Here is a table of some forty towns, showing how much the clock ought to be before or behind the sun if it is to agree with Greenwich mean time; which the railways had long adopted; and it is now, by an Act of 1880, the only legal time for all Great Britain, and Dublin time for Ireland. There is a longer list in Britten's 'Watchmaker's Hand-Book'.

GREENWICH TIME BEFORE LOCAL.					
	M.	S.		M.	S.
Westminster Palace	0	30	Carlisle	11	38
Peterborough	1	0	Liverpool	11	53
Hull	1	8	Edinburgh	12	43
Lincoln	2	4	Exeter	14	18
Doncaster	4	12	Plymouth	16	30
York	4	24	Glasgow	17	0
Portsmouth	4	24	Holyhead	18	36
Leicester	4	33	Cardigan	18	40
Oxford	5	1	Falmouth	20	12
Southampton	5	36	Dublin	25	22
Derby	5	52			
Leeds	6	4	LOCAL TIME BEFORE GREENWICH.		
Newcastle	6	24			
Lichfield	7	18			
Birmingham	7	33	Grimsby	0	0
Berwick	8	0	Louth	0	0
Aberdeen	8	23	Boston	0	0
Worcester	8	41	Cambridge	0	23
Manchester	9	0	Colchester	3	32
Bath	9	26	Ipswich	4	38
Bristol	10	12	Norwich	5	13
Shrewsbury	10	56	Dover	5	16
Chester	11	32	Paris	9	21

Though such contrivances are of no real use, for they would certainly not answer as well as independent sidereal and mean time clocks beating their own seconds, yet it may save calculators some trouble if I add here that the late Astronomer Royal in 1849 communicated to the R. A. S. three different arrangements of trains which he had received from a Dr. Henderson for enabling one clock to show sidereal and mean time on two different dials.

1. Let the 24-hour arbor of the mean solar clock carry a wheel of 247, driving one of 331 on an arbor which carries also a wheel of 43, driving one of 32 (only I should put these two smaller wheels first, the mean clock driving). The last of these three arbors will carry a hand showing sidereal hours, but, remember, hours only; and you must descend again to sidereal seconds by a train of the usual numbers, so as to get the 86,400 in a day. Of course the same train reversed would give mean time from a sidereal clock.

2. Let the 24-hour arbor carry a wheel of 96, driving one of 79 on an arbor which carries a wheel of 157, driving one of 133 on an arbor which carries a wheel of 72, driving one of 103. A hand on this fourth arbor will show sidereal hours.

3. Let the 24-hour arbor carry a wheel of 50, driving one of 30 on an arbor which carries a wheel of 182, driving one of 211 on an arbor which carries a wheel of 196, driving one of 281. This fourth wheel will show sidereal hours with an error of a second in rather less than a thousand years, the sidereal day being equal to 86164·0906 mean seconds. The errors of the other two trains are a little greater, but still quite insignificant. But Sir G. Airy and Dr. Henderson did not notice, or perhaps know, as I do by experience, that all attempts to run the time down from the slow hands or wheels of a train to the fast ones, instead of putting the fast ones nearest to the pendulum as usual, invariably fail and show very large errors; and therefore such a combination of dials would be quite useless except

for the roughest approximation, which could be of no value to any one who wanted to reduce one kind of time to the other accurately, or to use the secondary dial for observations of any kind.

Moreover, there is this fatal objection to all contrivances for showing sidereal and mean time by hands depending ultimately on one pendulum, that the secondary clock, whichever it may be, cannot in any way beat or show accurately its own seconds, as there is a difference of 16 seconds to be spread over the day, or something more than ·6 an hour. I am only surprised that Sir G. Airy, who must have been quite aware of this, though he might have no experience of the other difficulty, should have thought it worth while to make that communication to the Royal Astronomical Society, as if it could be of any use to astronomers, or to anybody else that I can see.

There are some other clocks, however, which may possibly be useful to telegraph-offices where they want to know the local time at which messages arrive, or the English time when they were sent. They consist of a single common clock movement driving as many dials as you please, say for the capitals of all the principal countries, the hands being set just as much before or behind English time as the respective longitudes require; and of course they always keep their distance, and so show the local time of each country corresponding to whatever is the present Greenwich time.

We shall see presently some of the methods of finding local time, or at least the time of noon at any place; and though it belongs rather to astronomy than horology, it is not out of the way to mention that the longitude of any place is best found by carrying thither chronometers with Greenwich time and comparing it with the local time, except where the Greenwich time can be telegraphed direct without the need of carrying chronometers.

INSTRUMENTS FOR MEASURING TIME.

At the head of these of course stand the oldest of them all, sun-dials. But it would be a waste of time to say much about them, except in the forms in which they may be made subservient to the more useful and accurate machines by which they have been superseded. The principal feature of all sun-dials which have to indicate other hours besides noon is a *gnomon*, with a straight edge, or pair of edges, parallel to the earth's axis, and therefore inclined to the horizon at an angle equal to the latitude of the place, with a plate for the gnomon to throw a shadow on. Thus an Indian sun-dial has its gnomon nearly level, and it should be elevated above the plate, while a polar one would be quite upright. If the plate of the dial is part of a cylinder of which the gnomon is the axis, all the hour divisions will be equal, but not otherwise. The divisions before and after noon are equal whenever the plate is equally inclined to east and west. Sometimes they are set in other fanciful positions; in whatever position the plate is, the same hour marks are right for all times of the year, if the gnomon is parallel to the earth's axis. But all that belongs to the art of *dialling*, which may be found in any of the encyclopædias, and is of no use to us, as sun-dial observations cannot be relied on except when the sun is near the meridian, by reason of refraction which bends the sun's rays.

Meridian dial. A solar meridian mark or sun-dial for noon only is intelligible and usable by many persons who cannot or will not undertake sidereal observations, and is so easily made that everybody ought to have one who cares about having accurate time, and has not the means of getting it from some other source. The simplest of all ways of fixing such a mark is to set up a plate facing the south with a narrow vertical slit in it, reaching down to the bottom, upon a horizontal slab of smooth stone, and mark the line of brightness on the stone at the time of solar noon by a chronometer or a good watch carrying the time from some accurate source. The best way of making these gnomons is to screw a pair of thin zinc or copper cheeks on a cast iron upright piece, without attempting to make a sufficiently narrow slit in the iron itself. It should have a thick polygonal lump at the bottom screwing on to an iron rod or 'lewis' let into the stone, and lead or cement run round it after it is set upright and as near the meridian as possible. The octagon in the drawing shows the shape of this base and the section across the middle of the gnomon. I fixed the first of these, without a chronometer, in the garden of a vicarage near Cambridge in 1838, by the following method; and as it was afterwards tested by Professor Challis and found as right as can be practically observed, the method may be safely followed. A C is the slit, which I have generally had about 9 inches high. About an hour before noon, on a fine day in summer, when the shadow is short enough to lie within the slab, mark with a pencil where the top of the bright line, or centre of the little hole at A falls, say at S, and draw the arc of a circle S *s* with radius C S. Mark again two other spots, S′ and S″, at about 11.15 and 11.30 A.M., and draw the corresponding arcs, S′ *s*′ and S″ *s*″. Then watch for the times when the end of the bright line again falls on each of the three circles,

Fig. 1.

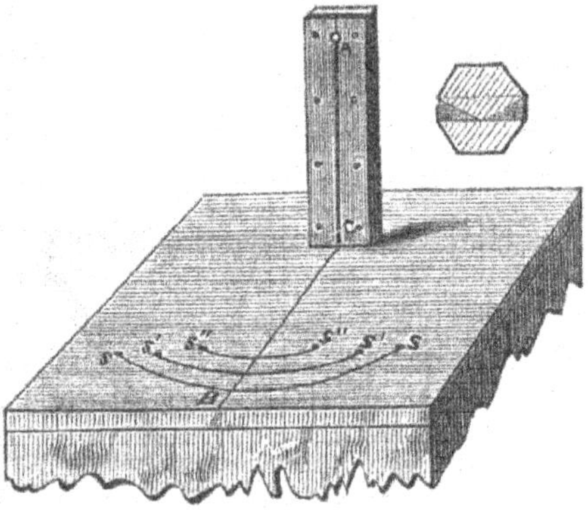

and mark the places *s s′ s″*. Bisect all the arcs, and if a straight line CB will run through all the bisections, you may be pretty sure that is the true meridian; if not, some mistake has been made, and you must try again another day.

As you will probably begin to fix the slab and gnomon by the compass, it may be as well to mention that the apparent magnetic north in England is about 18° west of real north at present; for it decreases now slightly with time, and is different in different parts of the world.

The Dipleidoscope is another meridian instrument, which was invented by the late Mr. J. M. Bloxam, a barrister in considerable practice, whose name I shall have to mention again with some clock inventions. He took a patent for it (long ago expired) and assigned it to the late Mr. Dent, by whom only they were made. Its name, compounded of διπλόος double, ἔιδος an image, and σκοπέω to see, indicates the principle of it; because in all positions but one you see a double image of the sun reflected in it; and if it is so fixed that the single image—*i.e.* the coincidence of the two—occurs at solar noon, it evidently becomes a meridian instrument. It has the advantage also of reflecting the sun when it is just too cloudy

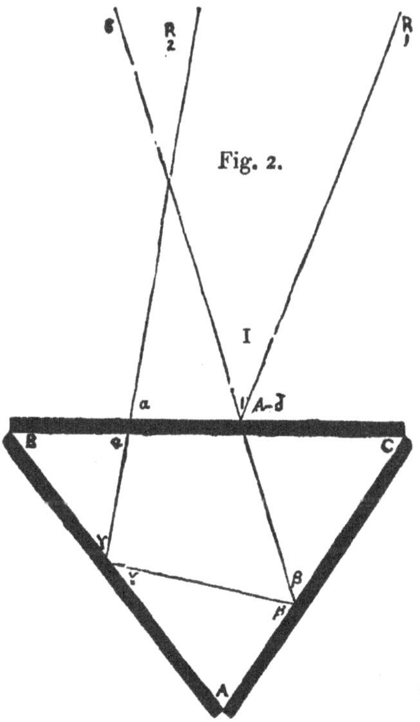

Fig. 2.

for a shadow to be distinct, and in fact you can only look at it through darkened glass when the sun is bright.

The instrument consists of three small plates of glass put together at their edges in a brass box about 2 inches wide and high, so as to form a hollow prism of any convenient angle, no precision being necessary in this. A B C in this figure is the section of it at right angles to the axis of the prism. The front glass B C is plain; the other two are blackened behind to form reflectors. But though the front glass is transparent, it also reflects, because there are dark ones behind it. S I and I R_1 are sections of the planes of incidence and reflection of the sun's rays from the glass B C, those planes being all parallel to the axis of the prism, which should be approximately parallel to the earth's axis. Part of the rays passes through that glass, and is reflected by glass A C to A B, and again reflected there and sent through B C to R_2. Now let the angle of incidence, and therefore of first reflection at I, be called $A - \delta$ (A being the angle between the two reflecting glasses), and let the other angles be designated as in the figure. It requires no mathematics beyond the knowledge that the three angles of every triangle = 180°, commonly called π, to see that the angle $\beta = \pi - (C + A - \delta)$ in the small triangle near C; and in the one near A, $\gamma = \pi - (A + \beta) = C - \delta$; and in the triangle near B, $\alpha = \pi - (B + \gamma) = \pi - (B + C - \delta) = A + \delta$. That is to say, the angle α, made by the plane of emergence of the twice reflected rays with the front glass, differs from that of the once reflected by $2\,\delta$. Therefore, if the prism is so placed as to make $\delta = 0$, which it will be if the angle of incidence = A, the twice reflected rays will come out parallel to the once reflected, and the two images of the sun will coincide. In fixing the instrument however, we have nothing to do with the angles; but simply to adjust it by trial with a chronometer (for it cannot be done without), so that the images do coincide at solar noon.

They were at first made so as to be fixtures on the stone where they were set, and so they were always exposed to the weather, and besides, if cemented in wrongly, they were wrong for ever. To avoid both these evils, I suggested the making of a brass plate to be fixed on the stone, with a raised slip adjustable by screws, against which the instrument is to be laid closely when used, but at other times it may be kept in the house. Some of them are made to turn on an axis parallel to the earth's axis, and then they can be presented to the sun at other hours besides noon, but only for the given latitude like a sun-dial. Some are made adjustable for latitude also. They are moreover made for star observations with the reflectors silvered instead of blackened, on account of the greater feebleness of star light. A table was published to be used with them, showing the time of first and last contact of the two images of the sun for every day in the year, as that is observable perhaps more accurately than the time of coincidence; at any rate it gives three chances of observation instead of one. But many people prefer my meridian slit.

WATER AND SAND-CLOCKS.

The earliest time-keeping *machine* is the clepsydra or water clock of the Greeks and Romans, which was no doubt made in various ways. Vitruvius mentions one made as a water wheel, which would probably be very irregular. The simplest in construction is a graduated cylindrical vessel with a hole in the bottom, and this appears to have been the most commonly used: but they must surely have discovered that the water in that case by no means runs out with uniform velocity, though they did not know, and some of the modern writers on antiquities apparently do not either, that the velocity varies as the square root of the height of the water above the hole. But if a trough is kept full by a stream, and a hole made anywhere in it, the water will then run uniformly into, and rise uniformly in, another cylindrical vessel, in which its height may be marked on the sides, if it is of glass, or else by a floating index, and that will make a very fair clock. I do not however find any notice of such a one in books on Greek and Roman antiquities. Various other forms of clepsydra are described in them, which it is not worth while to copy here.

The case is different with sand. If a column of dry sand ever so high stands over a small hole, the sand will run out no faster than if it is a very moderate height, for the same reason that a pile or cone of sand on a given base will stand up to a certain height and angle only, which depends on the amount of friction between the particles. The angles of the waist of an hour-glass ought to agree with that natural angle of the sand in order that it may run out uniformly to the end —if that is of any consequence, which it hardly is for the boiling of eggs, or even for the old use of limiting the length of sermons; which object might sometimes be advantageously accomplished now by a descending sounding-board, or as sounding-boards are gone out of fashion, by a pulpit bottom or 'drop' let down by clock work after 25 minutes.

The burning of graduated candles was another mode of marking time, and if the candles were of wax, as they probably were, and sheltered from wind, and the wicks uniform throughout, the measure would be accurate enough for any purpose for which it was likely to be used. The consumption of oil in a lamp might also be marked and used in the same way.

This seems the proper place to notice a genuine water clock, invented and used by Lord Rosse for the important purpose of driving an equatorial telescope so as to keep it pointed to a star, against the earth's rotation, for which there are various contrivances. It is described in the R. A. S. *Monthly Notices*, vol. xxvi., and the principle of it is that the water is let out slowly from a box by an india rubber tube carried by a float, so that its mouth is always at the same small depth below the surface of the water; and as the pressure depends only on that depth the outflow is uniform, and the descent of the float with it, and that has a cord attached to it, and at the other end attached to the wheel which drives the telescope. The great equatorial at Greenwich is driven by another kind of water clock, which will be noticed farther on, as it is combined with a revolving pendulum, and we will now proceed to what are properly called

CLOCKS.

The invention of clocks driven by a weight has been generally attributed to Pacificus, Archdeacon of Verona, in the ninth century; and also to Gerbert, afterwards Pope Silvester II., who made a clock at Magdeburg in 996, when he was an Archbishop. But there does not seem to be any contemporary proof that either of these were weight clocks; nor is it certain that Gerbert's was a clock at all; for it is described by an old writer quoted in Beckmann's 'History of Inventions' (4th ed. 1846) thus:—'Magdeburg horologium fecit, illud recte constituens consideratâ per fistulam stellâ nautarum duce'—*i.e.* 'he made a time-piece at Magdeburg, setting it by looking at the pole-star through a tube.' Beckmann thought William, Abbot of Hirshaw in the eleventh century, had the best claim to the invention; of whom it was said, 'Naturale horologium ad exemplum cœlestis hemisphœrii excogitavit;' and soon after, certain monks had the duty 'horologium dirigere et temperare, et signa pulsare;' which however looks rather like their directing the clock than the clock them, or as they say at sea, '*making it* ten o'clock,' rather than learning from the clock that it is so.

But he also says that in those times the day and the night were each divided into twelve hours, between sunset and sunrise; and if so, such hours could not be shown by any clocks; for we may be sure that no clocks of those days were able to adapt themselves to hours of such varying length. And all this seems to me very insufficient evidence of the existence of clocks, as we now understand the word; which, by the way, appears not to have driven out the older word 'horologe,' until the time of Henry VIII., having been long used for a bell (cloche).

But it may be safely concluded from the various allusions to horologia, and to their striking spontaneously, in the twelfth century, that genuine clocks had been invented before then, though there is no surviving description of the construction of any clock until the thirteenth century, when it appears that a certain 'horologium' was sent by the Sultan of Egypt in 1232 to the Emperor Frederic II. 'It resembled internally a celestial globe in which the sun, moon, and planets moved, *being impelled by weights and wheels*, so that they pointed out the hour, day and night, with certainty.' The oldest clock mentioned in England is that which was put up in a former clock-tower of Westminster, with some great bells, in 1288, out of a fine imposed on a corrupt Lord Chief Justice, of which a memorial survived near the same spot in a sun-dial which stood on a house in Palace Yard, pulled down within my time, with the inscription '*Discite justitiam moniti.*' In 1292 a clock is mentioned in Canterbury Cathedral, costing £30; and the old striking part of the original one in Exeter Cathedral was at work not long ago, and perhaps is still, though the going part had been replaced; and the same at Peterborough, probably of about the same date. That of Glastonbury Abbey, afterwards at Wells Cathedral, dated 1325, is now in the South Kensington Museum, going. One by Richard Wallingford, Abbot of St. Albans, made in 1326, is said to have been such as there was not in all Europe, showing various astronomical phænomena. There was also one at Dover

Castle, with the date 1348 on it. That also was exhibited going in 1876.

A description and picture of the clock made by Henry de Vick for Charles V. of France in 1370 may be seen in old editions of the *Encyclopædia Britannica*, whether authentic or made up from some other I do not know. I did not think it worth repeating in the eighth edition, for which I wrote the article on clocks. It was very like our common clocks now, except that it had only an hour hand, and a vibrating balance (but no balance spring) instead of a pendulum. It is impossible that such a clock could go well, and it seems strange that the apparently simpler contrivance of a pendulum should not have come till four centuries after clocks were first invented; and yet that is the general tradition. I suppose therefore that the pendulums in the old Peterborough, and Exeter, and other church clocks were added long after their original construction. Those clocks were wound up by long spikes or handles sticking out of the wooden barrel over which the rope goes which carries the weight. In many old church clocks the weight is only a large stone. It is not until quite modern times that church clocks had minute hands besides the hour ones, but in other respects there is surprisingly little difference in principle between the oldest of these machines and most turret clocks of the present day.

The going part of a clock is, and always has been, nothing but a train of some number of wheels and pinions, of which one turns in 12 hours, and another in 1 hour, if there is a minute hand. The first, or slowest, or 'great' wheel, is turned by a weight hanging by a rope wound round a barrel on that wheel's axis, or *arbor*, as it is called in clock-making; and the last or quickest wheel drives a fan-fly, or a fly-wheel, or a pair of vibrating arms, called a 'balance,' or a pendulum, to regulate the velocity of the train. A spring clock is merely a compound of a large watch and a common clock.

Any one who has looked at the inside of a clock—and it is useless to read this book until you have—knows that it consists of a few large wheels with a good many teeth, and that the teeth of every large wheel, except the highest, are engaged in or drive some much smaller

Fig. 3.

wheels fixed on the axes or arbors of all the large ones except the largest. These small wheels are called pinions, and their teeth are called 'leaves' for distinction. This figure 3 shows a set of pinions of from 6 to 16 leaves, the lowest and the highest numbers used. It also shows the kind of steel plate which is used for drawing *pinion wire*. For it is cheaper to make the arbors of such wire, turning off the projections all along the arbor except where they form the pinion itself, than to cut the pinions out of the solid. But they are not so good as cut ones, and I believe are not used in the best clocks.

Fig. 4.

The largest wheel in the train, which is called 'the great wheel,' *appears* to be stuck on the end of the barrel which carries the string and the weight attached to it—whether with a moving pulley or not, does not signify as a matter of principle, which is all we are considering at present. The only effect of the pulley is to take a double string of twice the length of a single one in the same clock case, which enables a larger barrel to be used and also prevents the string from untwisting.

But if the great wheel were really fixed to the barrel there could be no winding up, except by an endless chain contrivance, which we need not now consider. Accordingly the real arrangement is that the great wheel rides freely on the arbor of the barrel, and is connected with it by a ratchet and click; the ratchet being a saw-toothed wheel on the end of the barrel, and the click being on the great wheel with a spring to keep it in the ratchet teeth, which pass under it in winding, but cannot return, except with the great wheel itself. You will see a picture of it under 'Maintaining Powers,' farther on in the book. And it is all the same whether the barrel is turned in the old way by spokes, or in the modern way by a key or 'winder' put on the squared end of the arbor; only in the former case the barrel rides loose on the arbor, and in the latter is fixed to it and the wheel rides loose. Consequently the weight is always acting on the train, except at the time of winding; and in all good clocks there are contrivances for keeping the force on even then, called maintaining powers.

Now suppose the great wheel has 120 teeth, and the pinion which it drives has 10; then, if that pinion and its arbor and the wheel stuck on to it turn in one hour, the great wheel will evidently turn in 12 hours. The wheel which turns in an hour is called the centre wheel in house clocks, because that arbor comes through to the centre of the dial and carries the minute hand. Suppose that wheel has 64 teeth and drives a pinion of 8; then that pinion and the wheel on its arbor will turn in 8 minutes; and if that has 60 teeth and drives another pinion of 8, that pinion and its arbor will turn in a minute, and the wheel on that arbor is the scape-wheel, which drives the pendulum vibrating in a second, as we shall see presently. Not that there is any virtue in these particular numbers of teeth and leaves, or that the pendulum need vibrate seconds; most church clock pendulums are slower, and all short spring clock pendulums are faster. All that is requisite theoretically is, that the numbers of the teeth of all the wheels multiplied together, and divided by the numbers of the leaves of all the pinions multiplied together, should give the proper velocity-ratio between the slowest wheel and the quickest. Thus, if the scape wheel has to turn 60 times as fast as the centre wheel, and there is one between them, which may turn in any time, the product of the teeth divided by that of the leaves must = 60, and subject to that, you may distribute the numbers as you please—theoretically; but practically other considerations come in, such as that the slower wheels must be larger than the quicker ones, or they could not clear the arbors below them: that if the leaves of the pinions are very few they do not drive easily, and if they are many the teeth must be many and small, and more expensive to cut, and so forth; and the result is that, in the common long house clocks the numbers are usually what I gave just now; but in astronomical clocks or *regulators* they are higher, sometimes twice as much; in turret clocks they vary more, according to circumstances, as will be seen hereafter.

The simplest of all the methods of regulating the velocity of the train, and one which certainly existed before De Vick's time, is the fan fly, or a pair of arms with vanes which are resisted by the air. I think it by no means improbable, though it is never likely to be ascertained now, that some of those earlier clocks were trains of wheels with a fly to regulate their velocity, instead of a balance, which De Vick used in his going part, though he had a fly in the striking part, as the earlier English clocks have, and exactly as it is used to this day. So long as the force and friction of the train are uniform the velocity of the fly will be uniform, as the variation of density of the air is too small to affect it materially; and it should be observed, that a long fly with a rather slow motion is less affected by variations of force than a short and quick one; but as far more accurate methods are now used, it is unnecessary to go further into this.

A fly wheel, which is either a wheel with a heavy rim or a pair of weighted arms at right angles to the axis which carries them, is another method; but not so good, because, not being resisted by the air nearly so much as fans, it is much more affected by a change of force, which in clock work nearly always means a change of friction in the train. It acts simply by its moment of inertia, which is constant, and therefore the velocity cannot be constant if the force varies. In fact there is theoretically no limit to the velocity of a fly wheel driven by a weight, so long as the weight can go on falling, though practically a terminal velocity is soon reached, when the friction and the increasing resistance of the air balance the force; but of course this balance is disturbed and the velocity changes as soon as the force varies.

Conical Pendulum. A pair of weighted arms attached to a revolving vertical axis by horizontal hinges, so that they can fly further out as they go, will regulate the velocity more completely than a fly wheel or arms rigidly fixed, and still better if it has fans attached to it, but not completely enough to keep it uniform if the force varies much. They are like the 'governor' of a steam engine in appearance, but no further; for the governor arms work a lever which opens more or less of the throttle valve of the steam pipe, according as the engine is going too slow or too fast. A single ball or pair of balls hung in this way and driven by a clock train form what is called a *conical pendulum*, because each arm describes a cone, and the time of its revolution may easily be determined as follows, except so far as it is affected by friction and resistance of air:—

Let l be the length of each arm, and ϕ the angle at which it happens to be inclined to the vertical axis, which of course depends on the rate of revolution or angular velocity, which is usually called ω; then the centrifugal force of each ball $= \omega^2 \, l \, sin\phi$; and as that is the force which keeps the balls away from the vertical, it must balance the force which draws them to it, which is $g \; tan\phi$ (g being the usual symbol for the force of gravity, or twice the number of feet which a body falls in the first second of time, and g in this latitude is 32.2); therefore ω or the angular space moved over by the arms in one second $= \sqrt{\frac{g}{l \cos\phi}}$, and the time of a complete revolution through 360° or 2π, is $\frac{2\pi}{\omega} = 2\pi\sqrt{\frac{l\cos\phi}{g}}$. If you wish to know what that means in figures, you must express l in feet, as g is, and write the numerical value 3.14159 for π, and take the numerical value of $\cos\phi$ from a table of sines and cosines; and the result, after extracting the square root and dividing, is the number of seconds in which the revolution is performed. We shall see hereafter that it is just so much less than the time of a common vibrating pendulum of the same length as $\sqrt{\cos\phi}$ is less than 1. And as the cosine varies least when an angle is small, a clock of this kind will go better when the length of the arms and the weight of the balls are such that they make only a small angle with the axis when the clock weight is driving them. But again it must be remembered that these results are very much modified in actual working by the resistance of the air, which acts more strongly on the balls as they fly farther out, and thereby tends to regulate the velocity, as it does with a fan fly.

A clock of this kind is often used to turn the reflectors or coloured lenses of revolving lighthouses, and also for the more accurate purpose of driving large equatorial telescopes, to keep them pointed to a star notwithstanding the revolution of the earth. For the earth does not move by jerks, as a clock with a vibrating pendulum and escapement necessarily does. A revolving pendulum alone will not do without some contrivance to equalise the force upon it, or to check the pendulum itself by friction.

The simplest form of it is setting the revolving balls within a conical ring, which they can graze slightly, and it is better if they are furnished with slight grazing springs. And if the cone is also made raisable by a handle within reach of the observer, so that he can regulate the friction, probably this is enough in all ordinary cases.

But for telescopes of great importance superior contrivances are used. The chronograph at Greenwich, which drives a barrel covered with paper, on which the times of various observations are pricked by galvanic communication from the observers, is an ordinary clock with a revolving pendulum, driven by an arm which also carries round a kind of spade, dipped a little into an annular trough of water; and the farther out the pendulum swings the deeper it pushes the spade into the water, by a simple lever arrangement which can easily be imagined.

The great equatorial telescope there has a similar pendulum and trough; but besides that, the clock is itself driven by a 'Barker's mill,' or a pair of revolving horizontal arms on a vertical axis, all hollow and receiving water at the top of the axis. The arms have holes near their ends, on opposite sides, and the water flowing out there drives the arms the other way. The pendulum also works a throttle valve as the governor of a steam engine does, and so regulates the flow of water through the mill.

A revolving pendulum may be hung by a single wire, as

Fig. 5.

there is no tendency to twist, and they are so made in bedroom clocks, which have the advantage of being silent. But it is difficult to get a wire strong enough for a heavy pendulum which would not also be too stiff. The Greenwich ones are hung by a kind of universal joint, made of two pairs of suspension springs in stirrups, set across each other, one turned upwards and the other downwards, with a cross between.

We shall have to notice afterwards some other clocks for this purpose, in which a revolving continuous movement is combined with a vibrating pendulum and escapement; but I must describe escapements first.

It is perhaps worth mentioning that any point in a wheel revolving uniformly has always the same velocity in a horizontal direction as some point in a pendulum which makes a double vibration in the same time as the wheel revolves; and therefore, theoretically, a constant motion of a clock-train might be got by connecting some point in a 1 sec.-pendulum with a pin in a small disc revolving in 2 seconds, by a rod so long as to be practically horizontal. But it would probably be impracticable to keep the force constant enough to give just proper impulse to the pendulum; and if too much was given, there would be a jerk at the end of every beat, and if too little, the pendulum and the clock would stop.

BALANCE-WHEEL ESCAPEMENTS.

Before we go into the theory of pendulums we should notice the one other mode of regulating the motion of a clock-train which existed long before pendulums were applied. The earliest escapement of which there is any known description is that which De Vick's clock had, and which is called the crown-wheel escapement. The object of that, as of all the later escapements, was to let a tooth of the quickest wheel in the train escape past some stops called *pallets* at every vibration of the balance, and that wheel is thence called the scape-wheel, and a crown-wheel from its shape. The pallets A, B, in fig. 6, are pieces of steel fixed to the axis or arbor of the balance C in planes at right angles to each other, one of them set so as to be pushed one way by

Fig. 6.

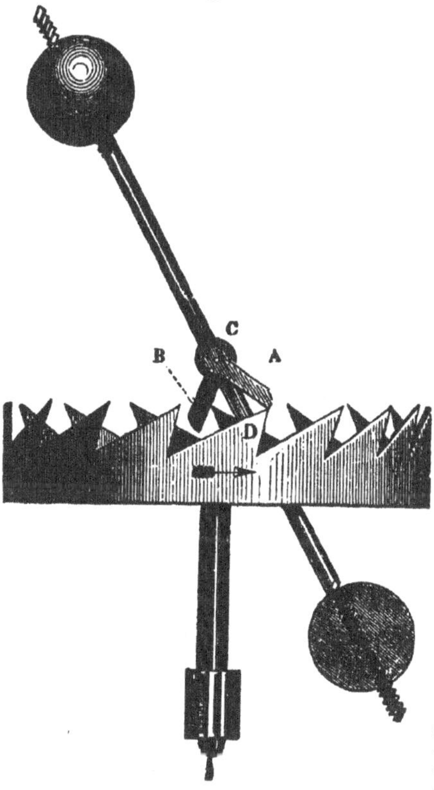

the front teeth D of the wheel, and the other the other way by the back teeth. As one tooth escapes past its pallet, the other pallet is in a position to receive and stop the opposite tooth. But as the balance has then acquired a swing in the direction in which it has been pushed by the escaping tooth, it does not stop immediately, but swings a little farther and so drives the wheel back again a little, producing what is called the recoil. This is just the same as the old 'vertical' watch escapement, which remained in use until a few years ago, with this important exception, that the time of vibration of a watch balance is regulated by a thin spiral spring fixed to it and to the frame, whereas this had none, and so its regulating power over the train depended solely on its moment of inertia, and on its swinging farther for any increase of force, which by no means makes it isochronous.

The same thing is still to be found in bottle-jacks; the piece of meat to be roasted forms the balance, and the noise at each change of its motion is the 'ticking' of the escapement. It is true the meat makes several turns for one tick, while De Vick's balance made only half a turn, but that is because there is a wheel on the pallet arbor in the jack, worked by a pinion on the meat arbor, as in the rack lever watch escapement; but otherwise the bottle-jack escapement is precisely the same as in De Vick's clock, in which the axis of the balance was vertical, and not horizontal as in fig. 6.

PENDULUMS.

Pendulums, like many other things, may have been invented several times over in different ages, or even in the same. In an old edition of the *Encyclopædia Britannica* it is said that 'the ancient astronomers of the east employed 'pendulums in measuring the times of their observations, 'patiently counting their vibrations during the phases of an 'eclipse or the transit of the stars, and renewing them by a 'little push of the finger when they languished; and 'Gassendi, Riccioli, and others, in more recent times 'followed their example.' If so, it is plain that this knowledge had itself languished and died, before the making of what has long been called Galileo's discovery in the church at Florence, that a chandelier, and therefore any other pendulum, vibrated different arcs in the same time, provided they were none of them large ones. When we consider the vast number of pendulums of various kinds that there are swinging about the world, it certainly is difficult to imagine that nobody ever made that observation before the sixteenth century. The application of it to the regulating of clocks however is a different thing, as that required invention as well as observation. To be sure, all that was needed was to omit one of the weights in De Vick's balance, and set it in a vertical instead of a horizontal plane, and it is strange enough that this slight but valuable alteration should have waited three centuries to be made. It would then assume this form (fig. 7), which is the same as the other in all but the position of the parts and the omission of one arm and weight of the balance. The bent end of the arm (called the *fork*) is substituted for a weight in this drawing, because it was afterwards found better to hang the pendulum independently, and connect it with that arm, called the *crutch*, by means of the fork. But I have seen small clocks of the last century, and even some modern French ones, with the crutch itself made into a pendulum by merely putting a ball at the end of it.

There seems no doubt however that the first person who investigated and established the mathematical theory and properties of the pendulum was Huyghens, the Dutch philosopher, in the seventeenth century; but it seems equally certain that the first pendulum clock was made for St. Paul's Church in Covent Garden, by Harris, a London clock-maker in 1621, though the credit of the invention was claimed also by Huyghens himself, and by Galileo's son, and Avicenna, and the celebrated Dr. Hooke, the undoubted inventor of the balance spring of watches, and the discoverer of its theory.

Fig. 7.

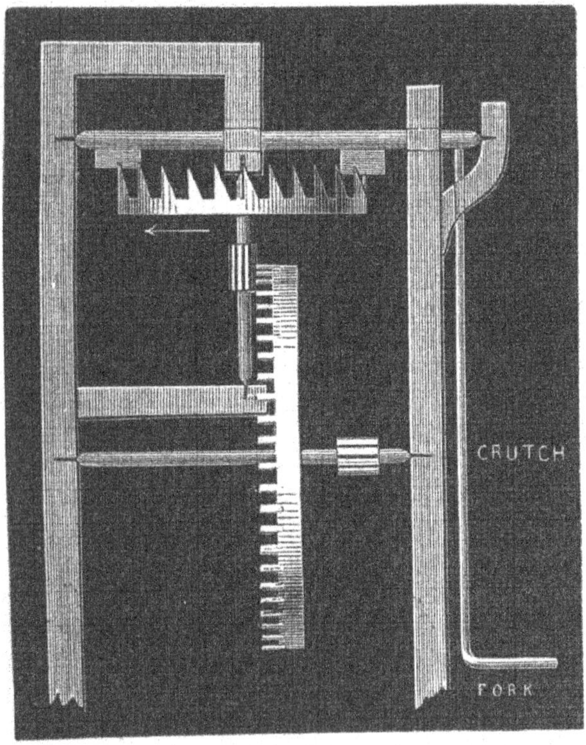

The main point of Huyghens's discovery seems at first sight a long way off any connection with what we now understand by a pendulum, viz., a weight or *bob* at the bottom of a long rod, which is hung by a string or a thin spring at the top, and the bob therefore swinging in a circular arc, or something very near it. For he proved that the curve in which a bob hung by a string of insensible weight must move in order to be isochronous in its vibrations, is *not* a circle, but a cycloid, or the curve traced out by a point in the rim of a circle rolling upon a straight line, *e.g.*, a nail in the tire of a carriage wheel rolling on a smooth road, or P in the circle D E P rolling on B G C in fig. 8; and he showed how a certain other property of the cycloid might be made use of to enable a pendulum bob to describe a cycloidal instead of a circular arc. It is not worth while to fill these pages with demonstrations which may be found in any mathematical treatise on mechanics, especially as I must assume the reader to have some of the knowledge which is only to be got from such books, in order to understand the demonstrations if I gave them. Therefore we may as well begin at this point, that a body moving by gravity in a cycloid (with the curve downwards, as B P F C in fig. 8) does describe

Fig. 8.

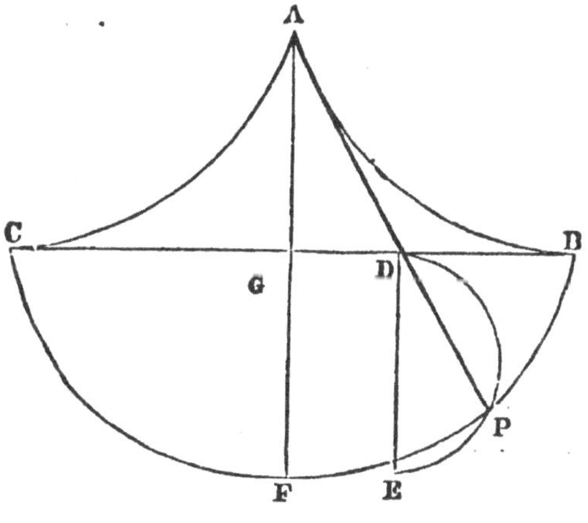

both large and small arcs in the same time; the reason of which is that the force may be proved to be always in proportion to the distance along the curve from its lowest point, F.

But how is a pendulum bob, hung by a string or wire from a fixed point, to be persuaded to describe a cycloid? It so happens that if a cycloid B F C is cut in two, and one half B F removed as a solid with a convex edge to A C (making A G=F G=D E), and the other half to A B, the end P of a string = A F, fixed at A, and moving between those 'cycloidal cheeks' will redescribe the old cycloid B F C. This is mathematically expressed by saying that the *involute* of a cycloid is another equal cycloid, and therefore also the *evolute* is; the evolute being the cheeks, and the involute the curve described from them: the proof of this belongs of course to geometry, not mechanics, and is of no consequence to us at present. Huyghens therefore proposed to hang clock pendulums by a string or a thin spring between cycloidal cheeks, and that was for some time thought a very superior method of making clocks. I have no doubt there are some still in existence, as I have seen them.

But after a time it was found that clocks went rather worse with these cheeks than without them; and then it occurred to somebody that the cycloidal theory is only true for what is called in mathematics a simple pendulum, or one in which not only the bob, but the centre of the bob, is alone supposed to have any weight; and of course there is no such thing possible, for the rod must have some thickness, or it is not stiff enough to work, or to be driven by the clock, and if you make the bob very heavy, with the view of rendering the weight of the rod insignificant, then the bob itself must be large, and differs considerably in its mechanical effect from a single imaginary heavy point at its centre. And besides that, the spring or string cannot be made to act against the cheeks without friction and other disturbing causes; all which things are said to have been proved to produce greater deviations from isochronism than a variation of several degrees in the arc. Indeed we shall see hereafter that the common clock escapements tend to produce an error of their own, which the deviation from cycloidal vibration (commonly called the *circular error*) is actually useful in counteracting.

Nevertheless the cycloidal theory is valuable to this extent: it shows why a common pendulum is very nearly isochronous for different *small* arcs; for the string A P will evidently describe very nearly the same curve near the bottom F, whether the cheeks are there or not: in other words, a bob vibrating in a small circular arc is almost identical with one in a cycloidal arc described by a string of the same length. The actual time of a circular vibration cannot be calculated without the aid of the higher branches of the integral calculus, and even then it can only be exhibited in the form of a rather awkward series, which would be little better than useless, except for small arcs; and for them it is quite sufficient to take only the two first terms of it. The whole calculation may be found in *Pratt's Mechanics*, or in the 8th edition of the *Encyclopædia Britannica*, under article *Pendulum*. The first term is simply the expression for a cycloidal vibration, $t = \pi\sqrt{\frac{l}{g}}$, the letters of which I have already explained at p. 25, for the conical pendulum, whose time of vibration you now see is less than that of a plane pendulum of the same length in the proportion of $\sqrt{}$cosine of the angle which the conical one makes with the vertical axis.

Circular error.—The next term in the series (omitting all the later ones which are still smaller) constitutes the circular error K, or the excess of the time of vibration in a circular arc over that in the cycloidal one belonging to the same length of pendulum; and it is accurate enough for all practical purposes to say that $K = \pi \frac{a^2}{16}\sqrt{\frac{l}{g}}$; and therefore for a whole day, $K = 5400\ a^2$ in seconds, whatever the length of the pendulum is; the amount of which you may easily calculate from the fact that if $a = 2°$, as usual in clocks, its numerical value is ·035, and therefore K = about 6 seconds. But this is more than what is called the circular error in a clock pendulum; for we need not care what is the difference between the time of a cycloidal arc and the actual circular arc which the pendulum describes, but only between such two small circular arcs, a and a_1, as the pendulum is likely to describe in different states of the clock. This quantity, which we may call Δ K, is only 5400 $(a^2 \sim a_1^2)$; and if we assume the larger of the two arcs to be as much as 2½° and the smaller 2°, the circular error between them will be rather less than 4 seconds a day. This is a very large variation of arc for a tolerably good clock, and when it is as small as it generally is, the circular error may be expressed by differentiating the expression 5400 a^2, and we may say (so long as both the arc and its variations are small, remember) that $\Delta K = 10800\, a\, da$ (da being the variation of the arc). Thus if $a = 2°$ and $da = 10'$, Δ K will be very nearly 1 second: *i.e.* the clock will lose a second a day for such an increase of the arc, independently of any other increase or counteraction of the circular error, which may be produced by the escapement at the same time.

For the common purpose of finding the length of a pendulum to beat seconds, or any other required time, we need not trouble ourselves with anything beyond the equation $t = \pi\sqrt{\frac{l}{g}}$, in which t is the number or fraction of seconds, and l is expressed in feet, because g means 32·2 feet (in this latitude), and π is 3·1416 or the numerical value of 180°. Therefore in order that t may be 1 second, you will easily find by a little calculation that l must be 39·1393 inches; and having got that fixed, you may find the length of pendulum for any other time of vibration very easily by multiplying 39·14 inches by the square of the ratio of the intended time to 1 second. But to save trouble I will put down a few of the lengths, omitting small fractions.

SECONDS.	FT.	IN.	SECONDS.	INCHES.
4	52	2	1¼	61
3	29	4	1	39·14
2½	20	5	¾	22
2	13	0½	⅔	17·4
1½	7	4	½	9·78

These seconds, and the second used as the unit of time in all mathematical formulæ, are seconds of mean time; and as a sidereal day is shorter than a mean one in the ratio of ·99727 to 1, a sidereal pendulum must be shorter than a mean one in the square of that ratio, which makes the sidereal seconds pendulum at Greenwich 38·87 inches. I have met with persons who could not understand how an increasing gravity can make pendulums go faster; and others, with a little mathematical knowledge, think that the force varying as the distance from zero (in small arcs) is opposite to the fundamental law of gravity varying inversely as the square of the distance. The answer to the second of these is, that the zero, or middle, or lowest position of a pendulum is not an attracting body like the earth (which attracts as if it were all condensed into its centre); and the tendency towards zero is only a mathematical consequence of that law of gravity. The stronger the tendency to zero, or the stronger gravity is, the faster the pendulum will evidently fall; and the sooner gravity will bring it again to rest in rising. The time of rising always = the time of falling, under any force which is equal at equal heights during the rise and fall.

Centre of oscillation.—It must be remembered that these are only the theoretical lengths of *simple pendulums* with all the weight concentrated in the centre of the bob, and that this theoretical length by no means coincides with the actual length down to the centre of gravity of the pendulum, but is always longer. This length may properly be called the *radius of oscillation*, as the lower end of it is always called the centre of oscillation; which is not a fixed point in the body, like the centre of gravity, but a relative one, every axis of suspension having a radius and centre of oscillation of its own. It is not always a simple process to calculate this length (which we may as well call l) for any given pendulum, as it requires either the integral calculus or some rules deduced from it; but it will be easy to explain the nature and meaning of the quantities on which it depends. Let m be the mass of each particle of the pendulum, which in these calculations must not be confounded with the weight, which is written mg (g being the force of gravity), r the distance of m from the axis of suspension, and M the mass of the whole pendulum; then the radius of oscillation = the sum of each particle multiplied into the square of its distance from the axis, divided by the sum of each particle multiplied into its distance simply; that is to say, $l = \frac{\Sigma\, m\, r^2}{\Sigma mr}$ (Σ being used to indicate this kind of summation, which can only be performed by integration).

The numerator in this fraction is called the *moment of inertia* of the body with reference to that axis of suspension. Of course there is some quantity k^2 which $= \frac{\Sigma\, m\, r^2}{M}$, and k is then called the *radius of gyration* for that axis, and Mk^2 is obviously the moment of inertia again. In like manner some quantity $h = \frac{\Sigma\, m\, r}{M}$, and h is then the distance of the centre of gravity of the whole pendulum (not of the bob, remember) from the same axis, which is easy enough to find in bodies of the shape commonly used for pendulums. It appears then that the effective length l of a pendulum always $= \frac{k^2}{h}$; and it is only when the rod is very thin and the bob itself small but heavy, that k and h can be assumed to be even approximately identical, and therefore both $= l$, as they are in a simple pendulum. But it does conveniently happen that the centre of oscillation in small pendulums of the usual forms is generally near the centre of gravity of the *bob*, and sometimes coincides with it, as we shall see.

This quantity Mk^2, the moment of inertia, always appears in mathematical formulæ as the resister of forces of rotation or vibration on an axis, and of all disturbances of such forces after they have set the body in motion, and therefore we see at once why long and heavy pendulums are better than short and light ones. There are a few other simple propositions relating to the centre of oscillation and the theory of pendulums, which it will be appropriate to notice here.

Draw a pendulum of any shape you like, supposed to be vibrating in the plane of the paper, and call its centre of suspension S, and its centre of gravity G, and the distance between them h. All bodies have this property, that their moment of inertia round any axis through G (which we call Mk_1^2) is less than round any other axis parallel to that, and we are not concerned with any but parallel axes; and further, if the other axis is distant h from the centre of gravity, the k^2 round that new axis $= k_1^2 + h^2$. Therefore if O, somewhere below G, is the centre of oscillation corresponding to S, l or SO $= \frac{k_1^2 + h^2}{h}$; and $l - h$, or GO $= \frac{k_1^2}{SG}$; or the centres of suspension and of oscillation are reciprocal. In fact, if the pendulum is wide enough, you may draw two circles round G with radii G S and G O, and if you stick an axis through the pendulum, at right angles to the same plane of vibration, anywhere in either of those two circles, it will vibrate in the same time as on the original S. Consequently, if you construct a pendulum (symmetrical on both sides of its middle plane of vibration, to prevent it swinging with a twist) with one fixed axis S, and another adjustable one for O, and a movable weight to adjust the time by, you can make it vibrate in the same time on both axes; and by adjusting the movable weight also, the pendulum can be made to agree with the vibrations of a clock pendulum which is beating seconds; and then the distance between S and O, or the two knife edges which represent them, being measured, we shall know that that is the length of the simple seconds pendulum. It has also been proved that if round axes are used instead of knife edges, the distance between them (not their centres) equally represents the radius of oscillation, provided the axes roll on a plane.

If all the standard yard measures in the kingdom should ever be lost, they could only be restored by this process, according to an Act of Parliament, 5 Geo. IV. c. 74, which declared that a yard is $\frac{36}{39 \cdot 1393}$ of the length of the pendulum which vibrates mean seconds in London at the level of the sea, in a vacuum. The force of gravity decreases so much towards the equator that an English pendulum would lose 2¼ minutes a day there. The following rule for the length of the seconds pendulum in any latitude has been deduced from observation, near enough for all practical purposes: $l = (1 - \cdot 0027 \cos 2 \text{ lat.})\ 39 \cdot 1156$ inches; that number of inches being the length of the pendulum at lat. 45°. When the latitude is more than 45°, cos 2 lat. becomes —, and so l exceeds its length at 45°, as of course it ought to do. Another still simpler formula is $l = 39 \cdot 017$ [the length of a seconds pendulum at the equator] $+ \cdot 2 \sin^2$ of latitude.

That Act however has been repealed, and some standard weights and measures are deposited in various public places under an Act of 18 & 19 Vic. c. 72. The late Astronomer Royal provided a set, open to anybody, at the Observatory gate. But if they all happened to be destroyed, new ones could be made from the old pendulum ratio. Indeed Mr. Johnson of Wilmington Square had one ready for that calamity. A simple pendulum 36 in. long would vibrate in a time which is to 1s. as 36^2 to 39.14^2, or .846 of a second, or 70.92 times in a minute; and he has made a clock with a reversible pendulum vibrating in that time, and consequently 36 inches between the two knife edges.

The length of a seconds pendulum so nearly resembles the French metre of 39.371 in., that some persons may fancy that that most ridiculous and mischievous revolutionary measure had an origin even as rational as being the length of a seconds pendulum in some latitude. But it has not. It was intended to be the 40 millionth part of a meridian of the earth—about as rational a standard as if we enacted that the yard should be the 420 millionth of the mean distance of the moon, which it is very nearly; and astronomers know the moon's distance within a less fraction than the difference of the metre from what it pretends to be, but is not.* Yet there are people who want to force on all the world this absurd, inconvenient, and useless measure, invented by a nation whose language is declining over all the world; while the English language, with that standard of measures which every man carries in his arms, his legs, and in his head, is spreading over all the world, so that it will soon be the only universal language to be found everywhere, if it is not so already. Doctrinaires of this kind may cram penny-school girls with French metres, and centimetres, and kilograms; but our yard grew and will remain as the natural standard of length until the stature of the human race alters. For it is the length of a good stride of a man of what is generally considered the best height, and that height is two such lengths, and so is the stretch of his arms, and a yard is the natural length of his walking-stick. A metre would be the yard of a nation of giants. With the yard too goes the equally natural and still older measure of

* See the essay 'On Celestial Weights and Measures,' in Sir J. Herschel's *Familiar Lectures*, condemning the metre strongly.

a foot, which all nations had, with such small variations as would occur in times when they had no scientific provisions for preserving exact standards. Some great authorities believe inches to have been the oldest measure of all; and the Egyptian cubit, which was unquestionably used in building the Pyramids, from the many simple multiples of it which occur there, confirmed by the discovery of one accidentally built up in a wall at Thebes, was probably 20 of their inches, being a little more than 20 of ours; and the 'sacred cubit' of the Jews was 25, according to Sir Isaac Newton. Probably the authors of the new 25 inch to a mile Ordnance survey of England little thought that they were reverting to that ancient standard. At the same time, I do not accept the theory of the Scotch Astronomer Royal,* that the 25 inch cubit was embodied in the Pyramid dimensions, besides the one of 20·7 inches, or 20 of some of the old continental inches. A variation of one thousandth is practically nothing in unscientific ages, and is less than the variation of many foot-rules now, and it is singular that if our inch were a 1000th longer, the earth's polar axis would be just 500 million inches.

Short and slow pendulums.—There is a kind of pendulum which is properly enough used in the instrument called a *metronome*, for counting the time in music lessons in a way that requires no particular accuracy, and occasionally, but very improperly, used for small clocks. It follows from the propositions I have been explaining, that if a pendulum with a heavy rod is set vibrating on an axis very little above its c. g. it will vibrate slowly, like a scalebeam; and consequently you may have a 2 seconds pendulum in the compass of a few inches. But even if the weight of it were equal to that of a large pendulum, its regulating power, or power to resist disturbance, would be very much less,

* See an amusing pamphlet on it by Mr. F. D. Wackerbarth, and the remarks on the Pyramid in my 'Book on Building.' (Lockwood & Co.)

because that is measured by the moment of inertia Mk^2, which is the sum of the weight of each particle × its distance ^{2} from the axis, and that of course must be small in a short pendulum, though its time may be long if h is very small. The time of a metronome pendulum is adjusted by a small sliding weight on the rod above the axis of vibration, the bob being of course below it. Moving the weight up brings the c. g. of the whole nearer to the axis, and also increases the Ml^2 of the whole, and so in both ways makes it go slower, and *vice versâ*. It is kept going by a roughly made watch movement and a common recoil escapement, such as I shall describe presently for clocks.

Shape of pendulums.—Whatever the shape of a pendulum is in other respects, it is essential that the back and front should be alike both in weight and shape; or to speak mathematically, that it should be symmetrical on each side of the middle plane of its vibration, or it will *wobble*, and vary its time in some irregular and incalculable way. It does not signify however, if one side of the bob is larger than the other in the direction of vibration; but as that would look ugly and the other does not, one never sees the ugly but innocent deviation, but frequently the other; for the pendulums of common clocks are often unsymmetrical in the back and front, and are therefore bad ones, though perhaps as good as the clocks they belong to. For this reason the old fashioned lens-shaped bob, or the flat cheese shape which some clockmakers use for church pendulums, are not good ones, because it requires great care to fix them with their own middle plane coinciding with the middle of the rod. It is true that a lens moves through the air with less resistance than any other shape; but then it must be very large to be of the same weight as a sphere or a thick cylinder with the axis vertical, and we shall see afterwards that nearly all clock errors are inversely as the weight (and length) of the pendulum. But there is a shape which would probably be found to combine some of the advantages of both, though Baily's experiments (in *Phil. Trans.* for 1832) showed some results which no one could have foreseen; I mean a cylinder of elliptic section, or one having the same horizontal section as a lens, but thicker; or of the section formed by two arcs of a circle of 120°, which would make the horizontal length and thickness as 12⅛ to 7 inches—a very convenient size for a large bob 13 in. high; which would weigh about 188 lbs. in cast iron and 288 in lead, and is equal to a round cylinder of 8¼ inches diameter. But such figures are only approximate, the weight depending on the density of the casting; and I have neglected the hole for the rod, as that would be compensated and more by the rod's own weight. A spherical bob is not so good as this shape, because a slight error or looseness in the hole would throw much more of the weight on one side and increase the tendency to *wobble*, a very serious defect. For all practical purposes a cylinder is probably the best shape, and is used now in all the best large clocks. I also introduced the plan of making the top slightly domical, to prevent bits of dirt resting on it, which would accelerate the pendulum, as we shall see presently under 'regulation.'

Bobs for large clocks are generally made of cast iron, partly because it is cheaper than lead, and also because very heavy lead bobs are liable to be bruised out of shape in moving about before the pendulum is hung. But there can be no doubt that lead is best, as its specific gravity is half as much again as iron, and therefore it is less resisted by the air, which we shall see afterwards affects pendulums by no means insensibly. For the same reason it is a mistake to put lead bobs in brass cases, as is usually done in 'regulators' or astronomical clocks, as it increases the bulk much more than the weight. The lead looks very well japanned. It is equally absurd to case lead weights in brass, though it does no actual harm as it does in the bob.

Fig. 9.

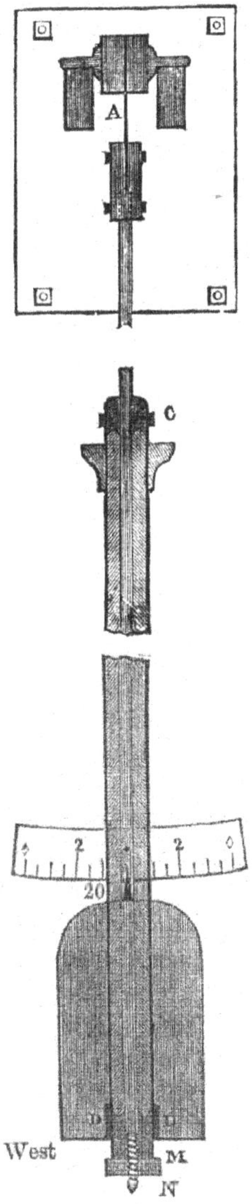

Suspension of pendulums.—Pendulums are generally suspended by a spring, for the obvious reason that they then swing without any friction—except of the air. Some French pendulums are still hung by a string, but that has some friction. The spring is not designed to act with any force as a spring, though it does very slightly, and therefore is not at all analogous to the balance spring of a watch, which is thus the substitute for gravity. The spring is enclosed in a slit, or between two pieces called chops at the top of the pendulum, which also carry the rod, except in the American clocks, where the wire itself is flattened out into a spring. The top of the spring is also riveted between two smaller pieces in common clocks, so as to make a lump which holds up the spring in the cock, which is a fixed piece with a slit in it just wide enough to hold the spring. In clocks of the higher class the two upper chops are much thicker and firmly screwed together, and a large pin put through them, and the chops themselves go through the cock tightly and are held up by the pin which lies in Vs. This will be better understood by this picture of a heavy pendulum and its suspension. We are not concerned with the lower part just yet. The nearly square plate is a cast iron plate bolted to the wall, out of which project the two thick pieces on which the pin lies. The pin itself is like a bolt with a tail projecting backwards through the head, and the piece corresponding to the head, on the right side in the figure, is the nut. The chops of the pendulum itself are shown a little below, and the way I introduced of making them for 'regulator' or astronomical clock pendulums which have the steel rod screwed into a solid piece, is to file out a slit wide enough to hold the spring and a piece of brass on each side of it, and the whole is pinched together by a screw. You cannot cut a slit thin enough for the spring alone. This will be shown in the figure of the four-legged gravity escapement hereafter.

A perfectly firm suspension is essential to good performance of any clock, and the heavier the pendulum the firmer its support must be. It is well known that clocks will influence each other when fixed against a wooden wall, however firm it may appear, and that one pendulum will thus actually set another of equal length vibrating. Consequently a free pendulum can be kept swinging from an apparently immovable support which has an invisible vibration or twist imparted to it by a clock below beating in the same time. All good regulators have the pendulum cock screwed to the back of the case, and that firmly screwed to a strong wall; or, better still, the cock is cast with a cast iron back or bracket which also carries the movement, and that is screwed directly to the wall, the whole front of the case lifting off. In the Westminster clock the pendulum cock is a large iron frame built in right through the wall with a flange behind, and the cocks of many other large clocks are fixed with bolts through the wall. With a firm suspension the pendulum swings farther than with a weak one, and we shall see afterwards that (other things being equal) the errors of clocks vary inversely as the square and sometimes as the cube of the arc.

It is hardly necessary to say that the whole of the pendulum suspension should be so adjusted that the bend of the spring may be exactly opposite the pivot of the pallet arbor, in order that the fork which embraces the pendulum, as described at page 30, may have as little friction as possible. It must not however be tight, for the reason just now given, viz. that no point in the pendulum describes quite a circle, and the difference is sensible enough generally to stop the clock if the fork fits the pendulum rod too tightly. It is equally necessary that the plane of vibration of the pendulum should be exactly at right angles to the pallet arbor, or else there will be a sliding motion of the pendulum backwards and forwards in the fork. But obvious as these things are, they are all constantly neglected, especially in large clocks, where the friction of all the parts is also the greatest. The fork indeed is seldom left too tight, because that mistake tells its own story directly; but, by way of making up for it, it is very often so loose that you hear the shake of it from one side to the other at every beat. There ought to be a drop of oil there, as that is just enough to keep a steady hold without either shake or tightness, since oiled surfaces are not really in metallic contact.

In old church clocks a long pendulum was often put for convenience a good way off the clock, but so as to swing in the same plane as the crutch, which is connected with it by a light horizontal wooden bar, no heavier than an organ 'tracker.' The pendulum top too was often higher than the pallet arbor, so that the pallets moved through a larger arc than the pendulum. There is no advantage in that; but if the pivots at the ends of the horizontal bar fit closely, there may be less loss of power by 'shake' than there generally is between the fork and the pendulum. Indeed I have seen old regulators made in that way, even with the pendulum in the usual place, by using two short horizontal arms thus, 

, C being the pivot in the crutch, P the pivot in the pendulum, and D the one at the junction of the two arms DC, DP. But this is decidedly inferior to the other plan of putting the pendulum on one side of the clock, since that requires only two pivots instead of three, and there is no twisting action on the pendulum. If the horizontal bar were attached by slight springs it would be better still, for there would then be neither shake nor friction, and the springs would no more impede the action than the pendulum spring does, for any moderate arc. There is yet in existence one great London clock of the last century with fans on the pendulum just above the bob (like the wings on the ankles of Mercury) to prevent it swinging too far, and they are actually placed obliquely, of which the effect of course is to make it swing with a twist. The late Mr. Vulliamy told me he removed fans from the pendulum of the Horse Guards clock. Sometimes heavy pendulums are hung by two narrow springs instead of one broad and thin one, to secure the vibration in the proper plane; but it is a bad plan, because it is difficult to get the two springs equal in all respects. The springs of 'regulator' pendulums of 14 or 15 lbs. are generally about ½ an inch broad and 2 inches long; I mean clear of the chops of course; that of the 6 cwt. pendulum at Westminster is 3 in. by 5 and $^1/_{60}$ in. thick.

There should always be a degree plate, with a pointer to it on the pendulum wherever it can be most conveniently seen. The length of 4° on the plate is always ·07 × its distance from the top of the pendulum; and as the only use of the degree plate is to see that the pendulum keeps the same arc, or to see how it varies, no very great accuracy is required in the degrees.

Pendulum springs.—Various opinions have been propounded on the proper strength for them, but none on any solid grounds, and some on altogether mistaken ideas of the duty of the spring. Probably the thinner it is the better, provided it is thick enough not to be bent too sharply or strained by the weight of the pendulum. Every now and then a clock is troubled with the disease of spring-breaking over and over again. Sometimes this proceeds from the lower edges of the upper chops being left sharp, which in time cut the spring; and perhaps more frequently from some unseen inequality in the fixing; and it is better to let the pendulum hang at first with the lower chops a little loose, and only to screw them up after the pendulum has got a square bearing, and to take care not to put it out in doing so. It is wonderful how much more trouble people will take to do things wrong than would serve to do them right, and it is almost incredible that some clock-makers send out their best clocks with the lower edges of the upper chops not horizontal but rounded into a circular arc. The slightest reflection would show anybody that springs so fixed must tend to 'buckle,' or bend not in a straight line, at every vibration. And others make the upper chops, which have the strain of vibration, thinner and weaker than the lower ones which have no strain. There should always be a block of wood close under the bob of a heavy pendulum, to prevent the top from breaking the crutch and pallets if it falls from the spring breaking.

A spring does not bend only at one point as a string does; and therefore a pendulum bob hung by a spring does not move exactly in a circle, but in something more like a cycloid described with that radius of curvature, as in fig. 8, p. 32; and it has often been attempted to make springs which would render the pendulum absolutely isochronous for all such arcs as it is likely to swing. Possibly the thing could be done, if it was worth doing; but both I and some other persons who have spent a great deal of time on experiments have come to the conclusion that it is not, for reasons which will appear when we come to consider the effect of the escapement on the time of vibration. Indeed I may say at once, with respect to the only escapement I have yet described, and all others on the recoiling principle, that the circular error, which it is the object of these spring contrivances to correct, is already more than corrected by those escapements: for the clocks never lose, but gain, when the arc of the pendulum increases; so that the circular error is actually useful for counteracting the escapement error, and the clock goes better than it would with a perfect cycloidal pendulum, or any equivalent contrivance.

Pendulum regulation.—Though pendulums can be made by calculation and experience very nearly of the proper length, they always require some adjustment afterwards, besides regulating from time to time, according to the state of the clock. The usual way of doing this is by a screw cut at the lower end of the rod, with a nut on which the bob rests, and by which it can be raised or lowered to make the pendulum go faster or slower. The lower part of the rod is always made into a square or some shape on which the bob cannot turn, except in certain compensated pendulums, as we shall see; for otherwise the bob would twist round with the nut. By that means you can hold the bob and the rod steady while you turn the nut. It is convenient to have the thread of the screw such that one complete turn of the nut will make a minute a day difference in time; and the better class of pendulums have a large round nut divided by marks which mean a second a day, with an index fixed to the bob.

It is easy to find what should be the width of the thread for one turn to do a minute a day. Let l be the length of the pendulum, and dl a small increase of l, and dt of t. Then we know that

$$\frac{t + dt}{t} = \frac{\sqrt{l + dl}}{\sqrt{l}}\text{; or, } 1 + \frac{dt}{t} = \sqrt{1 + \frac{dl}{l}}$$

There is a result of the binomial theorem which we shall often have occasion to use in pendulum problems, which may be stated thus:—When a quantity is of the form $1 + m$, m being a *small* fraction, $(1 + m)^n$ practically, or as astronomers say, sensibly $= 1 + nm$, whether n is $+$ or $-$, a fraction or an integer; and it may be as well to mention for those who do not know it that $(1 + m)^{-n}$ means $\frac{1}{(1 + m)^n}$. Consequently here $\frac{dt}{t} = \frac{dl}{2l}$; and multiplying by all the seconds in a day, 86400, and calling dT the daily increase, $dT = \frac{43200\, dl}{l}$ sec. So if we want dT to be a minute for one turn of the nut, dl or the width of the thread must $= \frac{l}{720}$, or about $\frac{1}{18}$ inch, a very convenient size for a fine screw.

But this mode of regulating is inconvenient for large clocks with very heavy bobs, and it is better to fix a collar on the rod at some convenient height, on which small weights can be laid to accelerate the clock, and taken off to retard it. The place where a given weight is most effective is at the middle of the length, but it does practically as well to put it somewhat higher, and the higher it is the less risk there is of shaking the pendulum in putting the weight on or off, by which any accurately going clock is seriously disturbed. The proportion and position of these regulating weights are determined as follows:—Let m be a little weight added to the pendulum at a distance d below the top, and b above O the centre of oscillation; then—

$$\frac{t + dt}{t} = \frac{1}{\sqrt{l}} \frac{\sqrt{Ml^2 + md^2}}{\sqrt{Ml + md}} = \frac{\sqrt{1 + \frac{md^2}{Ml^2}}}{\sqrt{1 + \frac{md}{Ml}}}$$

and since $\frac{m}{M}$ is very small this practically

$$= \left(1 + \frac{md^2}{2\,Ml^2}\right)\left(1 - \frac{md}{2Ml}\right)$$

which again for the same reason $= 1 - \frac{m}{2M}\left(\frac{d}{l} - \frac{d^2}{l^2}\right)$

From which you see at once, that if $d = \frac{l}{2}$, $-dT$ or the daily acceleration $= \frac{10800\,m}{M}$; or the 10800th part of M laid on the pendulum half way from the top will accelerate it a second a day; and if $d =$ either $\frac{l}{3}$ or $\frac{2l}{3}$, m must $= \frac{M}{7200}$ to do a second a day.

We shall get exactly the same result for b if we reckon from o; for if you substitute $l - b$ for d in this last equation, the result is that $\frac{dt}{t} = -\frac{m}{2\,M}\left(\frac{b}{l} - \frac{b^2}{l^2}\right)$; as might have been foreseen from the fact that the centres of suspension and oscillation are reciprocal (p. 38). You will easily see too that the worst mode of regulating is by a sliding weight near the middle of the pendulum; for the acceleration by moving the weight upwards from b to d is measured by $\frac{dl - bl + b^2 - d^2}{l^2}$, which is o when $d = b$, and very little while they are anything near equal. In short the middle of the pendulum is the place of maximum effect for the weight, and of minimum effect for moving it.

Temporary regulation is also sometimes required in the best clocks: *i.e.* they want putting on or back a few seconds or less, though the pendulum may not want permanently altering. Stopping the pendulum and setting it off again so disturbs it for some time that it must be always avoided if possible. One way of altering the clock a very little is to put on or take off a small regulating weight for a certain number of hours. Another mode now resorted to in very fine clocks is to apply a temporary magnet, which can be made and unmade for any required time, so as to act on the pendulum either with or against gravity, and so to accelerate or retard the clock, until it shows the proper time. Other methods have been tried which all disturbed the pendulum too much. The first is used at Westminster, where it is easy to put on small weights without disturbing that slow and heavy pendulum. I regulate my best regulator with a gravity escapement by bits of card laid on the bob of about 40 lbs.: that is, for permanent regulation.

Some common clocks are regulated by a nut at the top, above the cock, which pulls up the spring through the slit in the cock; but this is thought unfit for fine work, because the spring ought to be tighter in the cock than is possible if it is to be capable of being pulled down by the weight of the pendulum. But I rather doubt that, if it is made carefully.

Compensation of pendulums.—All substances of which pendulums can be made expand by heat, and consequently every pendulum naturally goes slower in hot weather than in cold; and though the lengthening of the rod is far too small to measure, except by most delicate experiments, it is enough to make a difference of a minute a week between moderate winter and summer heat (40° and 70°) with a common iron wire pendulum, and in five days with a brass one, and a minute in three weeks even with a wooden rod, which varies the least of all materials, but is subject to a little uncertainty from absorption of damp. The following is a table of the expansion of such materials as can be used for pendulums, *i.e.* of $\frac{dl}{l}$ for 1000° (F) of heat, which I use to get rid of unnecessary decimals, as we are only concerned with the proportions :—

White deal, and perhaps other woods, have ·0023 expansion, and specific gravity from ·5 up to ·9.

				lbs.
Flint glass expands	.	·0048 : a cubic inch	=	·1166
Platinum . .	.	·0056	,,	·76
Steel rod, tempered	.	·0064	,,	·28
Cast iron . .	.	·0066	,,	·26
Iron rod . .	.	·0070	,,	·28
Brass . . .	.	·0100	,,	·30
Copper . .	.	·0106	,,	·32
Silver . . .	.	·0115	,,	·38
Lead . . .	.	·0165	,,	·41
Zinc . . .	.	·0160	,,	·253
Mercury, in bulk,	.	·1000	,,	·49

its expansion in length depending on the vessel.

These differences make it easy to devise a pendulum which will always keep the bob at the same height. To take the most simple example, suppose we set a brass tube 6·4 ft. long on a nut at the bottom of a steel rod of 10 ft.; it is evident that the top of the tube, or a bob resting on it, will stay at the same height in all temperatures. This would be a very inconvenient sort of pendulum, but we can do the same thing by breaking up the long lengths into pieces, none of which must be more than 3 ft. for a 39 inch pendulum, and consequently a steel and brass pendulum must have three lengths of steel and two of brass. Accordingly the old *gridiron* pendulum, invented by Harrison, who became the first chronometer maker of his time from a carpenter, had a central steel rod with a cross at the bottom carrying two upright brass rods as columns, on the top of which was another cross piece from which hung two steel wires, again carrying a cross at the bottom, on which two more brass columns stood, and from the top of them hung two more steel wires carrying the bob.

Another form of it, invented by Troughton, was to substitute alternate brass and steel tubes for each pair of rods. If iron were used instead of steel, another pair of each would be required, making thirteen rods altogether.

But since the introduction of zinc-working brass pendulums have gone out, because zinc expands enough to do the compensation with only one zinc tube, and is stiff enough, which lead is not. It might also be done with platinum and brass, but that is too expensive, and zinc with steel or iron does just as well. The figure at p. 44 shows the best construction of large pendulums of this kind, and that of small ones is substantially the same, only short zinc tubes may be much thinner in proportion than long tubes, which have to carry heavy weights, acting as a column which must run no risk of bending. For small pendulums the zinc need not be more than ·1 inch thick, but for 2-sec. pendulums, with bobs of suitable weight, the zinc tubes are made three-fold, all drawn tight upon each other, and altogether $\frac{3}{8}$ in. thick. The tubes for somewhat lighter bobs and shorter pendulums may be half that thickness, or even less. No rule can be given for such things. For 1-sec. pendulums a steel rod $\frac{1}{4}$ in. thick is enough, but for a heavy bob I prefer $\frac{3}{8}$ in., as it has to be tapped for screws at each end. I will give some further directions about construction presently.

The calculations for the proper length of zinc are not quite so simple as you may imagine, except as a first approximation: in fact, if you attempt it all at once it is much too complicated to be practicable. We have seen at p. 51 in effect that the transfer of a very small weight from the lower part of a pendulum to the upper produces a great acceleration, and this happens in the rising of the weight of both tubes under expansion, yet it cannot be taken into account in any simple calculation. I suppose it is from this cause that the expansion of zinc, given in all other tables as ·017 for 1000° of heat, gives wrong results for these calculations, as is proved by experience. I find ·016 is the nearest value to produce right practical results. You must remember, too, that the point which we have to keep at the same level is no measurable one, such as the centre, or the bottom, or the top of the bob, but the c. o. of the whole pendulum, and we know nothing *a priori* of its relation to the top of the tubes or the total length of the pendulum, by which I always mean down to the bottom of the bob; for all these have to be found by trial. In the usual astronomical, or 1-sec. pendulum of this kind, it happens that the c. o. of the whole nearly coincides with the c. g. of a bob of 9×3 or 3½ or 4 inches; but in large ones with heavy tubes it comes much nearer to the top.

Again, an iron bob may be considered as attached to the iron tube anywhere, as they both expand equally; and therefore, as fixed at the c. o. And a lead bob may in fact be fixed at the c. o., and now generally is in the best pendulums, in order to avoid the slower changes of temperature in a thick bob than in tubes. But in commoner ones it is hung on a flange or collar at the bottom, and then you get the benefit of the expansion upwards of about 4½ in. of lead (up to the c. g. or c. o.) in addition to the zinc tube, which therefore has to be shorter than in the two other cases. And further, if the rod is steel its expansion is at a less rate than of the iron tube.

Bearing in mind, then, that any practicable calculation is only approximate, we may proceed with them as follows: Let l (as usual) be the length from the top of the spring to the c. o.; r the length to the bottom of the zinc tube, which rests on the nut at the bottom of the rod; z the zinc tube to be found; s the iron tube, which may or may not $= z$, as just now explained; c the height from the bottom to the c. o. Then for a steel rod, and a lead bob resting on the bottom, we must have—

$$\cdot016\, z + \cdot0165\, c = \cdot0064\, r + \cdot007\, s \quad . \quad . \quad . \quad . \quad \text{A.}$$

But now $s = z$, assuming all the bottoms to coincide, as they always do nearly; and we know by experience c to be about 4·5 in. for a 9-in. bob, and r about 44 in a 1-sec. pendulum, and we may introduce them at once, bearing in mind that the moment you translate one algebraical letter into inches all the rest become numbers of inches only. Then transposing all the unknown and known terms to opposite sides, equation A becomes—

$$(\cdot 016 - \cdot 007)\, z = \cdot 2816 - \cdot 0742 = \cdot 2074,$$

which gives $z = 23$; but it is safer to begin with the tubes a little longer, in case the compensation is not found quite enough, as they are easy to shorten, but impossible to lengthen.

When the bob is fixed, or may be considered fixed, at its c. o., the equations will be, first, for a steel rod and lead bob fixed at its middle,

$$\cdot 016\, z = \cdot 0064\, r + \cdot 007\, s = \cdot 0064\, r + \cdot 007(z - c) \quad . \; . \; . \quad \text{B},$$

and for iron rod and bob, $\cdot 016\, z = \cdot 007\, (r + 2 + c)$ . . . C. Taking the same values as before for r, &c., B will give 28·3 in. for z; which, however, is more than the length of zinc in the Greenwich sidereal pendulum, adjusted to mean solar time in order to compare it with others. I had my own made full 28 in., and I have no reason to think it is over compensated, but I cannot say I have tried it specially for compensation. The Greenwich one, with $z = 26$ more agrees with the tabular expansion ·017 of zinc, but not quite.

The Westminster pendulum, I know by its weekly rate for a year, is exactly compensated. It was made at first for the ·017 expansion, and was soon found to be under-compensated, and we had to make new tubes. If you insert in equation C the lengths given for that pendulum in the table below, you will see that ·016 almost exactly agrees with the ascertained results, c being of course $= r - l$. All the heights of bobs in this table are reckoned to the top of the dome, and my own, the third in the table, at Batch Wood, St. Albans, is slightly domed at the bottom also, and the weights are those of the whole pendulum approximately, as nothing respecting compensation turns on them. All the measures are inches.

sec.	*l*	*r*	*z*	*bob*	Weight of pendulum.
1¼	61	68·5	48·5	14 × 8	200 lbs.
1½	88	98	68	15 × 9	300 ,,
1½	88	96·5	61	14 × 8 lead	300 ,,
2	156·5	173	125	20 × 12	700 ,,

The tube should have some holes or long slots (which look better) on each side, to allow the air to reach the zinc, so that both tubes may change their temperature simultaneously. At Westminster the zinc weighs 67 lbs., and the iron tube 40 lbs., and the rod 62 lbs., out of the 700 lbs. total weight.

That is no longer the heaviest pendulum in England, for Mr. Godman, a clockmaker, organ-builder, and bicycle-maker at St. Albans, has put a 2½-sec. pendulum to the clock which he made for St. Peter's church there with a bob of 9 cwt., which he made in pieces called 'shifters' in clock weights, piled on each other like cheeses, to make it manageable without machinery, which is practically as good as one piece. And that has lately been exceeded by a 2-sec. pendulum of 12 cwt. in St. Nicolas's Church at Bristol, by Mr. Langford there. It is an odd coincidence that both those makers had the boldness also to put the gravity escapement a long way off the clock, because for local reasons it could not be near the best place for the pendulum, and with a gravity escapement that may be done very well. I need hardly say that thin iron rods on each side of the zinc tube may be used instead of the iron tube if you like.

In one or both those places there was this good reason for so separating the escapement from the clock—and there may be the same in others, especially where new gravity escapements are put to old clocks—that the clock itself could not be fixed firmly enough to carry the pendulum without yielding; the importance of which has been noticed at p. 45, and will be proved farther on. But there was no difficulty in fixing the escapement with the pendulum on iron brackets bolted to the wall.

Some persons appear to be misled by the well-known necessity for a secondary or auxiliary compensation in chronometer balances, which will be explained afterwards, into supposing that no *single* compensation of a pendulum can be complete.* But they depend on different principles, and the errors of an uncompensated balance are twenty times those of the commonest iron wire pendulum for the same variations of heat, being 6 sec. a day for 1° of heat, according to experiments. I may as well however explain a little more about the matter. The complete mathematical calculation for compensated pendulums does exhibit a very small and insignificant error, which is never taken account of in the ordinary calculations, and I only mention it to say that it is so small, and is not the defect which the persons I refer to have imagined. It is the effect on the moment of inertia, and therefore on the time of the pendulum, of the increase of length of all the rods and tubes, which of course have all some weight of their own. The moment of inertia of every such piece, say of length z inches and mass Z (which is z times the mass of one inch long of such rod or tube), of which the c. g. is h from the suspension, must be then $Z\left(h^2 + \frac{z^2}{3 \times 4}\right)$; and the moment of inertia of the whole is the sum of all those quantities in the numerator of the fraction $\frac{\Sigma mr^2}{\Sigma mr}$ or l. And to each of them has to be added its own increment for heat, viz. for the rod and the two tubes, and strictly even for the increase of length of the bob, and the resulting one for h in each case; and the

* See *Horological Journal*, April, 1882.

squares of all these appear in the numerator, and their first powers in the denominator. This is much too complicated for calculation except by successive approximations.

It is quite true that if a pendulum is not compensated at all, its errors increase more slowly than the temperature, because the time varies only as the square root of the length, while the length varies as the temperature. But that does not the least affect the question of how it behaves when it is compensated, so as to keep some point in the bob which practically agrees with the centre of oscillation, or is at l from the top at the same level. I say 'practically agrees,' because it need not be very exact. For if the c. o., which is at least approximately known, in the bob is kept at the proper level, it is clear that every other point pretty near it will be so too, since all the motions are absolutely very small, and of course their differences are smaller still. And that is the answer to this imaginary difficulty. I may anticipate the balance compensation so much as to say that besides the 20 times greater effect of a given change of temperature thereon, the rate of a balance depends on an external cause, viz. the varying strength of the spring, enormously more than on the mere variation of the moment of inertia, on which a pendulum rate depends, where also it is partly corrected by gravity acting farther from the vertical as the pendulum lengthens; *i.e.* both the numerator and denominator increase, but the numerator increases most, as it contains squares. Gravity has nothing at all to do with a balance, if it is balanced, as of course it should be.

Wood and Lead.—The same is the principle of the simplest of all the compensation pendulums. According to the tabular rates of expansion of lead and of a deal rod, it is easily calculated that a bob 14 in. high on a rod 46 in. long ought to make a compensated 39 in. pendulum; for c. o. is about ·4 in. below c. g. of the bob when it is 14 in. long, so that the rod must be 39·14 + 6·6, or nearly

46 in., and then we shall have $6.6 \times .0016 = 46 \times .0023$, which is right. Such a pendulum however was found to be overcompensated, and at first one is inclined to conclude that the expansion of deal has been overrated; but this was with a dead escapement, and I am satisfied, for reasons which will appear farther on, that such escapements contain a sort of irregular compensation of their own; and I advise no one to rely on a shorter proportion of lead bob than the above for a wooden rod for other kinds of escapements such as I shall describe.

Wooden rods are generally and rightly used for large clocks where compensated pendulums cannot be afforded; but such clocks are exposed to great changes of temperature —often as much as 40° between winter and summer, and to 30° on the average of each; which would make a difference of 20 sec. a week with an uncompensated wooden pendulum; whereas good public clocks ought to be guaranteed to vary not more than 5 or 6 sec. a week at the utmost—a much easier condition than 1 sec. a day, because one day may partially correct another. The Westminster clock only varies 1 sec. a week on the average, omitting casual disturbances by men and thunderstorms. Wooden rods should be as thin as will bear proper fastening at the ends, which may be thicker for that purpose. Any thickness which could be practically used and worked will bear ten times the weight of any bob that would be used. They should also be thoroughly dried and saturated with oil or some antiseptic fluid, and varnished to keep them from absorbing damp. Small ones are sometimes gilt; but even then they are not to be relied on, and I believe all the best makers agree that they are unfit for the highest class of clocks, though they will do for all but the highest. The late Mr. Vulliamy used to use mahogany and teak instead of deal, but I do not know whether it was from experience, which alone can determine such a question. I should think that the creosoting process would be a good one for wood pendulums; but that also wants trying.

Wood, zinc, and lead compensation.—There are places which invite the use of longer pendulums than can well be made of iron and zinc only, because a very long zinc tube requires to be very thick to be safe against bending. If wood can be made trustworthy by creosoting it in a vacuum (as they do railway sleepers), or in any other way, a very good long compensation pendulum may be made, with a zinc tube only about a third as long as an iron rod requires, and only a fifth of the whole length of the pendulum. Suppose we want a 2½ sec. pendulum, of which the simple length is 244·6 in., with a cylindrical lead bob, say 13 × 10 in., which will weigh 400 lbs. The lead itself will give us a little compensation due to the height of the c. o. above the bottom of the bob. The problem can only be solved by successive approximations, and they are too long to go through here; but if you try the following figures you will find them come as near as any pendulum can be made without adjustment, especially as there is some uncertainty about the expansion rate of wood, which I have taken according to the above table, and also taken its specific gravity at ·1 of iron for simplicity or a cubic inch = ·028 lb., for a heavyish wood, such as teak or mahogany. Take the total length to bottom of bob 253 in., of which about 240 in. will be wood rod, with a section of 1¼ in. diameter, and it will weigh about 7½ lbs. First let us see what length of zinc tube this will require, supposing the bob to be hung by a pair of thin steel rods from a cross head resting on the top of the zinc tube. There should however be a smaller collar fitting the top of the tube, with a slightly rounded top for the cross head to lie on, as there would otherwise be a risk of its not bearing square on the tube, but all on one edge. The expansion of so much of the bob as lies below the c. o. of the pendulum will give us some compensation; and as we have assumed that to be the difference between 244·6 and 253, we have 8·4 × ·0165 = ·138, for the expansion of the lead upwards; and 240 × ·0023 = ·5552 for that of the wood downwards. Calling the zinc tube z, we have ·017z upwards. The steel rods will be quite 2 in. longer than the tube, and there are 13 in. of fastenings and spring; so that we must add at least (z + 15) ·0064 downwards. Equating the downward and upward expansions, you will find that z must be 50 inches.

Smeaton's pendulum.—I have a clock with an old 1-sec. pendulum by Holmes, a celebrated clockmaker of the last century, with the following compensation, which was invented by Smeaton, the great engineer. The rod is of brass, 43 inches long + 2 inches of steel spring, and on a collar screwed to the bottom of it rests a thin zinc tube 12⅔ inches long, from the top of which is hung an iron tube of the same length, by the end being merely turned over, and on the bottom of that tube turned outwards rests a lead bob also of the same length, enclosing the tubes, so that the pendulum looks simply like a glass rod with a lead bob. The c.o. of the pendulum is 6 inches from the bottom of the bob, and therefore the expansion of 6 inches of lead + that of the zinc tube upwards has to balance that of the glass rod and the iron tube downwards, as you may calculate from the table that it will very nearly.

Another form of steel, zinc, and glass pendulum has been proposed, with a glass tube instead of the iron one, which would require a shorter length of zinc; but I see no advantage in it over the common one.

Compound Bar compensation.—Before going to the mercurial pendulum, which requires more detailed examination, we will dispose of a few others which require shorter notice. This one (fig. 10) is founded on the principle of the compensated balance in watches. W C W is a compound bar of brass and iron brazed together with the brass side downwards. As brass expands more than iron, the bar will bend upwards as it gets warmer and carry the weights W W with it, and they may be so adjusted both by magnitude and position as to raise the centre of oscillation as much as the elongation of the pendulum rod lets it down. I cannot give any details of the proportions, as this plan does not appear to have been used in England except in experiments,

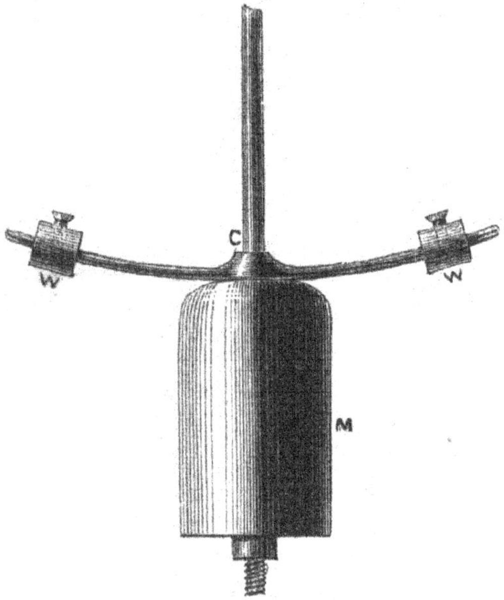

Fig. 10.

and old Mr. Dent told me it was found inferior to the other methods, over which I do not see that it has any advantage except the facility of adjustment. It is said in Kater's treatise to have answered in France, but the French seem to prefer complicated compensations to simple ones. It might however be used to complete the adjustment of another compensation left imperfect for that purpose, in which case the weights would only need to be very small.

Homogeneous compensation.—All the compensations I have yet mentioned require the use of at least two substances of different expansibility; but there is one which does not; and though it is practically of little or no value, in any form in which it has yet been made or suggested, it is worth while to describe the principle of it. If the pendulum spring is drawn up through a close slit in the cock as much as the pendulum lengthens by heat, its effective length will remain the same; and this may be done by various arrangements, of which the simplest is hanging the spring to another cock above the slit one, set on the top of a stiff bar long enough to expand upwards as much as the pendulum rod expands downwards. This bar may be either of the same or of a different metal from the pendulum rod, and its length will of course depend on its material; or the slit cock may be on one end of a lever, of which the other is pulled down by a wire of the same length and metal as the pendulum rod. But there is a serious objection to this plan in every form; if the slit is loose enough to let the spring slide up and down through it, it is too loose for the proper action of the pendulum in a clock good enough to require a compensation for temperature. There is no other kind practically worth describing.

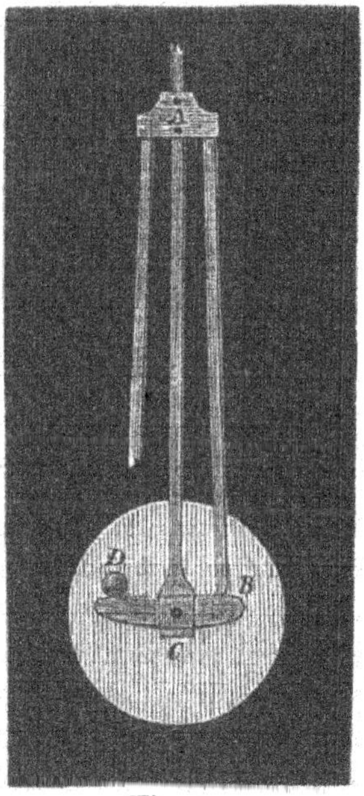

Fig. 11.

Ellicott's pendulum.—There is another bad one which is still used in small French clocks, though it has long ago been abandoned in England, where it was invented by Ellicott a clockmaker in the last century. A C the main rod is of iron or steel, and it has a pair of levers set on a cross pin at the bottom, of which only one B C D is shown here: at A there is a strong collar fixed to the rod, and between that collar and the short arm of each lever stands a stiff brass rod. The bob hangs by two pins, of which D is one, on the long arms of the levers; and it is evident that by a proper adjustment of the levers the bob may be made to rise under the expansion of the brass rods just as much as the expansion of the iron rod lets down C, the axis of the levers. But this action involves considerable friction at D, and the pressure on the ends of the brass rods must very much exceed the weight of the pendulum, and it is said to move by jerks, and is altogether inferior even to the old gridiron, and much more to the zinc or mercurial pendulum, besides being much more difficult to make properly. I suppose they are only made because they have a kind of scientific look to ignorant people, in clocks made to show and to sell.

Mercurial compensation.—The principle of this, which is used in all the best astronomical clocks (subject to a remark to be made afterwards), is the same as of the wood and lead; only, mercury being fluid requires a different mode of calculation. For it must be in a jar which has some sideway expansion of its own, and the rise of the mercury is only the excess of its expansion in bulk over that which the increased width of the jar allows it. The jars can only be made of either glass or iron, as mercury amalgamates with and so destroys any other metal that could be used. Although iron is the best, glass is the most commonly used, either from old habit or ignorance, and so we must consider both of them, and we will take the glass first, and in both cases neglect at first the weight of the jar itself: which by a strange oversight was forgotten altogether by Francis Baily, P.R.A.S., in a far more elaborate paper on this subject (in vol. i. of their Memoirs) than there is any need of, seeing that after all a compensated pendulum can only be accurately adjusted by trial, of itself or one exactly like it. It is almost equally strange that this mistake was never discovered for forty-four years, though the best clockmakers, and the late Mr. Bloxam, who most thoroughly investigated the theory of escapements in two papers in the 22nd and 27th vols. of the R.A.S. *Memoirs*, had practically found that the received height of mercury was insufficient. But they all attributed it to the pendulum spring being stiffer in cold than heat; which always appeared to me a most inadequate explanation, considering the weakness of the spring in proportion to the weight of a pendulum. In 1863 I had to calculate the height for the iron jar of a pendulum intended to weigh 40 lbs., and finding the result far beyond Baily's proportions, I looked carefully through his paper and then found out the mistake, and published the correction of it in the *Mechanics' Magazine*, of 5 Feb. 1864. Another reason why it had escaped discovery was that dead escapements, which nearly all astronomical clocks had, contain (as I said before) an irregular kind of compensation of their own, from the greater fluidity of the oil in warm weather. It is remarkable that Hardy, the inventor of another kind of escapement which I shall notice afterwards, had calculated the height of his mercury rightly before Baily's paper which everybody accepted.

Mercury expands ·1 in bulk while the glass jar increases ·0048 in diameter, and ∴ the sectional area increases as $(1{\cdot}0048)^2$ or 1·0096 to 1; ∴ the mercury will rise ·1—·01 (say) or ·09 of its own height. And since the c. g. of the bob is very nearly at O, let x be half the height of mercury; then ·09 x must $=$ ·0066 $(l + x)$ for the steel rod, which is usually carried down to the bottom of the jar in the form of a long 'stirrup,' on which the jar stands. That makes x $=$ 3·1 in. or the height of the mercury 6·2 in., which has to be increased ·1, or rather more for a reason which will appear afterwards, and that is Baily's result—without considering the weight of the jar.

The weight of the glass jar and rod may be a 6th of the ultimate or corrected weight of the mercury; and about so much more height must be added for it, as it is evidently much the same as if all the bob of pendulum were mercury, but its rise only $^5/_6$ of its real amount. And approximately this is near enough, seeing that pendulums vary in their proportions and must be finally adjusted by trial; and glass jars are going out of use for the best pendulums, which it is now agreed ought to be much heavier than the old two-inch wide cylinder allows.

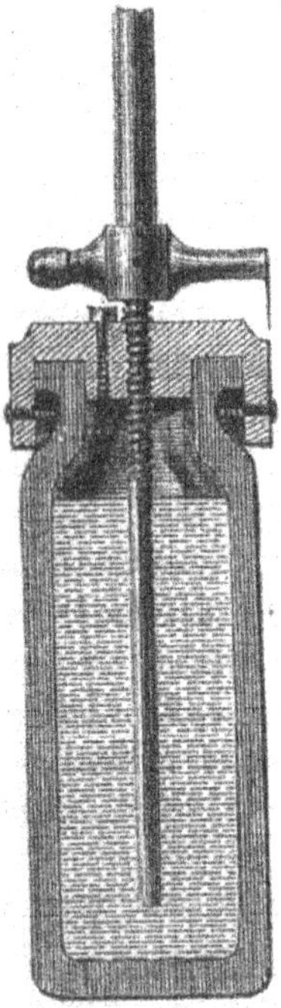

Fig. 12.

The late Captain Kater used for his experiments entirely glass pendulums, which of course require a less height of mercury than steel ones. There is however no advantage in that, but rather the contrary, as the jar must be inconveniently wide to hold any considerable weight, besides the difficulty of handling a heavy glass pendulum safely. Others have made the steel rod go through the bottom of the jar; but that is a bad plan, as it involves stuffing to make the hole mercury-tight, and allows no subsequent adjustment without risk of destroying that tightness. It is desirable however that the rod should plunge into the mercury, that they may take the same temperature, and the best way of doing this with a glass jar would be that shown in this drawing. The shape of the jar speaks for itself. The black ring round the neck is in two pieces of brass, put on with a piece of thin leather to improve the fitting, before the cap is dropped over it and screwed to them with two or three screws to each half of the ring. All other plans with the same object that I have seen involved some cementing to the glass or stuffing, which are not to be trusted. The small hole in the cap is for adjusting the quantity of mercury according to experience.

Cast iron jars are superior to glass, first because they are safer, and firmer on the rod, and the construction simpler when done properly; secondly, because the mercury can be heated in them to drive off any moisture or air bubbles between the mercury and the jar, as is done with barometers; thirdly, because (contrary to the common notion) the extra height required for the weight of the jar is an advantage, especially when you want a very heavy bob. Calling the jar about a third of the mercury in weight, a 40 lbs. pendulum will have a jar only 3½ in. wide outside; and one *holding* 40 lbs. of mercury will be 3½ in. inside, for we shall find the mercury to be nearly 9 in. high, and a cubic in. of pure mercury weighs half a pound, and common mercury weighs slightly less. The best construction, and much cheaper than that generally used till lately, is to cast the jar ¼ in. thick, and twice as long as it need be, in order to secure sound casting, as they do for cannon, cutting off the 'dead head;' turn the outside if you like for appearance, or else japan it, and 'enamel' the inside, *i.e.* coat it with glass, which stops up all the pores. The cap is partially dropped into the top of the jar just turned out to fit it and held by 4 or 5 screws through the jar. The rim is marked out into divisions of 1 sec. a day as described at p. 49, an index being fixed to the steel rod, which is tapped and screwed through the cap exactly as in the last figure, while the cross-head is to hold it steady when you turn the jar for regulating. A 40 lbs. pendulum made thus costs less than old Mr. Dent used to pay for the pendulum alone without the mercury, on the old construction. This is the construction now used by Mr. Brock of 64, George Street, Portman Square, who used to be with Dent, and has done all except church clock work for me for some years. Enamelling the jars was his idea.

For an approximate calculation like the last, we have only to use the expansion ·0066 of cast iron instead of glass for the jar, and that makes the rise of the mercury ·1—·0132 or ·087 nearly. And assuming the mercury to weigh 3 times the jar and rod, the rise must be $^4/_3$ times as much as if they had no weight, or we may treat the rise of the mercury relatively as ·065, or ¾ of its real amount. We know that the height must be something over 8 in., and the total length of the pendulum nearly 43½, allowing the c. g. of the bob to be a very little above O. It is not worth while to distinguish between the expansion of cast iron and steel for the short length of the cylinder, and we may say that the rise of the c. g. of the mercury, or the increase of x, or ·065x, must = 43·5 × ·0064 = ·2784, and ∴ x = 4·3 nearly. The assumption that x is something over 4 in., for the mere purpose of finding the total length of the rod, does not affect the result sensibly, as you will see if you try it as an inch more or less. But this is not quite enough, because it is not the c. g. but the c. o. that has to be kept at the same height, and we must find how much less h really is than l, for we have only yet stated it approximately. If b is the radius of the cylinder when the weight of the rod is insignificant, $l\,h = k^2$, or $= h^2 + \frac{x^2}{3} + \frac{b^2}{4}$, but b is so small compared with h that that term produces no sensible effect. Then putting 39·14 for l and either 4·3 or 4·4 for x, you will find by solving the quadratic equation that h = 38·98 or $l - h$ = ·16 in. = ·036x: ∴ we must say ·964 × ·065x or ·0626x = ·2784 as before, and ∴ x = 4·44, which makes the height of mercury just under 9 in. in a jar which weighs ⅓ of the mercury. I found mine not quite enough compensated with 8½ in., and it is perhaps a little overdone with 9—a widely different result from Baily's 6·6 for an iron jar. The calculation of compensation otherwise than by this method of approximation runs into equations of a high order and becomes unmanageable, and after all it is better to adjust the height of the mercury finally by trial. The longer the bob is, or the heavier the rod and compensation tubes, the more the c. g. of the bob (not of the whole pendulum) is below c. o. A very long bob is not desirable on account of the increased resistance of the air, and therefore it is not expedient to make the jar more than about a third of the weight of the mercury, which is in every way a convenient size.

It is now thought at Greenwich that zinc and steel pendulums are as good as mercurial ones. The normal sidereal clock there has a zinc pendulum, and I am told that its daily variation of rate is reckoned by hundredths of a second. If this is so, it will be a considerable saving in expense, inasmuch as there is no doubt that pendulums ought to be much heavier than used to be the fashion. I must say that the performance of the great Westminster clock goes far to confirm the Greenwich opinion, for it has no discoverable error of temperature.

Barometric compensation.—It is evident that the larger the bob is for its weight, *i.e.* the less its specific gravity, the more it will be resisted by the air, and the smaller arc it will swing under a given force. We shall see afterwards that all the escapement errors involve $\frac{Wh}{Ml}$ divided by either a^2 or a^3, Wh being the force that actually drives the pendulum for a day, or weight × daily fall, after deducting the friction, and Ml the same as before, and a the arc from zero as usual. Therefore the larger a is *for a given weight* the better a great deal. Moreover, the greater the density of the air is, the more it diminishes the effective specific gravity of the pendulum, viz., by an amount = sp. g. of the air, which is variously estimated at something near the 840th of that of water, and therefore about a 9000th of an ordinary lead pendulum with deal rod, or a mercurial one with a glass jar and the usual appendages, which make its mean sp. g. about 11, water being 1. And this reduction is considerably increased in a swinging pendulum by the air which it drags along with it. Baily found, by comparing the times of free pendulums of various kinds in vacuo and in air (see *Phil. Trans.* of 1832), that the stationary loss of sp. g. is sometimes doubled by vibration. One specially curious result should be noted, that a thin flat rod with a very elliptical section was more affected in this way than a round one three times as thick, although a lens-shaped bob was less affected than a sphere of the same diameter and of course much heavier in proportion to its surface, which so far gives it an advantage. But we have no information whether the air affects a lens more or less than a sphere of the same weight.

Bloxam said in a note to his paper in the R.A.S. *Memoirs* of 1853 that Baily overlooked the fact that the current produced in the descent of the pendulum prevents it from being retarded in the ascent as much as it would have been if the air had been at rest. He also always found the circular error to be less than its theoretical value, and the resistance of the air doubtless tends to produce this effect.

Without going through Baily's various results it is enough to say that he found pendulums of sp. g. about 11 gain nearly 13 sec. a day by being put into a vacuum; as that was the daily increase of time (+ Δ T or — 'rate') for an inch rise of barometer, which is shortly called the barometric error, apart from any circular or escapement error which may accompany it. A small platinum ball hung by a wire gained 5 sec. in vacuo, against 9½ for a lead one of the same size (for a sphere is less affected than a cylinder); the gain was nearly 14 for a brass and 56 for an ivory ball, and 19 for a round copper rod ·41 in. thick and 5 ft. long; which last was 3 times as much as its stationary loss of specific gravity. But this effect on the rod as a whole pendulum is insignificant when it carries a bob much heavier than itself, as it usually does.

Baily found also by calculation that with an arc of 2° 45′ the barometric error ought to be neutralized by the circular error, *i.e.* by the diminution of arc produced by the increased density of the air. That result differs from the one arrived at by Bloxam, but was verified by the performance of the Westminster clock during the year 1872, in which I ascertained that the barometric error does not exist, *i.e.* is exactly neutralized, by the fact that applying any correction, either + or —, bearing a constant proportion to the actual variations of the barometer, would have increased the small average variation of the clock. I mention as a fact, without professing to account for it, that Bloxam found the barometric error of his clock less with an arc of 90° than with a considerably larger one. But I am decidedly in favour of large arcs, especially for large clocks exposed to greater variations of force than small ones, and I adopted one at Westminster before I knew that it agreed with Baily's calculation, which it has so remarkably verified.

In order to avoid the barometric error, Mr. Carrington adopted the plan of a perfectly air-tight clock case, the winder going through a 'stuffing-box,' and then he exhausts the air down to some given pressure a few inches below the usual height of the barometer (see R.A.S. *Notices* of Nov. 1872). But this is much too troublesome for common use. Dr. Robinson, of the Armagh Observatory, long ago adopted the much simpler plan of compensating the error by attaching a pair of very thin barometers on the right and left sides of the pendulum rod, above the bob (R.A.S. *Memoirs*, vol. 5). This evidently has the effect of raising a small weight of mercury from near the bottom of the pendulum to near the top when the density of the air increases, and *vice versâ*. But I shall presently point out an inconvenience of this mode of doing it, and Dr. Robinson expressly says that the adjustments were troublesome. The barometric error also varies considerably even between clocks of nominally the same kind; so much that it is not safe to assume any given amount beforehand, but it must be ascertained from observation of that particular clock before the calculations are made for correcting it. In the best kind of astronomical clocks, with detached, or with gravity escapements, but not dead ones, we may take the daily loss ($+ \Delta$ T) for the present at 0·3 sec. a day for an inch rise of barometer, or that is the barometric error so far as it is unconnected by others which attend it.

In a paper on this subject in the R.A.S. *Monthly Notices* of 1873 I remarked that two barometers are unnecessary, because a slight want of symmetry in (not across) the plane of vibration cannot affect the pendulum. Moreover, with a bob of the length required for an iron jar mercurial pendulum there is not room to put the barometer tube entirely above the jar. What we have to do is to find the position and the diameter $2x$ of the barometer which fixed at some given height will thus compensate a 39 inch pendulum of given weight, say 40 lbs. It is tempting at first sight to make the great mercury jar serve as a basin for the barometer; but there are various practical objections to it, and it is much better to use a syphon tube bent over the side of the jar, in which therefore a 'rise of barometer' of 1 in. will be only an absolute rise r of ½ in. in the longer leg, with a fall of ½ in. in the shorter. Let the mercury in the long leg reach to d below the top of the pendulum, and in the short one to b above the centre of oscillation: ρ, the specific gravity of mercury, is such that a cubic inch of mercury is very nearly half a pound. Then, by the same mode of calculation as at p. 50, the little weight transferred from b to d being $\pi\rho x^2r$,

$$-\Delta T = \frac{43200\pi\rho x^2 r}{M}\,\frac{ld - d^2 + b^2 - lb}{l^2}\text{ sec.},$$

or putting ¼ for ρr and 40 for M we must have (nearly)

$$850\,x^2\,\frac{ld - d^2 + b^2 - lb}{l^2} = {\cdot}3.$$

Now d must be substantially $>b$ for this to keep its sign, and also to keep tolerably constant for different degrees of rise. For simplicity let $b = 0$, and $\therefore d = 9$ in.: this gives $2x$ or the diameter of the tube ·03 in. If the barometric error is 3 times as much, $2x$ must be $\sqrt{3}$ times as much or about ·05; in either case rather a thermometer tube than a barometer one. The tube, after being bent over and down either the right or left side of the jar, had better be brought up to the rod again, and the two branches joined there by heat for strength, and tied to the rod by waxed thread rather than any metal fastening. If the compensation is found to be overdone the tube has only to be untied and raised, and if insufficient it may be lowered, if room enough has been left for it in the bend.

As the barometric error never lasts long in one direction it will very seldom be thought necessary to apply the compensation to large clocks; and we have seen that it can be otherwise materially reduced, if not absolutely corrected, as it is in the Westminster pendulum, by making it swing a large arc. If it were necessary to apply the barometric compensation to a long pendulum the best way would be to put an annular basin for the mercury on the top of the bob, round the compensation tubes, and have a straight barometer dipping into it. It may easily be calculated that the same barometric error of ·3 sec. a day with a 1¼ sec. pendulum of 200 lbs., and also with a 1½ sec. pendulum of 300 lbs., would be corrected by a barometer tube of about a sixth of an inch diameter. The reason why the same tube is enough for the heavier pendulum is that the rise of the mercury is in a more effective place, *i.e.* near the middle of the pendulum. A 13 ft. pendulum of 700 lbs. with the lower surface of the mercury 6 in. above c. o. requires a tube of ·3 diameter for the barometric error of ·3 sec.: another odd coincidence of figures. If the barometric error in any particular clock is found to be 2 or 3 times as much as this, $2x$ or the diameter of the tube must be $\sqrt{2}$ or $\sqrt{3}$ as much as the above figures, and it must vary in the same way with the $\sqrt{\text{weight}}$ of the pendulum. We shall have to consider this further when we come to escapements.

The normal sidereal clock at Greenwich has a barometric compensation of a more complicated kind. An independent fixed barometer raises and lowers a magnet which attracts the pendulum more or less according to its position. I must say that seems to me a roundabout way of doing the business, and I should have been inclined first to try a large arc of $2^\circ\ 45'$ like that of the Westminster clock, and if that is not sufficient, then the simple barometric tube compensation, with the bottom of the tube near the bottom of the bob.

RECOIL ANCHOR PALLETS.

The next important invention which followed that of pendulums, and that very soon, was a pair of anchor-like pallets moving in the plane of the scapewheel, instead of the 'vertical escapement' with pallets set across a crown wheel, which pallets being very short required a long swing of the pendulum to let the wheel escape. It is not indeed absolutely necessary that crown wheel pallets should be very short, and they would go with less friction if they were long and the teeth of the wheel few; but the recoil would be more violent, and they would require more careful adjustment, and as a matter of fact they always were short. Anchor pallets in the form in which they were first invented, either by Dr. Hooke, whom I have already mentioned as one of the claimants of the pendulum, or by Clements, a London clockmaker of his time, had a recoil, no less than the crown wheel pallets, but they could be made to escape at as small an arc as you please. Fig. 13 is a drawing of the recoil escapement, as it is always called, which is still used in all the common clocks in the world, though it has long been abandoned in all that make any pretension to great accuracy.

Fig. 13.

In this figure the tooth *a* has just escaped from the left pallet A, and *b* has dropped on to pallet B; the pendulum is therefore moving to the left, and it will not stop immediately but will go on a little farther and so make the wheel recoil a little, as you may see clearly enough in any old-fashioned house clock with a seconds hand: as it returns, the wheel urges the pendulum again to the right and so gives the impulse which is necessary to maintain its motion against the resistance of the air and the friction of the escapement itself; and then tooth *b* escapes, and the tooth below *a* falls on A and the same action takes place there. You observe that I have drawn the acting faces convex. For some theoretical reasons they ought to be concave; but, as very often happens in clockwork, the one practical reason of friction preponderates the other way. Even if they are made flat, the teeth always wear holes in them, though the teeth are of brass and the pallets of steel, made as hard as possible, and it is evident that the friction at the recoil is much greater if the pallets are concave than convex. Moreover it is always found that as the pendulum arc decreases from any decrease of force in the clock, it loses, and *vice versâ;* and concave pallets would not diminish this error, but increase it. Some considerable persons stuck to this escapement for some time after the dead escapement was invented, being apparently misled by the fact that variations of force produce less variation of the *arc* in this than in the dead escapement, because the friction of the recoil checks the arc; but it does not follow that the variations of *time* are less: in fact it has been proved both by experience and calculation that they are not.

I may as well mention here that a 'steady rate' means, not necessarily that the clock is going *right*, but that it is going *uniformly*, or regularly gaining or losing exactly the same number of seconds a day or a week. The rate is always written + when the clock gains, and — when it loses; which you must remember is just the opposite of the signs appended to the expressions for the various errors of the time in the mathematical formulæ; for when t the time of vibration of the pendulum increases, dt is +, and the clock loses and the *rate* is too little or —.

Harrison's recoil escapement had scarcely any friction. It was invented by him when he was a working carpenter in Lincolnshire. Any one who wants to see a description of it will find one in the 7th edition of the *Encyclopædia Britannica*; but I did not think it worth while to repeat it in the 8th; nor here, as it is a mere obsolete curiosity, and nobody else ever made it to answer, even before better things were invented. If such a thing were worth doing, it might be done much more simply by a three-legged scapewheel, such as I shall describe as one form of the dead escapement (at p. 102), but without the horizontal or dead pallets there, and with the impulse pallets set deeper and not in the same line, though parallel to the pendulum. The impulse would be given with nearly as little friction as it is there, and the recoil would also have very little; and if the force on the scapewheel were made uniform, as it may be by a contrivance which I shall describe for turret clocks, there is no reason why such an escapement should not go very well, though the dead one would go better.

Clock out of beat.—Most people seem to know that the beats of a clock ought to sound equal in time, but most people have a very erroneous notion that this depends on the clock being set on a perfectly level surface, or standing vertically; whereas that has nothing at all to do with it, unless the crutch has been so adjusted that the pallets do escape at equal angles when the clock case stands upright. No doubt they ought to be so adjusted, because a clock looks better standing straight than crooked. In the best clocks the beat is made adjustable by beat-screws at the bottom of the crutch or fork. In common ones it is simply bent by hand till the beats sound equal. Mr. Dent made the fork pins in turret-clock escapements to open with a spring, to prevent the teeth being damaged in case they should be caught by the escaping corner of the pallets when the clock is put back, or the pendulum set going without the clock being wound up. Each 'prong' of the fork must have a separate spring, both set against a stop between them. The crutch and everything attached to the pallets ought to be kept as light as possible, because they are in fact a pendulum, moving on pivots instead of a spring, and therefore with much more friction than the real pendulum. But a long crutch is better than a short one, because less angular motion and force is lost in the looseness or 'shake,' which, as I explained at p. 46, must be left between the fork and the pendulum. The proper way to try whether a clock is in beat is to let the pendulum swing only just far enough for the escape, and then you will easily hear if the beats are unequal.

When a clock with any kind of anchor escapement (which all clocks may be assumed to have, unless the contrary is known) sounds 'out of beat,' it wants either one side lifting or the crutch bending—which you please. If the right-hand beat of the pendulum (*i.e.* the blow on the left-hand pallet) comes too quick, the right pallet escapes too soon or does not go deep enough. Therefore the right side of the clock wants lifting; for that is equivalent to moving the pendulum and pallets to the left. Or it wants the fork or bottom of the crutch bending to the right, which, remember, is making a bend *in* the crutch to the left, for that carries the pallets to the left. Or if there is a beat screw in the fork, it wants turning to the right or 'setting up,' so as to move the crutch to the left. And *vice versâ*, if the left beat comes too quick, the left side of the clock wants raising, or the fork bending to the left, or the beat screw turning to the left.

I shall notice afterwards, under French clocks, a contrivance which some of them have to make the beats gradually adjust themselves. Recoil pallets—and dead ones too—should only just clear the teeth, and when the points get worn so as to allow more clearance the pallets should be renewed.

DEAD ESCAPEMENTS.

These are also of the anchor kind; but in fig. 14, as in fig. 13, a tooth of the scapewheel has just escaped from pallet A and another has just fallen on to pallet B; but the face on which it falls is very different, as the shape of the teeth also is, but that is only to give room for clearance, and in both cases it is only the point of the tooth that acts. Here you see the face of the pallet is divided into two, and one part is an arc of a circle of radius C B while the other part is much the same as it was in the recoil pallets. The consequence evidently is that there is no recoil here, and the tooth lies *dead* on the circular face B E (which is therefore called the dead face) however far the pendulum may swing, until it returns to B, and then the tooth begins to act on the impulse face and gives the impulse to the pendulum exactly as in the other escapement. In the same way the

Fig. 14.

dead face D G of the other pallet is part of a circle with radius C D: the dotted line being the old recoil shape.

This is the dead escapement, which was invented early in the last century by Graham, who invented also the mercurial pendulum and the horizontal watch escapement; and the great advantage of it is that although a slight increase of force on the scapewheel increases the arc of the pendulum, it does not sensibly increase the time, if the escapement is properly made. This was first demonstrated mathematically by Sir G. Airy, in the *Cambridge Philosophical Transactions* in 1827 (Vol. III. p. 105), and it may be made tolerably clear without any mathematics, as follows. If each tooth could drop exactly on the corner dividing the dead and impulse faces of the pallet, as at B, it is clear that the impulse on each pallet would begin and end when the pendulum is at the same distance on each side of zero or the vertical. In other words, as much of the impulse would be given while the pendulum is falling as while it is rising, and therefore its gravity would be increased on one side and diminished on the other through equal arcs and during equal times, and therefore on the whole the accelerating force on the pendulum, and therefore its time of vibration, not altered. Neither would the friction on the dead faces, if it were constant throughout, and if it acted through the same arcs before and after zero, affect the time directly; for while the pendulum is falling the friction acts contrary to gravity, and with gravity while it is rising after the escape has taken place. As it is always resisting the motion of the pendulum it tends to diminish the arc; and on the other hand, the impulse always tends to increase it; so that here also there is counteraction to some extent; but as the friction on the pallets does not vary in any definite proportion to the force of the train, but sometimes one way and sometimes the other, no useful relation of this kind can be established. All we can say about the arc is that it increases under an increased force or a diminished friction, until the remaining friction on the pallets and the resistance of the air stops the increase.

Mr. Bloxam came to the conclusion, in his elaborate papers on escapements in vols. 22 and 27 of the *Astronomical Society's Memoirs*, that so far from the dead friction being a thing to be disregarded as constant and not materially affecting the rate of the clock, as Mr. Airy had assumed, it probably affects it more than all the other escapement errors together in an astronomical clock with even a moderately good train, and in a way which it is impossible to calculate on account of the different circumstances under which it varies. It is evidently very improbable that the friction can be made the same while the point of the tooth is (as we may say) ploughing its way up the pallet during the ascent of the pendulum and sliding down it in the descent. Having said this I shall not attempt to deal any farther with the dead friction, but its existence must be borne in mind as capable of either mitigating or increasing the other errors, as the case may be; and some idea of the magnitude of its effects may be formed from this, that I remember the arc of a new church clock of Mr. Vulliamy's increasing from 2° 30′ to 3° 30′ in the first year, from no other visible cause than the self-polishing of pallets. And it afterwards fell off again from the pins of the heavy scapewheel wearing hollows in the pallets where they dropped.

Now let us apply the same reasoning to the recoil escapement, and we shall find that the result is just the opposite from that in the dead. That part of the impulse which is within the points where the teeth fall on the pallets is the same as in the dead escapement, and therefore may be taken not to affect the time; but in the ascending portion of the recoil the force is acting with gravity, and so it is in the descending portion. Whenever the force of gravity is increased the time of a falling or swinging body is diminished; and therefore on the whole any increase of the force of the clock in the recoil escapement tends to make the pendulum go faster. But the recoil resists the pendulum in rising as much as it impels it in falling, and by means of the friction resists it much more; and as the force through the other portion of the arc, corresponding to the impulse in the dead escapement, tends to increase the arc, they may so nearly balance each other that an increase of clock weight may produce no visible alteration in the arc at the very same time that it is, as we may say, knocking the pendulum backwards and forwards more rapidly between the same limits; whereas the dead escapement just sends it so much faster as to make the whole vibration take very nearly the same time while it has to pass through a longer space—subject to the following modifications.

The first of them is this: the teeth cannot safely be made to drop exactly on the corners of the pallets, but must have a little of the dead face to fall upon; and therefore the angle or arc of impulse before zero must be rather less than that after zero, and therefore the tendency to increase the time preponderates. For the same reason there is necessarily rather more of the dead friction in the descent of the pendulum, where it acts against gravity, than in the ascent, and so that also tends to slowness. The greater this difference is, or the higher up the dead faces the teeth drop, the greater these causes of error are; and yet it is very seldom that one sees a dead escapement whose maker appears to have had the least idea of this fact; for I suppose if they had, they would not make them as if they thought the right thing was for the tooth to fall as far over the dead face of the pallets as possible, instead of falling as near the corner as possible.

Another modification is the circular error, which I have already explained, and which acts in the same direction as the one last mentioned, increasing the time with the increase of the force and therefore of the arc.

But there is another of the opposite kind; for when the

whole arc increases, that portion of the arc and of the time during which the impulse or disturbance of the natural time of vibration takes place becomes less in proportion to the whole, and that diminishes the increase of time which would otherwise be caused.

And yet further, we have hitherto assumed that nothing varies except the force of the scapewheel teeth on the pallets, and the friction which is due thereto. But the friction on the pallets may and constantly does vary even more than the force of the clock, and generally in just the opposite direction; for as the clock and pallets get dirty together, the force on the pallets decreases, which accelerates the time, while the friction increases, which retards it; and so on the whole it is by no means certain which result will preponderate in the natural state of the clock, or by how much; and the only certain way to get a steady rate out of a dead escapement clock is to take as much care as possible to keep the force and the friction constant; which is only to be done, and can be done in small clocks with light wheels, by very accurate workmanship, highly finished and hard acting surfaces, and keeping them clean and just sufficiently oiled, and above all a good pendulum, properly fixed. It is necessary for this also that the two pivots of the great wheel should be as nearly equal as possible, even at the cost of making the back one larger than it otherwise need be. I have frequently observed the arc to be sensibly greater at that end of the week at which the string is at the back end of the barrel, and therefore the weight acting principally on the thin pivot. When a pivot has to be used for winding, it must be made much thicker than is required for a mere pivot, and I find it an improvement to let both the pivots rest on friction wheels as large as there is room for. They may both be on one arbor, but the front one must ride loose, as they will not have quite the same velocity under the different-sized pivots. The holes should be lengthened a little downwards to make the pivots rest on the friction wheels.

Something more precise than the above general reasoning is of course necessary for actually measuring the different elements of variation of rate. This is what Mr. Airy did, or rather laid the foundation for doing, in the paper I have already referred to; and taking it up at the point where his calculations ended and his inferences began, I carried it farther in two papers in the *Cambridge Transactions* in 1848 and 1852 (vols. 8, 9). His calculations are rather too long and complicated to insert here, and they may be found in *Pratt's Mechanics* and perhaps some other Cambridge books; so I will take them up at the same point here.

The first important mathematical result arrived at by Mr. Airy was this: If ϕ is a disturbing angular force on a pendulum when it is at the angle θ after zero (reckoned + when it tends to increase θ), and a the extreme arc, then the increase of time of one vibration due to the disturbance, which we will call

$$\Delta = \frac{l}{\pi g a^2} \int \frac{\phi\, \theta\, d\theta}{\sqrt{a^2 - \theta^2}},$$

this integral being taken between the limits through which the disturbing force acts. He also found a corresponding formula for the increase of arc; but that is of no use towards ascertaining how much the arc really will be increased by the continual action of the disturbance, as it is soon limited by friction and resistance of air in a way which cannot be calculated.

Before we can make any use of this value of Δ we must see what ϕ is in the particular escapement. In order to do this let us call the angle which the impulse face of each pallet makes with the dead face δ; then, since the tooth, taken as a prolonged radius of the wheel, ought to be a tangent to the dead face, δ will be also the inclination of the tooth to the impulse face at the beginning of the impulse; and for this purpose we may assume it to continue the same throughout, though in fact it increases a little.* Let p be the distance of each pallet from their arbor, and Pg the moving force of the clock-weight as it arrives at the points of the scapewheel teeth, Ml the mass and length of the pendulum (which we may treat as its equivalent simple one for this calculation); then the equation of motion will be

$$\frac{d^2\theta}{dt^2} = -\frac{g\theta}{l} + \frac{Ppg \tan\delta}{Ml^2},$$

since g tends to decrease θ after zero where it is $+$, and ϕ, which represents the other term in the equation, does the contrary. And as this term is independent of θ, we have

$$\Delta = \frac{Pp \tan\delta}{Ml\,\pi a^2}\int\frac{\theta\, d\theta}{\sqrt{a^2-\theta^2}}.$$

The limits between which the integration is to be taken are from $\theta = -\beta$, where the impulse begins before zero, to $+\gamma$, some rather larger angle where it ends after zero; and the result will be

$$\Delta = \frac{Pp \tan\delta}{Ml\,\pi a^2}\left(\sqrt{a^2-\beta^2} - \sqrt{a^2-\gamma^2}\right)$$

Now before going any farther we may see at once from this, that if β could be made $= \gamma$, *i.e.* if the impulse could be made to begin just as much before as it ends after zero, Δ would $= 0$, and we might save ourselves all further trouble, and pronounce the dead escapement perfect, or capable of being made perfect, so far as the impulse is concerned. But it seems impossible to make $\gamma - \beta$ much less

* If κ is the motion of the wheel on the pallet during each beat, the angle of the tooth with the pallet face will be $\delta + \kappa + \beta + \gamma$ at the end of the impulse on the down pallet, and $\delta + \kappa - \beta - \gamma$ on the up pallet: κ cannot be more than $5°$ in a 30-toothed wheel, and $\beta + \gamma$ is never more than $1\frac{1}{2}°$; and as δ is generally about $60°$, the variation is too small to affect these calculations materially.

than 30′, and in fact it is seldom made so little. And it will not do to say that as 30′ only = ·00085, that still leaves Δ very small, and so the clock must go very well; for it must be remembered, first that Δ only means the difference of time of a single vibration out of the whole number T in a day; and T is 86400 for a seconds pendulum; and further, that Δ is not after all the error of the clock between one day and another, but only the difference between the time of a pendulum swinging freely, and one kept going (or in mathematical language *disturbed*) by a clock escapement; and therefore we shall have to go one step farther and find the variation of Δ itself before we can know anything about the going of the clock, since Δ cannot be made = 0. But before we do that it will be convenient to examine the value of it as it stands.

Let h be the daily fall of the clock weight Wg. The drop of the tooth at each beat, or the space through which the moving force Pg acts, ought to be nearly = the thickness of the pallet $= p\,(\beta+\gamma)\tan\delta$; and this × T (the number of beats in the day) × Pg would $= Wgh$, but for the loss in the friction of the train and the slight difference between the actual drop of the tooth and its theoretical drop, which is the thickness of the pallet. For the present purpose it is of no consequence whether W is a little more or less, and therefore we may neglect this difference and consider

$TPp\tan\delta = \dfrac{Wh}{\beta+\gamma}$; and therefore the excess of time of the clock pendulum over the free pendulum in the day, or

$$\Delta T = \frac{Wh}{Ml\pi a^2}\,\frac{\sqrt{a^2-\beta^2} - \sqrt{a^2-\gamma^2}}{\beta+\gamma}.$$

This is by no means a pleasant expression to deal with in a general way; but it is not difficult to draw the necessary conclusions from it, by assuming some particular values of the different quantities, in accordance with what they usually have in clocks. Although the error of the clock is not represented by ΔT, but by the variations of it, still if we can make ΔT very small, the variations of it in the same escapement will be smaller still; although it might happen, and in another kind of escapement does happen, that the variations are the least when ΔT is at a maximum. But there is no fluctuation of that kind here; and before going any farther we may observe at once the advantage of a long and heavy pendulum, seeing that Ml is always fixed in the denominator of the expression for the rate of the clock due to the escapement. Moreover it is a fact that a long and heavy pendulum requires very little (if any) more force Wh to drive it than a short and light one; the reason of which is that the chief impediment to the motion of the pendulum is the resistance of the air, and the resistance to the surface does not increase in anything like the same proportion as the weight, if the bob is of a good shape. Therefore Wh in the numerator need not be materially increased for a very great increase of Ml in the denominator, which is directly equivalent to a very great increase in the accuracy of the clock; and you will see that the same is true in the other escapements.

The next point to consider is the length of the arc of impulse $\gamma+\beta$. As we have already seen that $\gamma-\beta$ ought to be made as small as possible, but cannot be made less than about $30'$, let us take that difference as fixed, and see whether the whole angle of impulse $\gamma+\beta$ ought to be large or small compared with α. For simplicity of calculation, let us try the effect of making α, γ, and β in the proportions of 8, 7, 5, and 8, 5, 3, and 8, 4, 2, keeping $\gamma-\beta$ constant, you observe. Substituting those values then in the above formula for ΔT, you will find that it comes out in the proportions of 20, 14, and 13 in the three cases respectively. In other words, it is better *not* to make γ nearly $=\alpha$, or the impulse to last nearly to the end of the arc; and it is satisfactory to know that this agrees with the conclusion which old Mr. Dent came to from observation, though it is contrary to the practice of most other clockmakers, who seem to prefer that sleepy-looking kind of escapement in which the second-hand moves very slowly and the 'excursion' of the pendulum beyond the impulse is very little. He altered the transit clock at Greenwich from a long to a short impulse accordingly with good effect.

Taking it then as proved that γ the angle of impulse after zero ought not to be nearly so large as a, we shall be able to simplify the above value of ΔT by assuming that $\frac{\gamma}{a}$ is so small that all higher powers of it than the square may be omitted. And then $(a^2-\beta^2)^{\frac{1}{2}}-(a^2-\gamma^2)^{\frac{1}{2}}$ may be expanded by the binomial theorem, and only two terms of each expansion need be taken, and the equation will assume the simple form.

$$\Delta T = \frac{Wh(\gamma-\beta)}{2\pi Mla^3}.$$

Mr. Airy concluded that the smaller $\gamma+\beta$ is, the better, because it appeared in the numerator of his expression for ΔT. But he left the expression for the force unreduced into the terms involving the shape of the pallets, in which we saw that $\gamma+\beta$ was lying hid, ready to come out in the denominator of the fraction; and therefore, as it is also in the numerator, it disappears altogether, on the assumption that γ is moderately small, as it ought to be; and beyond that the above expression for ΔT gives no information as to the proper size of γ or the length of the impulse.

But then another consideration comes in. If you make the impulse very short, the pallet will slip away before the tooth has time to catch it; and the shorter the angle of impulse is the more of it is lost by the inertia of the wheel (which therefore ought to be light): in fact this is another reason, besides the necessity for leaving a little of the dead face for the tooth to fall upon, why the angle β at which the impulse really begins cannot help being sensibly less than the angle γ at which it ends. Therefore also there is no use in making the pallet corners sharp, for the tooth cannot follow the pallet immediately, and it had better slide off a slightly rounded corner than drop on to the impulse face with a kind of jump off a sharp corner. On the whole the result appears to be that the escape had better take place at something near 1°, and consequently the impulse should not begin later than 30′ before zero, assuming a the extreme arc to be 2°, which will make

$$\frac{\gamma-\beta}{2\pi} = \frac{1}{720}, \text{ and therefore } \Delta T = \frac{Wh}{720\ Ml\ a^3}.$$

The value of $\frac{Wh}{Ml}$ varies very much according to the quality of the clock: in the best astronomical clocks it may be taken to be as little as $^1/_{30}$ to make the pendulum swing 2°. As the force which arrives at the pallets is now represented by W, we must treat it as variable together with the arc; and so, differentiating ΔT, we shall have

$$d\Delta T = \frac{1 \text{ sec.}}{21600\ a^3}\left(\frac{d\,W}{W} - \frac{3\ da}{a}\right) = 1^{s}\!\cdot\!2\left(\frac{d\,W}{W} - \frac{3\ da}{a}\right).$$

If there was any definite relation between the ratio of increase of the force and the arc, this would give a very easy calculation for the variation of rate so far as the impulse is concerned. But there is not, as it depends on the state of the different parts of the clock. Sometimes it may happen that the proportionate decrease of the arc from increased friction is just ⅓ that of the force which arrives at the escapement, and then there will be no variation in the rate. Sometimes you may increase the clock-weight considerably without making much impression on the arc, if the pallets are dirty; and generally in an artificial experiment of that sort, except while the arc is smaller than 2°, $\frac{da}{a}$ is likely to be less than $\frac{d\,W}{3\,W}$, and then the clock will lose, even independently of the circular error which tends the same way, and which we know would be 10800 $a\,da$ if it were not in great measure corrected by the pendulum spring, though it is very difficult to say how much. On the other hand if you clean and oil the pallets alone the arc is sure to increase, and yet the clock will generally gain, because the increase is chiefly due to the diminution of the dead friction, which (as I explained before) would diminish the time, independently of the term $-\frac{3da}{a}$ belonging to the effect of the impulse, which retards less on a long arc than a short one.

Half-dead escapement.—In order to counteract the disposition of dead escapement clocks to gain as the arc decreases under ordinary circumstances (which, you remember, is the opposite of what happens in the recoil escapement), Berthoud a celebrated French clockmaker invented the plan of making the dead faces not quite dead, but with a slight recoil, so as to get a sort of compromise between the effects of the two escapements. Large clocks, which are subject to great variation of train force, are distinctly better when so made than with quite dead pallets. Moreover the variations of the arc are rather checked by the half-dead pallets. The largest variations of arc I ever saw in a good clock, were in one of Mr. Vulliamy's, who used to take particular pains to make his pin-wheel pallets quite dead by cutting them out of a turned cylinder of radius equal to their distance from the arbor. A very slight recoil, such as you can hardly see in the motion of the wheel, is enough. But the best authorities are of opinion that a purely dead escapement is better in astronomical clocks, where the friction and variations of force are much less than in turret clocks.

Loseby's isochronal spring.—Another plan for isochronising the long and short arcs was invented by Mr. Loseby a chronometer maker in London, and exhibited in 1851. A large circular loop of very thin steel wire is set on a stud from the back of the clock case, say on the right side of the pendulum, so as to embrace the rod nearly half-way down, just catching it as it swings to the left side of the loop. The farther it swings of course the more it has to stretch the loop, and the resistance increases in a high ratio with the degree of elongation; and it seems that this can be adjusted so as to isochronise the pendulum in a dead escapement under great variations of the force of the train. So at least the Astronomer Royal reported after trying some experiments on it. But it at once occurred to me that such experiments proved nothing as to the effect of such a spring on an astronomical clock in its natural state, in which the variations of the pallet friction are generally greater than those of the train and produce the opposite effect, as is evident from the second term of the equation at page 90, and then the spring would make it worse. As chairman of the Horological Jury of the Exhibition, I wrote to this effect to Mr. Airy, who then made a different class of experiments, this time by artificial variations of the pallet friction, and he issued a fresh report in 1853, that 'Mr. Loseby's invention was *not* perfectly successful.'

Large and small arcs.—There is one more point in the theory of dead escapements which requires particular attention. You observe that a^3 appears in the denominator of the expression for the variation of impulse rate, and so it would in that belonging to the dead friction. That is, the three resisters of disturbance of the rate are the weight of the pendulum, its length, and the cube of its arc. But the arc in any given clock in its normal state of friction varies in some irregular way with the force of the train, *i.e.* the clock-weight and its fall; and an increase of the weight in any given ratio may or may not increase the arc in the cube root of that ratio: but so long as the arc is small, its increase will most likely be a great deal more than that; and if so there is a clear advantage in increasing the weight; subject always to this memento, that the circular error also increases with the arc. I have a note of having once cleaned a regulator and oiled it with sperm oil, and the arc increased from 1° 45′ to 2° 15′, and yet I found no material alteration in the rate. I am satisfied that the very small arcs which used to be the fashion are a mistake, and that from 2° to 2° 40′ is much better. A writer in *H. J.* xix. 125 said that, in consequence of this advice, he had increased the weight and arc of an Austrian clock, which had almost reduced its errors from minutes to seconds.

Importance of firm fixing.—Whatever increases the arc without increasing the weight is obviously a great advantage; and the principal things which do that are diminution of friction and inertia of the train, and steadiness of suspension of the pendulum. I cannot give a better proof of how much the arc depends on that, than the effect of hanging the Westminster pendulum on its proper cock, which is a large cast iron bracket built into the wall; the arc increased full 45′ over what it had been in the factory, where it was hung on what seemed a perfectly firm support, a strong timber frame built up from the ground. Even smaller pendulums generally increase their arc from about 2° in the factory to 2° 30′ as soon as they are properly fixed to a good wall on stone corbels or iron beams. This shows the extreme badness of the common way of fixing large clocks on a stool or timber frame set upon a wooden floor in a tower, and common clocks by a single nail through a thin back of the case (see p. 45).

The friction is of course only to be diminished by proper shaping of all the acting surfaces and making them of the best metals for working together. Brass wheels and steel pinions, and also brass teeth and steel pallets seem to be the best in small clocks, although there are other cases where steel and steel act better together, as in the horizontal watch escapement. In large clocks cast iron wheels and pinions suit each other better and wear less than anything else, as has long been known by the great machine makers, though scarcely any clock-makers choose to believe it, and of course refuse to try, being what they call 'practical men,' who understand by the word *experience* the constant use of one thing or one way of doing it and absolute ignorance of any other. Mr. Vulliamy used to think steel scapewheel teeth or pins better than brass ones, and they have been occasionally used by other people, but I think are now generally disused. At the same time it is certain that steel pins do best in my escapements which will be described presently. In the best clocks the pallets have jewels, generally sapphires, let in for the teeth to act upon, and it is quite ascertained that brass teeth suit them best.

When the pallets are steel it is scarcely necessary to say that they ought to be as hard and as smooth as possible; especially the former, for if they are not smooth at first the teeth will make them so in time, but soft ones will never get hard. They are hardened like files, by being heated red hot and cooled suddenly, and not tempered at all. The sharp hollow corners, which are considered by ignorant people a sign of fine work, are apt to crack in hardening, and as such a corner is always a weak place besides, they ought not to be so cut out; and the same remark applies to every hollow corner in every part of a machine, unless something else has to fit into it. Probably it is best to heat the pallets in lead melted red hot, and cool them in oil, which is now adopted for some larger steel things, as I have seen pallets twisted in the ordinary mode of hardening. The same may be said of pivots and pinions, except that they are tempered and not left quite hard.

Aluminium bronze.—The alloy of copper and aluminium, to which this name is given, seems to me very superior to either brass or gunmetal for many horological purposes. It is stronger, much more elastic, smoother, far less liable to tarnish, *i.e.* to decay; and for small articles, such as the scapewheels of clocks, and all the wheels of chronometers and watches, the excess of the cost over brass would be insignificant. It solders well with either common 'silver solder,' or another with less silver in it. Mr. J. F. Stanistreet, of Liverpool, an amateur clockmaker, told me he had used it; and it is used for cheap watch cases.

Weight of scapewheel.—It is important to keep the upper wheels in the train, and particularly the scapewheel, as light as possible. It is certain that every blow you hear in the working of a machine indicates some loss of force, and wearing out of surfaces, and that the machine would be better without it—unless it is a hammer; and the heavier the blow is, of course the worse it is. In clock escapements a sudden stop and a blow of some amount is inevitable; but there is no reason why it should be increased by making the scapewheel three or four times the necessary size and weight, and the drop of the teeth more than is necessary to clear the pallets. I have already mentioned also that the greater the inertia of the scapewheel, the longer it is in effectively catching the pallet, although you cannot see the interval. I have seen church clocks, not merely old ones like that of St. Paul's Cathedral, but new ones, with scapewheels a foot in diameter, and weighing several pounds, and you may sometimes hear the thump of every beat in the churchyard. Bloxam calculated that a pendulum of 15 lbs. does not require half as much force as a marine chronometer, if a great deal of it were not wasted in having to start a heavy train from rest at every beat; otherwise its weight would be of little consequence. Mr. E. D. Johnson has made regulators with very light and small trains, in fact, little more than chronometer trains, and he says they answer perfectly.

I may observe here that the rims of wheels (in which most of the inertia lies), and indeed the whole wheel, may be made materially lighter with 5 spokes than with 4, because the arcs are shorter. Sometimes, in the most expensive clocks, they are made with 6; but 5 spokes leave very little more than $^1/_6$ of the rim open, on account of their own thickness, as you will see in p. 99; and they seem to me quite close enough for clock-wheels; and of course every unnecessary spoke adds unnecessary work and expense; that number is used throughout the Westminster clock and many others now. Here too, as in the pallets, and indeed in every possible place, the modern workman who is taught to think 'high finish' the perfection of work, or in other words, to display as much finger work and as little head work as possible, files or 'crosses out' the corners as sharp as possible, instead of leaving them rounded a little, which would make the wheel stronger with no appreciable increase of weight. I believe it would be a very good rule that a sharp hollow corner ought never to be allowed anywhere, unless something has to fit into it, as it always makes a weak place, in which, if anywhere, things crack in casting, fly in hardening, or break in working, and moreover is so easy to do, that as a proof of good workmanship it is contemptible, even if it were not really bad besides.

Length of pallets.—A French clock-maker in the Exhibition of 1851 had an apparatus for illustrating the superiority of moderately short pallets over long ones. It does not require much apparatus to prove that; for assuming the scapewheel to be of any given size, it is evident that the farther the pallets are from their arbor the longer is the run of the teeth upon them, and the more friction there is affecting the pendulum. The usual proportion seems to be to make the distance of the pallets from their centre = the wheel's diameter (generally 1⅞ inches in regulators), and embracing 10½ teeth, *i.e.* from one dead face to the other.

This seems to me rather an unnecessary length, and I should prefer 9½, or half a tooth under instead of over one third of the number in the wheel.

Various rules and calculations have been given in books for the length and position of the pallets, according to the conditions which their authors thought most important. I think the most important of them all is to keep the dead friction a minimum, consistently with a sufficient length of pallets to prevent much force being wasted in clearance. The way to do this is to make the dead faces always perpendicular to the pressure of the teeth, *i.e.* to the circumference of the wheel. Assuming 10 tooth spaces, or ⅓ of the wheel, to be embraced from one under face B to the other A (fig. 14) as usual, *i.e.* 9½ spaces from one dead face B to the other D, the pallet centre C will be the intersection of two tangents to the circumference at B and D. That makes the distance of centres $d = 1{\cdot}84\,r$ (r the radius of the wheel), and p or C B and C D = $1{\cdot}54\,r$, which is quite enough. As one dead face is above and the other below, the lengths of the pallets as a whole are unequal, the appearance of which the clock-makers dislike, though it is of no consequence. If you determine to have them equal, and make $d = 2\,r$, as is usual, and B C and A C = $1{\cdot}73\,r$, the pressure will be perpendicular to the dead face B, but not quite so to D: if you make $d = 1{\cdot}72\,r$, and F C and D C each = $1{\cdot}4\,r$, the dead face D will be right and the other wrong; if d is anything between those sizes, one pallet will be rather better and the other rather worse, supposing the lengths to be equal.

The unimportance of equality of length of impulse on the two pallets is evident from the fact that in some kinds of escapements, as we shall see afterwards, there is only one impulse pallet, and there is no loss of force thereby. The only way in which force is lost, apart from friction, is in the drop of the teeth from the end of one pallet to the face of the other, or in other words, the clearance, which is necessary to prevent the point of the returning pallet from catching the top of the tooth which last escaped from it. Consequently the thickness of the pallets has to be rather less than half a tooth space. The slopes or impulse faces may be adjusted thus: Fix an index C P to the pallets, and put them and the wheel on centres without shake at the assumed distance on a board. First file down B till there is only just enough dead face to receive the tooth when the escape takes place at A, and leave the corner F at first obviously too sharp, *i.e.* rather too long, and A too blunt, or the corner G nearly square. Mark where the index points at the moment of escape, and mark 2° from that, which = ·035 C P = ¼ in. if C P = 7 in. File away F till the next escape takes place at that 2°, and then reduce G till that pallet has only just enough dead face for the tooth to fall on. This will make each escape take place at 1° from zero of the pendulum.

The teeth of dead scape wheels are often made of a most absurd shape, the weakest and heaviest possible, viz., with the sides radial, and therefore thinnest at the root, and with a heavy shoulder, and also often too long. The best form is a simple taper. What are called *club* teeth, with a lump at the end, were supposed to hold the oil better, which is apt to run away from the points, but they seem to be generally abandoned. In the best clocks the pallets are jewelled with sapphires, which require scarcely any oil.

Pin-wheel escapement.—There is a very convenient form of the dead escapement for large clocks, which goes by this name. It is said to have been invented by Lepaute in 1753, but also by Whitehurst of Derby. The teeth are pins of brass wire set in the face of the wheel, and the upper half of each cylinder cut off, as it could not act and would only waste room in the drop. But I introduced the plan of cutting off a small slice of the under or acting side also, as shown in fig. 15 (next page), because, unless that is done, you must either have the wheel very large, or the pins very thin, or long pallets, or a large angle of impulse, which are all objectionable. The advantages of this escapement are,

Fig. 15.

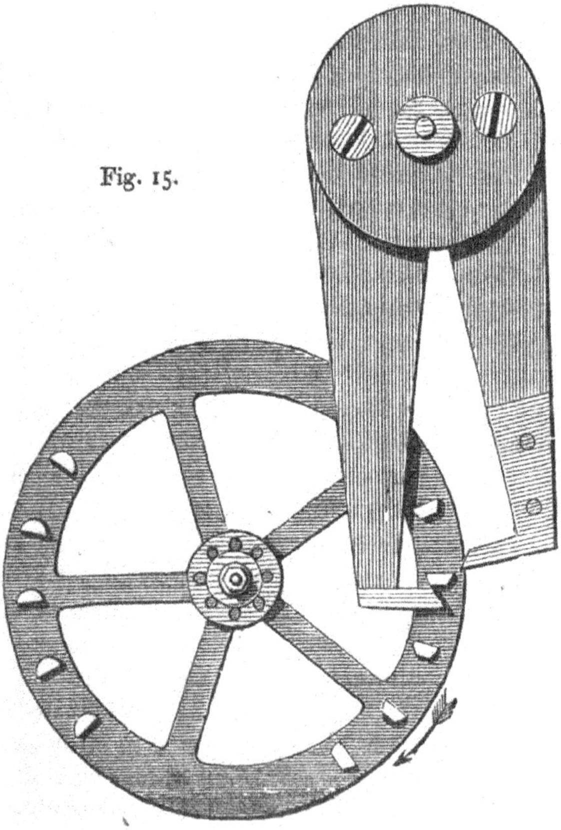

that it does not require so much accuracy of construction as the other, and less is lost in the drop, and therefore you can get many more pins than teeth to act in a wheel of given size, which often saves one wheel in the clock. If a pin gets damaged it is easily replaced, whereas if a tooth is damaged the wheel is ruined. The blow on both pallets being downwards, the action is more steady than it sometimes is in the other. The pallets are best made with their cross section rather convex, and also 'half dead.' The scape-wheel of the large clock at King's Cross, by which the Great Exhibition time was kept, and of many others made from my design, is only 4 inches wide, with 40 pins in it. The lower pallet should be the inner one, and the higher one outside the wheel, because this makes the action of the teeth on both of them more direct. If the pallets are on opposite sides of the wheel with two sets of pins, they may be alike. The pins must then be at alternate places.

Pin pallets.—Some of the best small French clocks now have an escapement which at first sight may be confounded with the pin-wheel escapement, but is really quite different. The scape-wheel is like a common dead one, and it is set (merely for show) in front of the dial; but the pallets are made of semi-cylindrical ruby pins; the effect of which is that the half-dead part of the action is not on the points of the teeth but on their faces. The impulse is the same as on common jewelled pallets, only with the faces round instead of flat. I have seen an old clock with similar pin pallets, made of a thick bristle held at both ends, for the sake of silence, and the fork also.

Single-pin escapement.—This very simple and neat-looking escapement has been several times re-invented. But I believe the only person who ever made it really succeed, from better attention to the proportions, was the late Mr. C. Macdowall, a very ingenious clockmaker, who died in 1872 at the age of eighty-two. There is an interesting memoir of him in the *Horological Journal* of Sept. 1873. But, like many other uneducated inventors, he was very difficult to convince that an *independent* inventor is not allowed by the world the credit of a *first* inventor unless he is so. He also invented that most useful instrument—the spiral drill; or rather I should say, he invented the practical mode of making it by twisting a piece of pinion wire, for the thing itself turned up in wood as an old Indian invention in the 1851 Exhibition. Another of his independent inventions was the 'helix lever wheels,' as he called them, but they also appeared in some German clocks in the same Exhibition; besides some other things which had been published in well known English books. He also persuaded himself somehow or other, that he invented the three-legged escapement which I shall next describe; but no one ever saw or heard of it until it had been made from my design, though he had shown me all his inventions, and I had helped him to get both a patent and an Exhibition medal for his single pin escapement, and old Mr. Dent, under my advice, had bought the patent for a considerable sum, and gave him an order for 500 watches on that plan if he could make or get them made, which he could not; and Mr. Dent accordingly got some made in Switzerland. I wore one of them long enough to see that it answered very well, but the expense of the two extra wheels which it requires overbalanced the advantages, though I hear they have

Fig. 16.

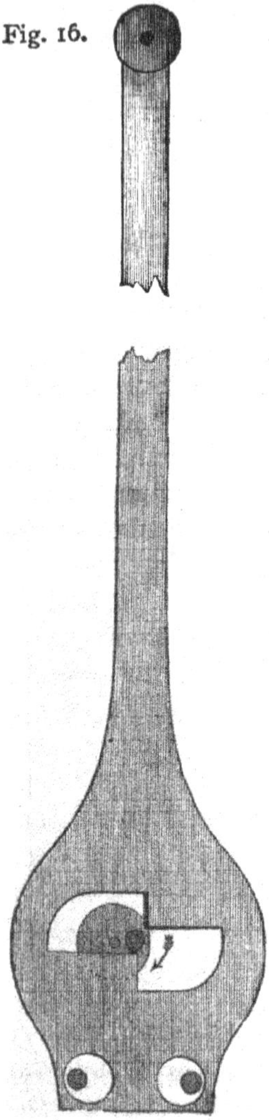

been also made in Paris from Macdowall's instructions, the patent not having been taken there, and being perhaps difficult to maintain here if it had been disputed; for the thing had certainly been published, as we found afterwards, in a French book. Its action is evident enough from this picture (fig. 16). It has the advantage of giving a great part, you may practically say half, of the impulse directly across the line of centres (of the wheel and pallet arbor), and therefore with very little friction. The pin is made of a ruby, and should be set very near the arbor. The older ones were put too far off. Macdowall made them with a 'neck-bearing,' *i.e.* the pivot was behind the wheel, or the arbor came through the frame, and had the wheel pinned or squared onto it. The pallet piece itself forms the crutch for the pendulum, the scape-disc being set behind the clock frame. It should have eccentric fork-pins to adjust for beat.

Three-legged dead escapement.—It occurred to me in 1851 that all the best or most direct part of the impulse in the single-pin escapement might be kept, the more oblique

Fig. 17.

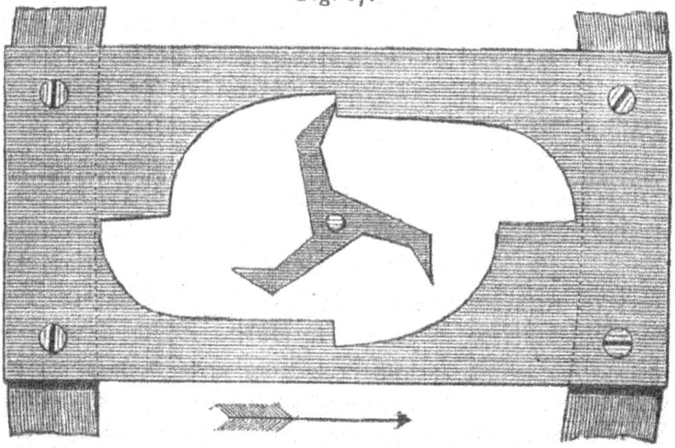

part got rid of, and one of the extra wheels saved, by using three pins or teeth instead of one; and the result was this escapement (fig. 17) (for clocks only, not watches), in which the upper tooth is shown in the act of giving the impulse. This is a *full-sized* view of the escapement which drove the Westminster pendulum of 6 cwt. for half a year, until it was superseded by another modification of it invented for the purpose of equalising the force of the impulse. That scapewheel was of steel and weighed only $^1/_6$ of an ounce or 73 grains, and the clock weight required for it with a common turret clock movement was only 18lbs. falling 6 feet a day, and less than a quarter of what a dead escapement had required; and considering that more force is lost by the inertia of such a train than in small clocks, we may say that the fraction $\frac{Wh}{Ml}$ was less than a third of its usual amount for the best astronomical pendulums swinging the same arc. I found also that it was possible to isochronise the long and short arcs—at least for such variations as actually occurred, by making the stopping or horizontal faces half-dead, as I have drawn them. The distance of the scapewheel from the pallet arbor should be about 24 times the radius of the wheel, to make it escape at 1°, allowing a little for clearance. It requires careful adjustment, for the greatest depth on the lower pallet is something less than $^1/_8$ of the radius of the wheel.

It is necessary in this and prudent in all dead escapements, of large clocks especially, to have a spring-fork to the crutch; *i.e.* the fork-pins attached to the crutch by springs, so that if the scapewheel is not turning from any accident while the pendulum is moving, and the pallets jam against the teeth, the fork springs may give way and let the pendulum go on without breaking a tooth, as it inevitably will unless so relieved. The pin-wheel escapement is specially liable to this, because the edges of the pins and pallets are not sharp as in the common toothed wheel escapements.

The friction might be still more reduced and the adjustment of the pallets made easier by making the escapement with long stopping teeth, as in fig. 18. It would also give room for a longer swing of the pendulum, which cannot safely be made above 2° in the other form, or the stopping faces will reach the scapewheel arbor. An escapement of this kind clearly reduces the pallet friction to the smallest amount possible in any dead escapement. It is however not free from the variations of force in the train.

Fig. 18.

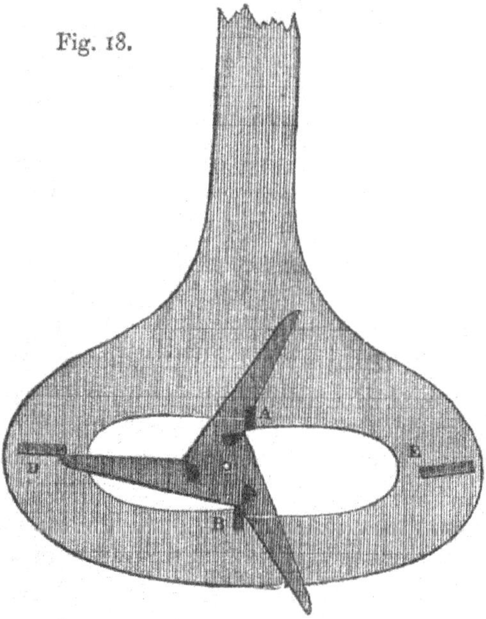

Constancy of force is only to be got by an entirely different class of escapements, or by the addition of a train remontoire, of which I shall speak hereafter under Turret Clocks.

Detached escapements.—There have been various contrivances for clock escapements on the chronometer principle of leaving the pendulum free or detached from the scapewheel except at the time of receiving the impulse and of unlocking the wheel. There is an old French one described in Rees's Cyclopædia, and Sir G. Airy invented another, which is described in his before-mentioned Cambridge paper, and has been lately adopted in the normal sidereal clock at Greenwich. A few of them had been made before by old Mr. Dent, with half-second pendulums, for which this escapement does not answer, as Mr. Brock, of George Street, Portman Square, has also told me. The Greenwich clock however with a seconds pendulum of about 30 lbs., appears to answer remarkably well, and therefore I copy the original picture, from which its action will be easily understood. The single pallet C P is exactly like the right hand or down pallet of a dead escapement. A B D is a detent attached to a fixed block at A by a slight spring, B being the catch which holds the scapewheel teeth or pins, for it does not matter which they are. (B D is the flat view of it, omitting the stop at B.) At D, on the underside of A B D, there is a very slight 'passing spring' projecting a little backwards, as shown in the separate sketch; so that when the pin at D on the pallet goes to the left it pushes the passing spring down, and so passes it; but when it comes to the right it lifts up the detent, because the passing spring being underneath cannot yield upwards. That unlocks the tooth at B, and lets another tooth at P fall upon that pallet, and give the impulse down its oblique face. The advantages which Sir G. Airy contemplated were, the abolition of the dead friction, and the power of making the impulse begin as much before zero as it ends after zero, the value of which we saw at p. 86.

Fig. 19.

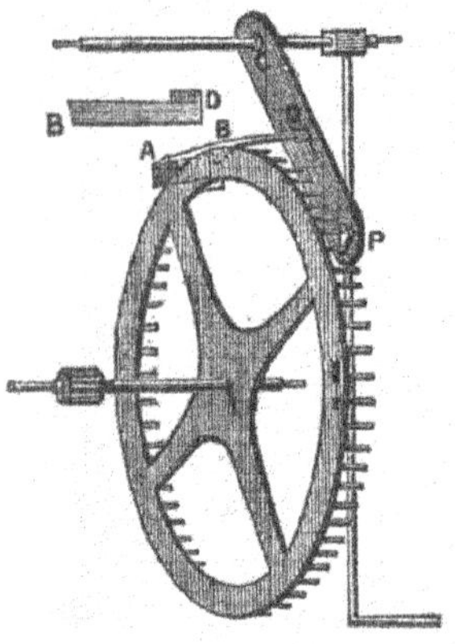

But I cannot help thinking the following a better plan, because it gives the impulse without any sensible friction, as in the three-legged escapement just described, and involves no passing spring, which is always a delicate arrangement. I had this clock made and set going in November 1865, and it went till August 1873, nearly 8 years, without being touched, even with oil, as I wished to see how long it would go. The arc decreased a little, and it had a slightly increasing gaining rate in the last few years; but I never had any clock go so well for anything like that time, though my gravity escapement clock, to be mentioned presently, has gone as well for shorter periods. The train is not a very fine one: with a fine train, or a coarse one with a remontoire for a turret clock, I have no doubt this escapement would go at least as well as any other. But since 1873 it has not equalled my gravity escapement clocks, and the gravity escapement is both cheaper and safer for large clocks than any train remontoire; which may therefore be considered obsolete.

Fig. 20.

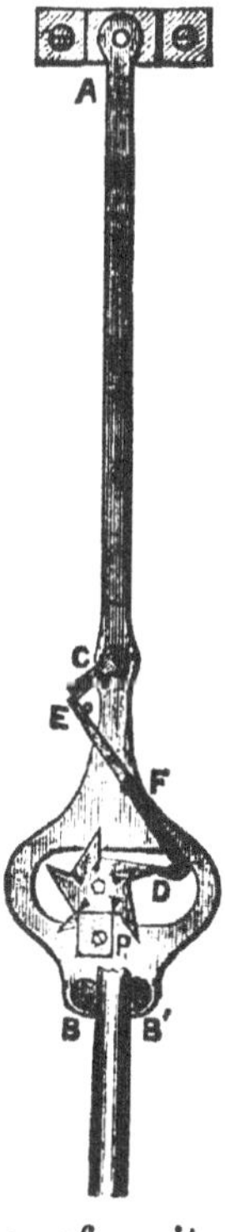

This figure shows the construction clearly enough, except as to size. The locking teeth of the scapewheel reach about ·6 in. from the centre, and the impulse pins ·25 in. Their centre is 7 in. below the top of the pendulum and crutch, and the impulse begins about 1° before zero, and ends 1° after. For the scapewheel turns 36° before and 36° after zero; and therefore, if r is the distance of the pins from the centre, and p the length of the crutch, $r \sin 36° \text{ must} = p \sin 1°$ for an impulse of 2°; which makes $p = 33{\cdot}7\, r$. Consequently the dimensions just now given are practically right, allowing a little for clearance, as you must. The pallet piece P is a separate piece of quite hard steel screwed on to the aperture in the crutch A P; and this requires care in adjustment, as the depth is less than $^1/_5$ of the radius of the wheel, being *r vers. sin* 36°; which however is more than in the three legs, for this turns 72° at each beat, and that only 60°, this being a single-beat escapement, which, it seems necessary to inform some persons, wastes no power if the train does not move at the return beat of the pendulum.

But for that it would be better to have no crutch, but make the action take place on the pendulum itself; it is hardly possible however to manage that behind a clock of regulator size, with the pendulum hung to the wall as it should be. The wheel lies behind the clock in a long cock embracing it and the crutch. The detent D E F is on a stud F in the back plate, though it looks here as if it were on the crutch, as the click C E really is. When the pendulum goes to the right (in this drawing) the click trips over the top of the detent, and when it returns to the left the click pushes the detent aside, and sets a locking tooth free, and then the scrapewheel turns and a pin gives the impulse. The shape of the pins should be observed. It will easily be seen that the scape-wheel should be at the bottom of the clock, and the great wheel at the top, as in my gravity escapement, and also in the dead three-legs described at pp. 102 and 104. B B′ are eccentric beat pins, by which the impulse can be made to begin as much before zero as you like, so as to make the angle which we called β greater than γ if necessary.

REMONTOIRE OR GRAVITY ESCAPEMENTS.

These are not to be confounded with the thing called a train remontoire, which I alluded to just now but shall not explain till we come to turret clocks, as it belongs only to them. A gravity or remontoire escapement is one in which the impulse is not given to the pendulum directly by the clock-train and weight, but by some other small weight lifted up, or a small spring bent up, always through the same distance, by the clock-train at every beat of the pendulum. And the great advantage of them is that the impulse is therefore constant; for the only consequence of a variation in the force of the clock is that the remontoire weights are lifted either faster or slower, which does not signify to the pendulum, as the lifting is always done when the pendulum is out of the way. If this can be managed with certainty, and without exposing the pendulum to some material variation of friction in the work of unlocking the escapement, which it must perform, its motion and therefore its time must be absolutely constant, since there is nothing to disturb it. It does not look a very difficult problem; and yet it puzzled the clockmakers to solve it in a satisfactory way for about a century, in consequence of certain small difficulties which nobody would guess until he had the opportunity of seeing them in action; and after all it was not done by the clockmakers, but by two lawyers, in different ways.

But first it may be asked, is the gravity escapement problem worth solving? Is not the dead or the detached escapement proved to be good enough both by experience and by mathematics? The answer is, that in science nothing is good enough when it can be improved upon: that both mathematics and experience prove that only by the most careful and delicate and therefore expensive work, and only on a small scale and with light weights, can clocks with 'impulse escapements' (*i.e.* receiving direct impulse from the train) be made to go as well as the best of them do; and although it is possible, by the addition of a train remontoire, to make even large and heavy dead escapement clocks go as well as astronomical ones, yet that apparatus involves not only some extra expense, but what is far more difficult to provide, some extra attention and intelligence in the people who have the care of it. We may however dispose of a great number of the contrivances for gravity escapements by the remark, that they require not less, but more delicacy of construction and careful handling afterwards than the finest dead escapement, even if they 'perform' any better, which scarcely any of them do.

Before we can appreciate the merits and the difficulties of this class of escapements, it is clearly necessary to understand the theory of them; which I shall be able to exhibit more briefly than that of the dead escapement, as some of the ground is common to them both, and especially that useful formula of Sir G. Airy's for the variation of the time caused by any escapement, from which I shall start again. But it will help our conception of the process if we first take some simple form of gravity escapement as an illustration of their principle; and it does not signify that that form was neither the first nor the best, on account of certain mechanical objections which we are not concerned with yet.

Mudge's gravity escapement.—The pallets A C, B C, are no longer fixed on one arbor, but on two, as close to the bend of the pendulum spring as possible. The acting faces are so shaped and placed that whenever the wheel moves from B towards A, a tooth will lift one of them until it is stopped by the nib *a* or *b* at the end of each acting face. Here the pallet A has just been lifted and is holding the tooth that lifted it; as soon as the pendulum comes, moving to the right, it will evidently push that pallet out of the way by means of the fork-pin P, and so free the tooth, and the wheel will begin to turn, and the opposite tooth will immediately lift the pallet B till it likewise is stopped by *b*. The pendulum is all the time going on rising to the right and carries the right pallet with it as far as it likes to go; when it begins to fall the pallet falls with it, not only to the place where it was taken up, but to a lower place, corresponding to that of the left pallet in the picture; and the fall of the weight of the pallet from the place where it is taken up by the pendulum to the place where the pendulum leaves it, or the difference between its rise and fall with the pendulum, constitutes the impulse, and that difference is evidently constant, however far the pendulum may swing.

Fig. 21.

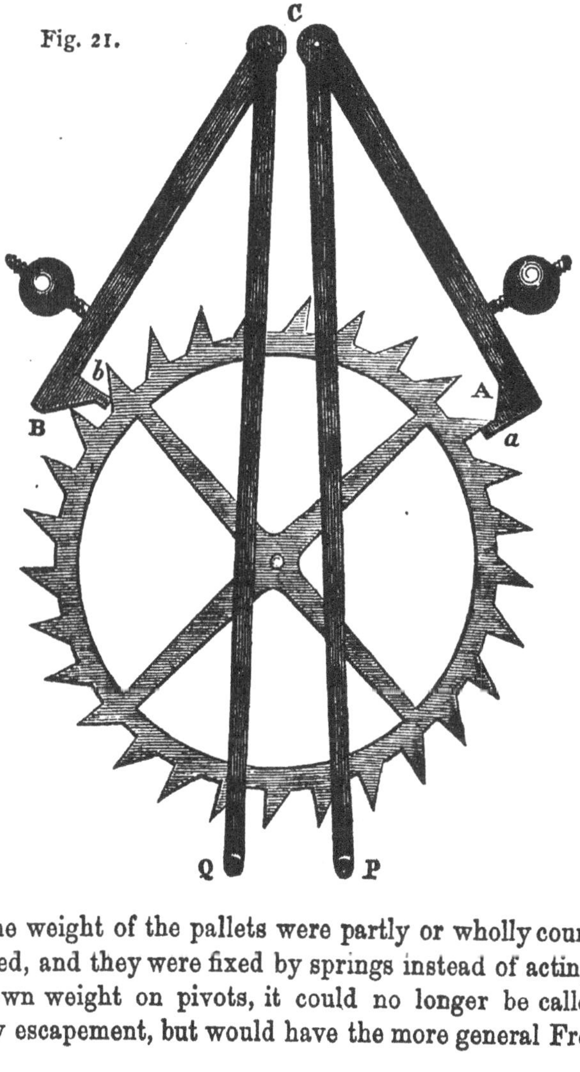

If the weight of the pallets were partly or wholly counterbalanced, and they were fixed by springs instead of acting by their own weight on pivots, it could no longer be called a gravity escapement, but would have the more general French name of a remontoire; but the principle would be the same, except that the force of the springs has a law of its own, and is more variable than that of gravity.

It is easy to show that the effect of a gravity escapement is to make the pendulum go faster than a free pendulum, by exactly the same reasoning as was used to contrast the dead and the recoil escapements; or still more simply, by considering that the remontoire weights are in effect so much weight added to the pendulum far above its centre of oscillation, and therefore they accelerate it. But this tells us nothing about the variation of the rate, in case the arc increases or decreases a little under any change of friction; and we shall see presently that a very curious result comes out with respect to this, which it is impossible to arrive at except by calculation.

To do this we must find out again what that quantity called ϕ, in Sir G. Airy's formula at p. 85, is for this escapement. Let the angle after zero at which the pendulum begins to lift the pallet be called γ, and the angle at which it leaves the other pallet which has been giving the impulse $\pm\ \beta$, according as that takes place after or before zero: so that, as in the dead escapement, the angle $\gamma + \beta$ is the angle of impulse, if the descending pallet is left after zero, which we shall see is the best arrangement. Let Pg be the weight of each pallet, and p the distance of its centre of gravity from its axis at C, and δ the angle which p makes with the pendulum when they are in contact. Ml^2 represents the moment of inertia of the pendulum as usual; strictly speaking we ought to add to it the moment of inertia of the pallets while they are in contact with the pendulum; but as that makes no difference in the nature of the result, and in the best escapements one pallet or the other is always in contact, we may either consider that as included in Ml^2 or neglect it altogether. Then the equation of motion will be

$$\frac{d^2\theta}{dt^2} = -\frac{g \sin\theta}{l} - \frac{\text{P}pg \sin(\delta + \theta)}{\text{M}l^2}.$$

We may put θ for sin θ, and 1 for cos θ (θ being small),

$$\therefore \frac{d^2\theta}{dt} = -\frac{g\,\theta}{l}\left(1 + \frac{Pp}{Ml}\right) - \frac{Ppg \sin\delta}{Ml^2}$$

This is of the same form as in the dead escapement; only here the constant term involving simply θ, which always indicates isochronism, has its coefficient increased; which shows that the pendulum will regularly go faster than if its own gravity alone acted upon it, exactly as it is accelerated by those small regulating weights which I described at p. 50. The other term constitutes the disturbing force ϕ, from which we are to learn what will be the disturbance of the isochronism if the arc varies a little; and this result will also be of the same form as before, except that the limits of ϕ are different, as it does not act now through the middle of the arc only, but from $-a$ to β and from γ to a; and even if β should be identical with γ the action is not continuous, for the falling pallet pushes the pendulum from $-a$ to β, but the pendulum lifts the pallet from γ up to a. Therefore the result of the integration will be (putting Wh for the sum of Pp sin δ for the whole day):

$$\Delta T = -\frac{Wh\left(\sqrt{a^2-\gamma^2}+\sqrt{a^2-\beta^2}\right)}{Ml\,\pi\,a^2(\gamma+\beta)};$$

which differs from the dead escapement formula only in the signs; but we shall see that that produces another very important difference. First however we may remark that if β is $-$ (that is, if the falling pallet is left before zero), or even if it is smaller than γ (and it cannot be larger with safety to the locking), ΔT is increased, and so will all its variations be when there are any. Therefore let us assume that β is as large as it can be, *i.e.* that one pallet is taken up just when the other is left, or $\beta = \gamma$; and then the formula becomes much more simple:

$$\Delta T = -\frac{Wh}{Ml\,\pi\,a^2}\sqrt{\frac{a^2}{\gamma^2}-1}$$

We must differentiate this, as before, to see what the variation of rate will be; but this time we need not consider W or the force of the impulse variable, because we know it is not—if the escapement is what it pretends to be. Then

$$d\,\Delta\,T = \frac{W h\,da}{M l\,\pi\,a^3} \frac{\dfrac{a^2}{\gamma^2} - 2}{\sqrt{\dfrac{a^2}{\gamma^2} - 1}}$$

From which this remarkable result appears,—that if the weight of the pallets is so adjusted that the pendulum swings through an arc $a = \gamma\sqrt{2}$, the rate will not vary at all, even when the arc does, except what may be due to the circular error. Unfortunately they both have the same sign, and therefore the escapement cannot be made to correct the circular error. But there happens to be a mechanical difficulty in making γ as large as $\dfrac{a}{\sqrt{2}}$ or $\cdot 71\,a$, which would be $85'$ if $a = 2°$; and moreover Mr. Bloxam came to the conclusion, as stated in his papers, that from other causes, especially the variation of density of the air, it is better to make γ considerably smaller than $\cdot 71\,a$. Let us see then what the variation of rate from the escapement will be for some such value of γ even as far from the theoretical value as $\dfrac{a}{4}$.

In the Westminster clock I know by trial that $\dfrac{Wh}{Ml}$ at the escapement is not more than $1/45$; taking it at that, and a at $2°\,40'$, as it is, and $\gamma = \dfrac{a}{4}$, you will see, if you take the trouble to make the calculation, that the clock will only gain about a second a month for a decrease of arc of $5'$, which is larger than is likely to happen, besides what is due to the circular error so far as it is uncorrected by the spring.

In those escapements where the falling pallet is left before zero, the expression for the variation of rate would be

$$d\,\Delta\,T = \frac{Wh\,da}{Ml\,\pi\,a^3\,(\gamma - \beta)} \left\{ \frac{a^2 - 2\gamma^2}{\sqrt{a^2 - \gamma^2}} + \frac{a^2 - 2\beta^2}{\sqrt{a^2 - \beta^2}} \right\}$$

in which you observe $\gamma - \beta$ instead of $\gamma + \beta$ is in the denominator, and therefore the variation is much larger than in the other form of the escapement. Theoretically indeed this might also be made $= 0$ by making the three angles satisfy this condition,

$$\sqrt{a^2 - \gamma^2}\ \sqrt{a^2 - \beta^2} = \frac{a^2}{2}.$$

Thus $\gamma = 90'$ and $\beta = 78'$ would be right for $a = 2°$; but these small differences would be even more inconvenient mechanically than $\gamma = 85'$ in the other case. That construction therefore is decidedly the worst, notwithstanding the tempting appearance of the pendulum being left free through the middle of its arc; a fact which would probably never have been known with certainty without this kind of investigation, as the errors would have been sure to be attributed to any but the right cause.

Reid mentions, at page 139 of his old book, a curious fact bearing on this. He says he cut a hole in the side of the case of a gravity escapement clock, which increased its arc from 1° 22′ to 1° 37′, and at the same time made it gain 42 sec. a day, and a larger hole increased the arc 12′ more, and I suppose the rate. Therefore the diminished resistance, from the pendulum being able to drive the air before it through the hole, much more than counteracted both the circular error and the escapement error, making the clock gain this great amount instead of losing a little under the increased arc. Moreover such an enormous variation of rate as this shows that the motion of the air in the clock-case may affect the rate materially. But I must add that I have twice tried the experiment myself in a still stronger way, by taking off the clock-case altogether from my own gravity escapement clock, and I could not observe any increase of arc at all—certainly not of 5′, nor any alteration of rate. The arc was however nearly 1° larger than in Reid's clock before he cut the holes; and this is another proof of the unsteadiness of very small arcs. I have since found that a small enclosure for a turret clock pendulum bob did diminish the arc a little.

Probably no one would foresee, without experiment, that the simple form of escapement in p. 110 would fail. Several of the most elaborate French turret clocks in the Exhibition of 1851 were on that plan, and people were very much surprised when I showed them that you could make all those clocks increase their arc visibly and speedily by increasing the clock-weight, which is directly contrary to the fundamental principle of a gravity escapement. This form of the escapement was invented by Mudge, a celebrated watchmaker whom I shall have to mention again, and it had long been known here that it would not answer, on account of its liability to *trip*, or to have the pallets jerked out so far by the motion of the wheel that the nib fails to catch the lifting tooth, and so three or four teeth run past and the time is altogether lost. The only way of avoiding this liability is to use a very highly finished train with high numbered pinions to keep the force uniform, and then to make that force only just enough to raise the pallets; but that is inconsistent with the clock being able to work anything but a small dial, and if it is not kept very clean and the oil fresh, it will be sure to stop. In short, the clock becomes too expensive and delicate, and requires too much attention to be tolerated in common use.

Cumming's escapement.—But even if all these risks were got over, there is still another radical defect in all such escapements, which does not appear to have been ever noticed before I pointed it out with reference to those clocks in the Exhibition. The force may easily be enough to raise the pallets a little too high, without jerking them over the tooth altogether, and then the pressure on the nib is quite enough to keep them there, and so the pallet is taken up by the pendulum at some angle greater than the proper one γ; and as it falls down with the pendulum to a constant place, the impulse lasts longer than it ought, and of course the arc is increased. I gave the name of *approximate tripping* to this defect, and any escapement which is liable to it is evidently worth nothing. It seems capable of happening even where the impulse or sloped portion of the pallet is put on one arm and the nib on a separate one which is not lifted at all by the wheel, but only by the pendulum, as in Cumming's escapement, which was invented very nearly a century ago, and for a time was thought to answer well. I suppose this is in consequence of the force with which the tooth strikes the stop pallet, sometimes throwing it a little out of its place, unless it is undercut or given a slight recoil the wrong way; and that is objectionable too, because it resists the unlocking, and does not always resist with the same force.

Hardy's escapement.—One of the objections to Cumming's escapement was the friction of no less than 8 pivots of the 4 arms which had to move with the pendulum in the course of each vibration. Hardy avoided this by setting the arms on springs instead of pivots; but that introduced another and probably a worse evil, because the stiffness of the springs varies with the temperature, which of course disturbs the rate of the pendulum. That escapement was consequently removed from the transit clock at Greenwich many years ago, and a dead one substituted. There were some very good rates of three Hardy's clocks published in *Pearson's Astronomy*; but they are practically extinct.

Kater's escapement.—The late Captain Kater, who paid great attention to the theory of pendulums, invented a gravity escapement, which is very fully described in the 130th vol. of *Philosophical Transactions*, on the principle of making the weight of the descending pallet unlock the scapewheel by falling upon an anchor like a pair of dead escapement pallets without the impulse faces. He supposed that as the inertia of the anchor would stop the pallet for a moment, the pendulum would leave it there, and so be itself free from the friction of unlocking. But here again it was found that the force of the wheel was apt to displace the anchor unless its pallets were undercut, and then the resistance was sometimes too great for the gravity pallets to overcome, unless they were too heavy for the pendulum; and after many attempts to make it go, that escapement also was taken out of the only clock to which I know of it being applied, and came into the hands of Mr. Bloxam, who showed me it. The failure of it is described in his paper before mentioned.

Gowland's escapement, which was in the Exhibition of 1851, was on the same principle as to the unlocking only. He had not even pallets for the impulse, but a pair of small weights, which hung on long arms or spikes projecting horizontally from the locking pallets, except when they were lifted off by similar arms projecting from the pendulum. This prevented the friction of any pallet pivots affecting the pendulum. The locking pallets were of Mudge's form, and were prevented from being driven too quickly and tripping, by paddles descending from them into a pot of oil—not a very elegant contrivance certainly, and requiring a good deal of extra force in the train. I heard no more of it after the Exhibition, at which I was not surprised.

M. Gannery, of Paris, had an escapement there also, on the same principle as to the small weights, which were hung by strings from the pallet arms and received in cups on the pendulum arms, or *vice versâ*. But instead of the oil pot, he made the scapewheel of the usual size with only 9 teeth of very slow rise and a nib at the end, which therefore lifted the pallets slowly and seemed to obviate the tendency to trip, so far as I could judge by merely looking at it. I should think however that the rigidity of the strings would be quite enough to affect the pendulum, and the friction in lifting was considerable. Moreover I doubt whether the stopping of the unlocking weight in any of this class of gravity escapements is decisive enough to make the difference of the angles of lift and of drop quite constant; and if it is not the escapement fails. Of that escapement also I heard no more, and the French members of the jury evidently thought very little of it. Nevertheless, the scapewheel with 9 teeth instead of 30, which reduces the pressure of the stops in that proportion, was a great advance in the right direction, though not absolutely new, because the same thing had been done much better some years before in

Bloxam's escapement.—This is so superior to the others that it deserves a more particular description. This drawing (next page) of the full size in Mr. Bloxam's own clock, is copied (with a little alteration for distinctness of exhibition) from his account of it in the Astronomical Society's Memoirs of 1853. The pallets are lifted alternately by the small wheel or pinion with 9 teeth, and with scarcely any friction, as the action is only for a short distance across the line of centres. The stopping is done by the long teeth, and the pressure there is less than the lift in the proportion of the radii of the small and large wheels. The stops are A and B: E and F are the fork pins which embrace the pendulum. The pallets above at C are cranked, that their centres of motion may be identical with that of the pendulum; which is perhaps an unnecessary refinement, especially as the pendulum spring has no one centre of motion. The size of the wheel determines that of the pallets, thus: if the radius of the wheel is 1 in. the length of each pallet down to the stop must be 2·8 in., to make the angle between the locking teeth and each pallet 90°. Mr. Bloxam made the angle γ, at which the pendulum leaves one pallet and takes up the other, only 20′ in his clock, and $a = 1.°40′$, these being the proportions which he concluded were the best to counteract the effect of variations of density of the air.

Fig. 22.

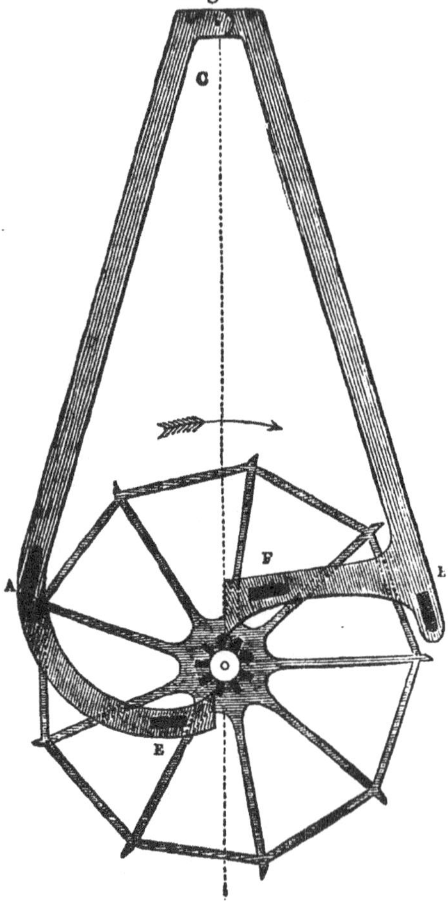

The objections to this escapement are, that it is delicate and expensive to make, both in itself and in the rest of the clock, and if a pallet got accidentally lifted the wheel would run with great velocity and probably break a tooth when the pallet stopped it again — an evil to which most of the previous escapements are equally liable, and it is a very serious one. It is also liable to trip unless the train is very fine, so that no more weight need be used than is absolutely necessary. But unless a gravity escapement will enable the clock to do with a coarse train, it is of no real use; at any rate it is quite certain not to come into use; and Mr. Bloxam said that he considered a fine train essential. That in his own clock was the finest and had almost the highest numbered pinions (18) I ever saw. And therefore after all, this can only be regarded as a theoretical or scientific solution of the gravity escapement problem, but hardly a practical one. I expressed this opinion of it in early editions of this book, and stated that old Mr. Dent and I had agreed that it would not do for the Westminster clock, even if protected from variations of force by a remontoire in the train. Somebody however thought he knew better; for in the 1862 Exhibition there appeared for a short time a small turret clock with this escapement, and a train remontoire, which was said in the newspapers to have been made under the direction of the Astronomer Royal. But it could not be made to behave properly, and after a few weeks was withdrawn.

At the same time, Mr. Bloxam deserves the credit of having first discovered, though he did not publish the discovery till long afterwards, that the practical conclusions of Sir G. Airy's Cambridge paper were erroneous; and that the errors of a dead escapement cannot be made either insignificant or constant; that the dead friction does not correct itself before and after zero; that the acceleration of a gravity escapement pendulum over a free pendulum does not signify the least; but that what does signify is the constancy of that acceleration; and that the variation is least when the acceleration is greatest, and may be made practically nothing by a particular arrangement of the angles of impulse. The Astronomer Royal afterwards in effect assented to all this, first by admitting in 1852 that my gravity escapement, which will be described presently, answered perfectly, and by himself communicating to the Royal Astronomical Society Bloxam's two papers of 1853 and 1858 sent from Madeira, the latter of them after he was dead; and then actually going too far and assuming that

Bloxam's escapement was a practical solution of the gravity escapement problem.

Sir E. Beckett's Gravity Escapements.—I have now to describe the only escapements of this kind which have ever come into real use; and that both for large and astronomical clocks, especially the former. The earliest form of them need no longer be described, as it has been superseded by two others, one with a four-legged scapewheel for small clocks, and the other with a double three-legged wheel for large ones.

The four-legged escapement is shown in this drawing of a regulator of this kind seen from behind, half the real size. To avoid confusion, only the centres of most of the wheels are shown, and you see the train is inverted, to save height in the frame and clockcase. G is the great wheel arbor, and the barrel is shaded; C the centre wheel, close to the top of pendulum and pallet arbors, B second wheel, A seconds-hand wheel, and E scapewheel. DD are the brackets which come out of a large cast-iron plate or back, from which also comes the pendulum cock, shown by partly broken lines, to leave the pallet cock visible. PP are the pillars, CST, CS'T' the pallets, and their inside pivots (the arbors being very short) run in a flat piece or cock between the centre wheel and the clock plate or frame. The scapewheel speaks for itself as to the long locking teeth; but it has two sets of lifting pins near the centre, pointing alternately backwards and forwards, one set lifting one pallet and the others the other, the pallets being in different planes, before and behind the wheel, with one stop S pointing forwards and the other S' backwards. This cannot be done otherwise with an even-numbered wheel, and though it can with a five-legged one, that is an inferior construction in other respects, and is much more difficult to make rightly, and has not one advantage when it is made. I say this from experience now, as I did before

Fig. 23. Back View.

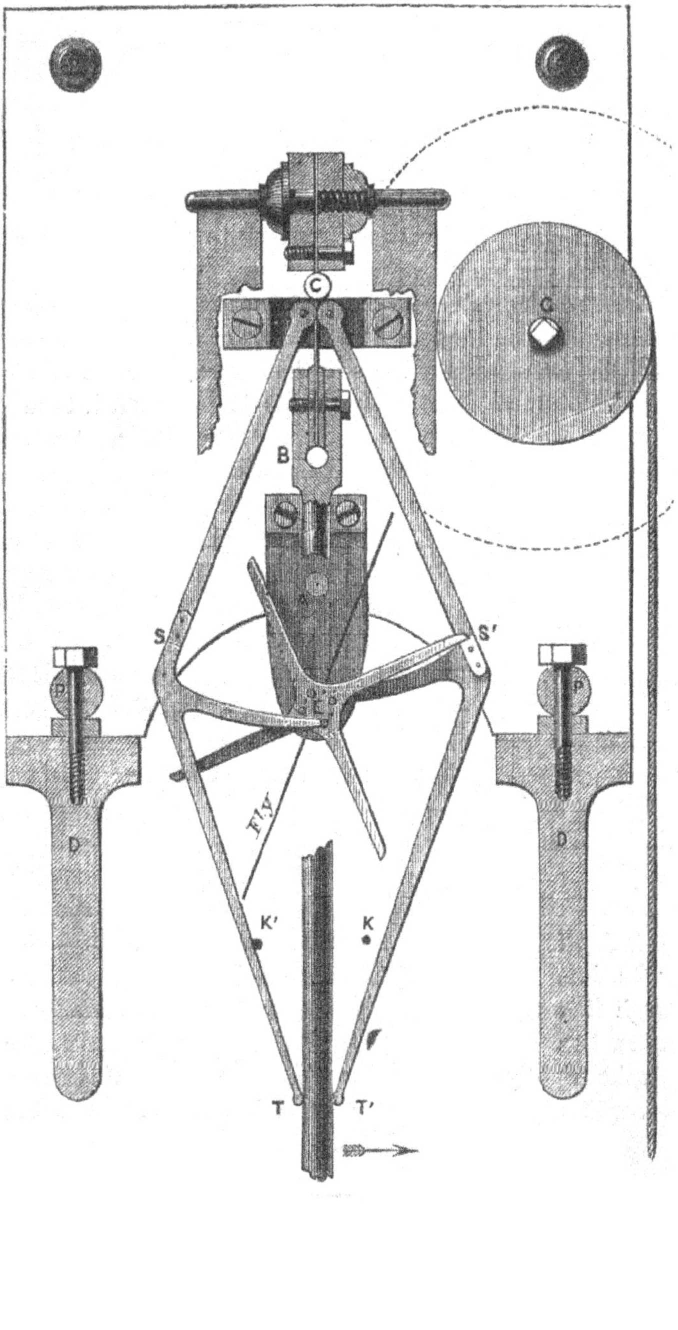

from theory. For with the strange propensity of mankind for taking more trouble to do wrong than it would take to do right, some persons would persist in making the escapement with five legs. I thought it might possibly save a little force or weight on the train, but it does not even do that; nor anything else, except give more trouble to adjust: whereas the four-legged one is so easy that the very first I had made went at once perfectly without any alteration.

The great feature of them is the regulation of the velocity and the avoidance of the banging on the pallets and the risk of tripping, either actual or approximate, by putting a common fan-fly on the scapewheel arbor. And this becomes possible or effective only by reason of the large motion of the wheel and fly at every beat: viz. 45° in this escapement, and 60° in the three-legs. A fly would be of very little use in Bloxam's escapement, to say nothing of its other difficulties. In order to leave room for the fly the seconds-wheel arbor is cut short and set in a cock within the frame, which leaves space enough for a fly above 2 in. long in each arm, from E to B.

The pallets require no beat screws, as they are only thin pieces of steel like square wires, which can easily be bent for adjustment; and even for a 40 lbs. pendulum they have to be made very light; they must be cut out of steel plate, and the lifting faces hardened. The pins are found to do better also of steel than of brass, and one set of them should be tapped into the wheel with left-handed screws, so that the action of lifting may not tend to loosen them. The wheel is either screwed on the arbor up to a shoulder with a left-handed screw, so that the action tends to tighten it, or is 'squared on' a hexagon and pinned. The scapewheels are also generally cut out of thin steel plate; but some have been made of a harder kind of gun-metal or softer kind of bell-metal, and I am not certain which is best. But in either case the stops must be as hard as possible, either steel quite hard or jewels. I think steel hard enough, and I do not believe in leaving any place where there is friction totally dry, though the oil may be the least possible.

Another great advantage of these escapements is that the length of the teeth or legs, and the largeness of their motion, make the pressure on the stops, or the work of unlocking by the pendulum, insensible; and therefore also they are incapable of holding up the pallets so as to cause 'approximate tripping' by any force that you can apply to the great wheel, provided the escapement is made properly. But with that same genius for doing things wrong when it is as easy to do them right, some persons have made the angle CSE, at the stop which is struck upwards, less than 90°, and then have said the escapement failed because it sometimes tripped, as it was pretty sure to do. For safety it is as well to put the up stop a little higher, and the down stop a little lower than their proper theoretical places, which are where both the angles would be exactly 90°.

That determines the theoretical distance of the pallet arbors C from E the scapewheel centre. If its diameter is 4 in. that distance is 5·2, and that *is* the proper distance of the top of the pendulum spring above E. The pallet arbors must evidently be a little lower. Mr. Bloxam made them cranked (see p. 119) in order to get their common axis in a line with the top of the spring; but that is an unnecessary refinement, at least for these escapements, and is never done. And as no point in the pendulum really swings quite in a circle, I doubt if the friction of the pallets on it would be sensibly less for their being both made to describe the same circle by cranking their arbors: at any rate it is too insignificant to care about.

The distance of the lifting pins from the centre should not be more than a 40th of EC, or else the angle of impulse 2γ will be larger than is found expedient. It is difficult to make the pallets light enough even with a small γ, and the larger it is the lighter they must be. The length of their tails down to the beat pins is arbitrary, but I found the Westminster clock perform decidedly better with the pallet tails long than short. The length here shown does very well, and it looks neat to make the two parts reciprocally parallel. The pins should be placed so that the lifting may take place equally across the line of centres CE, because then it is done with the least friction. For this purpose the pins which lift the lower pallet must be set on the radii which run along the acting faces of the teeth, and the other set of pins half way between them, with reversed screws, as I said before.

Any gravity escapement requires a heavier weight than a dead or detached one, other things being equal, because it must be strong enough to lift the pallets promptly and firmly always; but the superfluous force does not reach the pendulum, and therefore does no harm, provided the train is good enough not to waste much force generally in order to get over occasional weak places from bad wheel-cutting. Nothing tests the defects of a train like a gravity escapement. If there are bad places in a common clock train they must be very bad indeed for the teeth not to follow the pallets, though they may be giving no effective impulse for some seconds; but in a gravity clock the pallets have always to be lifted; and I have never got a train yet in which I could not hear some weak place, recurring always at the same time of day, and requiring at least an extra pound in the weight beyond what was generally wanted. For that reason, though a high numbered train is not, or ought not to be requisite for these clocks, it should be a thoroughly good one; and as defects of cutting affect low numbers more than high ones, it is better to have them rather high, though they need not be anything like what are used for first-rate dead escapements. I have the upper pinions of 10 and the centre one of 12, and if these are well cut, and still better if they are lantern pinions, they are enough; if they are not, you will soon hear and see it by the escapement, and the train should be rejected as a bad one.

In gravity regulators, for the same reason, the wheel must have a little run at the pallets before it begins to lift them, just as many clocks will not begin to strike if the hammer tail lies on the pins: *i.e.*, there must be banking pins K K′ for the pallets to rest on just clear of the lifting pins. And in turret clocks the banking pins are useful to reduce the arc without making the pallets too thin. They may be simply a thin piece of metal adjusted for the beat pins to fall on.

It is better to have a full-sized barrel with three lines and a fixed pulley for a 'three-quarter length' clock than two lines and a smaller barrel. A clock of this kind with a 40 lbs. pendulum swinging 2° requires a weight of from 20 to 24 lbs., according to the train, to lift the pallets promptly and firmly, and loud enough for an observer.

These clocks have also been made to strike a small bell every minute by a pin on the minute wheel, which enables an observer to go on for some time without looking at the clock. Of course this requires rather more weight, and no such extra friction could be tolerated in a dead escapement. My clocks strike one at the hour on a rather large bell, but that makes no sensible difference in the weight. When there is a full striking part, the train must be arranged in the usual position, which is inverted in the mere going clocks, to save unnecessary height of the frame and the pendulum being several inches above it, on account of the extra wheel and the length of the pallets. The great wheel is also put on the left side to keep it near the side of the case and out of the way of the pendulum. When a pendulum imparts vibration to the weight, as they do sometimes, the simplest cure for it is to put a board down the side for the weight to graze when it is near the pendulum. It has never happened in my clocks, and three lines tend to prevent it.

Unfortunately I have no daily rate of any of these clocks regularly taken in an observatory, until we come to the largest of them all; but judging of my own for long periods by the Westminster clock, which is tested daily, I have no difficulty in pronouncing it superior in steadiness of rate to any dead escapement whose rate I have ever seen. And they are made by Mr. Brock, of George Street, Portman Square, and I dare say by other makers, for half the price of the old-fashioned best dead escapement regulators with only 12lbs. pendulums.

Some persons have taken a great deal of unnecessary trouble to modify these escapements in order to avoid the fly, as if that did any harm; and have added train remontoires, as if that did any good to any gravity escapement which really is a gravity escapement—*i.e.* which has no tendency to trip, and gives a constant impulse clear of the friction of the train. One of these attempts to get rid of the fly was exhibited by Dr. Clark in 1862; but the wheel was stopped with a bang that thrilled all through the clock, which only tends to knock it to pieces, and keeps everything in a state of vibration. It was also a much more delicate construction, and I should prefer Bloxam's if I had to choose between them; for if you have not a fly, the less the motion is at each beat the better. Another contrivance for the same purpose, invented by a French workman here, was used by his employers in a few church clocks, and then abandoned, and my original form resumed by the makers in all their turret clocks. The fly should of course be as light as possible, either of thin steel or aluminium; and the best way to put it on is with a piece of watch spring pushed in through an oblong hole so as to press always on the arbor, which should *not* be very thin. Some persons do not seem to know that a long fly is more effective than a wide one of the same area.

Another mistake made by several of these inventors,

including Dr. Clark, was that of supposing it was better to let the pallets unlock the wheel in falling with the pendulum, forgetting that the pendulum is then just as much affected by the friction of unlocking as if it did it in rising, if that friction is sensible; and if it is not, still less can it signify whether it is borne directly or indirectly by the pendulum. It is still worse to keep the pallets moving in contact with the stops during each 'excursion' beyond the angle of unlocking, for that reduces it so far to a dead escapement. But the worst of all the 'improvements' was by the late Mr. Cooke of York, the eminent telescope-maker, who also made these clocks, and rounded the stops 'to make the unlocking easy;' which degraded it to an impulse escapement, the pallets being then driven away by the teeth with a force depending on the clock weight and friction of the train.

The double three-legged escapement (fig. 24 next page) is so called because it has two three-legged wheels, A B C and *a b c*, in different planes, with one set of 3 lifting pins between them. Here the two wheels must be squared on the arbor, and the lifting pins need only be shouldered between them. The pallets also lie in one plane between the wheels, but one stop S points forward to receive the A B C teeth, and the other S′ backward to receive the *a b c* teeth alternately. The reason for two wheels is that with one three-legged wheel you cannot have the pallets far from upright, which is evidently the worst position for them, as it requires much more dead weight to be moved at every beat in order to have weight enough for effective impulse. It should be understood that there is no particular mechanical advantage in the two wheels being set with the alternate teeth equidistant, appearing like a six-legged wheel. They may be set with one set of legs at 90° and 30° to the other set—or at any other angles to get any greater inclination of the pallets if desired. However the

Fig. 24.

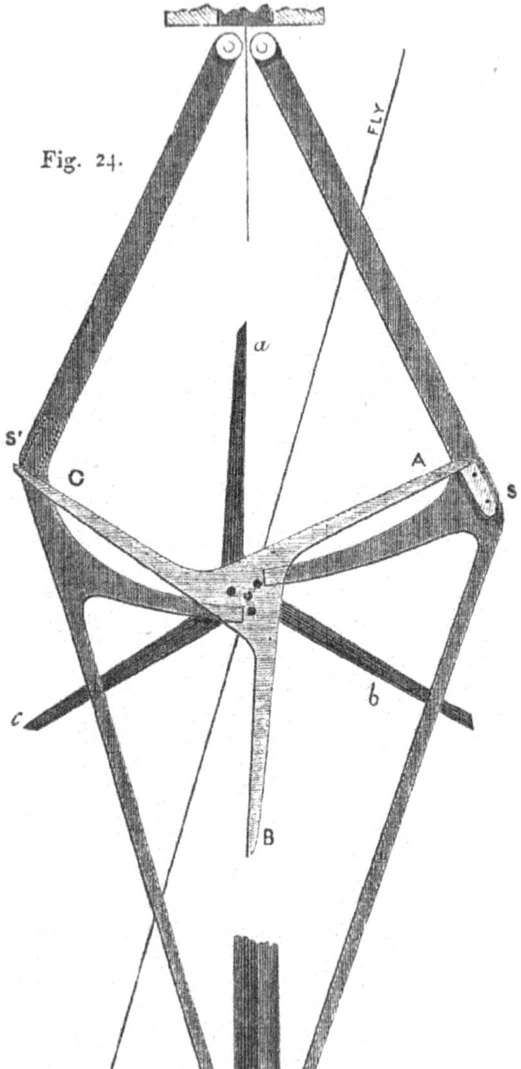

equidistant arrangement is the natural one, and is generally used, though not invariably. I need hardly say that a pair of wheels of this kind are very different from a six-legged wheel, which would only move 30° at each beat, while this moves 60°, besides other differences.

In this case the distance of the pendulum top from the scape-wheel centre evidently = diameter of scape-wheel. The lifting pins should not be farther from the centre than a 36th of this, or the pallets have to be inconveniently light and thin: the pins and arbor may be solid as a three-leaved pinion. They should be so placed that the one which is holding up a pallet and the one which is going to lift next may be vertically over each other, the third being on a level with the centre—*i.e.*, they will stand on the radii which form the acting faces of the teeth of one wheel, as you see here.

This escapement is the best for large clocks, which must have plenty of superfluous force to drive the hands in all weathers (which superfluous force all reaches the pendulum in clocks of the old kind, varying immensely); but the fly must, on that account, be much larger in proportion than in regulators, and care must be taken in planning the clock to leave room for it. In the smallest turret clocks I have always had the fly a foot long altogether and 1¼ in. wide, and in large ones considerably more. When the fly is very large, as at Westminster, the friction of a spring on the arbor is not enough, and there must be a larger 'roller' or blank-wheel pinned on the arbor for the spring to act on. At the same time it was very satisfactory to find that when the men had once forgotten to screw up the spring after doing something to the fly, the clock had never tripped in the days which had elapsed; and I have tried the same experiment elsewhere; but the four-legs will trip if the fly is loose. The greater obliquity of the pallets, 30° against 22½°, is the cause of this superiority of the three-legs; and this obliquity may be increased still more if you like by altering the relative position of the two scapewheels. But I by no means advise the omission of the fly, even then.

In very large clocks the pallet tails are too thick to bend for adjustment of the beat, and then eccentric beat pins are used, which require no description. They are usually made of brass, even in small clocks; but I think it would be better to cover them with hard wood. The finest clock of this kind I have seen was a regulator with four-legged escapement made by an amateur, Mr. George Salt, of Saltaire; and it had ivory beat pins, which certainly had less chatter than brass ones. The pendulum weighed nearly 40 lbs., and yet the clock weight was only 15 lbs., with about 4 ft. fall. This shows that a gravity escapement really requires very little more force than a dead one with as good a train as possible, though practically I should always make the weight abundant. One thing must be specially attended to, as a distinction between these and dead escapements: the beat pins must on no account be touched with oil or grease of any kind, but left absolutely dry, whatever they are made of; for the slightest adhesion to the pendulum is fatal; though in dead escapements the fork should always have a drop of oil, to keep it as close as possible without being tight. Moreover, care should be taken to make one pallet begin to lift simultaneously with the resting of the other, and neither before nor after.

The best evidence of the performance of these escapements is the annual report of the largest of them all at Westminster, of which I shall have more to say afterwards.* My own have often gone for months together without any difference from Westminster for which I could venture to correct, or to regulate the pendulum. I have never found on close inquiry that the rate of the very best dead escapement regulators

* And I have read in the *English Mechanic* that one under the charge of Professor Waldo, of Yale College Observatory, had gone for several months on a rate of only ·47 sec. a week.

approached that of Westminster and several other public clocks of this kind, of which I have occasionally had reports. I have had accounts of some that had been altered from other escapements to this, with the effect of reducing errors of minutes to seconds. The last was from a gentleman who said that his turret clock, made from this book by a man who had never made one before, has a better rate than any astronomical clock he had ever known.

CONSTRUCTION OF THE GOING PART OF CLOCKS.

Fig. 25 (p. 134) shows the arrangement of the going part of a common regulator, or a house clock of superior character, except that the pendulum in a clock of that kind ought to be hung by a cock on the back of the case; but I have already given a drawing showing that at p. 44; so I show another mode of suspension here. In the common recoil escapement clocks the pendulum only weighs a pound, or less, and is hung from a cock merely screwed to the back plate of the frame, which is a much weaker plan than this. A*a* is one of the pallets on the arbor *a*, and E*f* the crutch and fork; which generally embraces the pendulum, but sometimes goes through it, especially when it is a wooden rod. The weight is not hung by the single line, but by a double line going through a pulley, sometimes in the weight itself, but more frequently hung to it by a hook. This prevents the string from untwisting, and enables you to do with a thinner string, and it requires a barrel of twice the diameter which a single string would, and that is worth something when the pivots of the barrel are as large as they usually are, of which I have spoken already at p. 84. But it must be remembered that every pulley in a machine, and especially every moveable pulley, wastes power very sensibly by the friction and stiffness of the rope, especially when the moving power is at the slow end of the system of ropes. Theoretically it makes no difference whether a given weight with a given fall in a week is hung by one rope without any pulleys or by half a dozen; but practically you will find the weight has to be increased in a very high ratio for every additional pulley you add. You might soon reach a number at which no weight whatever would do the required work, but would all be wasted in overcoming the friction and stiffness of the ropes. This remark is chiefly important with reference to turret clocks, since not one architect in a hundred ever consults a clockmaker before he plans either dial-holes or clock-room or place for the weights to fall, and then people think it is the clockmaker's fault if the clock does not perform as well as if it had plenty of room for the weights to fall and other things to act properly.

The barrel is fixed to its arbor, of which the back end is a common pivot, but the front is carried through to the dial K and is squared for the key to take hold of it. The great wheel G rides loose on the arbor between the barrel and a collar shown just above G, and it is connected with the barrel by the ratchet and click, of which a front view and description has been already given at page 21. (There are in fact two ratchets R and *r* in fig. 25, but we need not inquire into the functions of the second at present: it will be explained at p. 149.) The great wheel G drives the centre pinion *c*, which always turns in an hour, and its arbor goes through to the dial and carries the minute hand in the way I will describe presently. The centre wheel C drives the second pinion *d*, on whose arbor is a wheel D which drives the scapewheel E by its pinion *e*. In moderately good clocks the pinions have all generally 8 teeth or leaves, and the wheels in that case have 96, 64, and 60 teeth, if the scapewheel turns in a minute as usual: in the best clocks the pinions are occasionally as high as 16; Mr. Bloxam's and Mr. Salt's are the only ones I ever saw with higher numbers. Of the scape-wheel, pallets, and pendulum, I

COMMON HOUSE CLOCK.

Fig. 25.

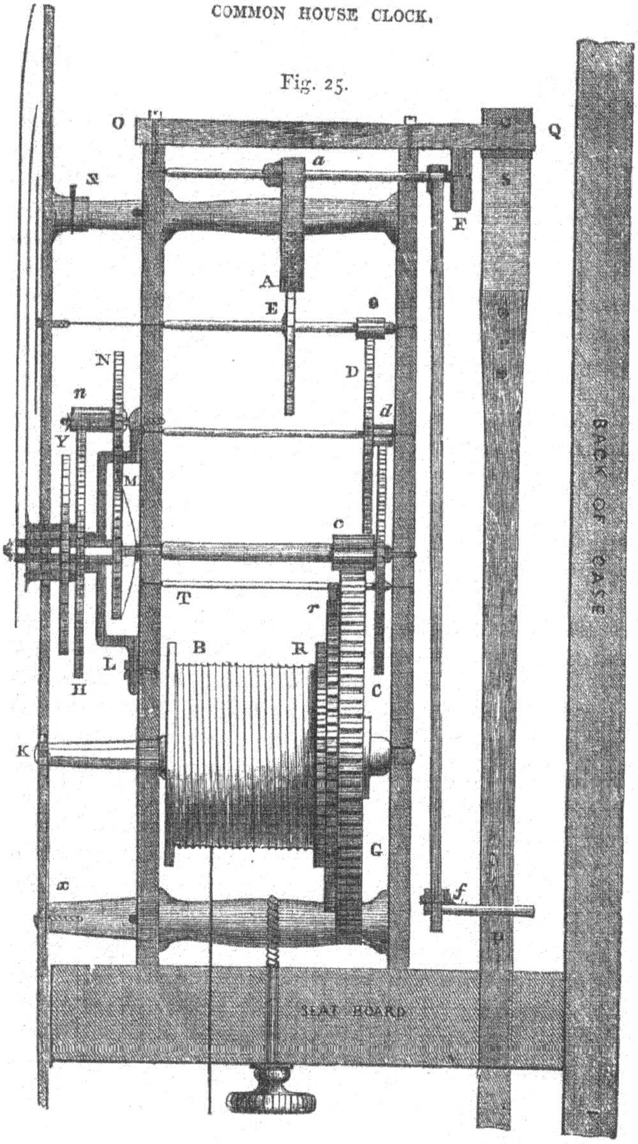

have said enough already. In the best clocks, and sometimes in common ones, the scapewheel arbor comes through the dial and carries a seconds hand.

Short clocks with half second pendulums are best made with a scapewheel of the usual number of 30 teeth, as either a large and heavy wheel, or teeth very closely set, are objectionable. In that case the scapewheel will of course turn twice in a minute, and the product of the numbers of the teeth of the centre and second wheels must = 120 × 8 × 8 if the pinions are of 8, which is best satisfied by 96 and 80. In very inferior clocks the scapewheel pinion is only 7, and then 84 and 80 teeth will do. The American clocks have lantern pinions, in which 6 pins are as good as 8 or 9 common teeth, and therefore the centre and second wheels need only have 72 and 60 teeth for a half-seconds pendulum. It is not worth while to go on with the calculation for ⅔ seconds pendulums or others, as the principle of it is perfectly obvious.

Dial work.—If the minute hand were fixed rigidly to the centre arbor, the clock could never be altered; and therefore the way in which it is fixed and yet alterable is this: there is a wheel M (in fig. 25) set on a hollow arbor or pipe which fits easily on the centre arbor, and has its own end squared for the hand to fit it. A small bent spring with a hole in the middle fits on the centre arbor, and the hole rests against a shoulder which is turned on the arbor just in front of the clock frame; the ends of the spring press against the back of the wheel M when it is pressed back, as it is after the hand is put on, and it is kept there by a collar and a small pin put through the end of the centre arbor.

And in this small matter there is room for a very common mistake: the hand is kept steady by the friction of this spring at one end and the collar at the other, and it will evidently be much steadier for the same amount of pressure if the spring fits the arbor tight, and so the friction is between the ends of the spring and the back of the wheel, than if it is loose on the arbor, simply on account of the difference of leverage at which the friction acts when you turn the hand with your own finger. In mere regulators without any striking part, this does not matter, because there is nothing to disturb the hand; but when the wheel M, or the equal wheel N which is driven by it, has to lift a lever every hour to discharge the striking part, it matters a great deal, because a great deal more friction is then required to hold it in its place against the pressure of the lever; and yet it seems to be the fashion in London to make the hole in the spring round instead of squaring it on to the arbor; which would take about ten minutes to do. The consequence is that a clock will sometimes take to losing unaccountably in this way, which I only first discovered by seeing the minute hand gradually lag behind the seconds hand; and that defect can only be permanently cured by making the spring very much stiffer than it need be if it is put on properly, *i.e.* with a square instead of a round hole.

Over the minute-wheel M and its hollow arbor there is fixed a thing called the *bridge*, which is shown at M L (fig. 25), and has another pipe enclosing that of the wheel M but not touching it; and the hour-wheel H with another hollow arbor still larger rides upon the bridge pipe, and is driven by a pinion *n* of 1-12th its own number of teeth, which is fixed to the wheel N of the same number as M. That wheel N is generally set upon a stud or pin screwed into the front plate, but is better with pivots in the frame and a cock. The hour hand is set on the end of the hour-wheel socket either with a small screw or pins. The thing marked Y in front of the hour-wheel has nothing to do with the going part of the clock, but is the *snail* which regulates the number of hours struck by the striking part, as will be explained therewith.

The hour hand in astronomical clocks generally has a

small circle to itself in the lower half of the dial, to prevent its hiding the seconds hand in the upper half, which it is important that an observer should always be able to see. Besides it moves with much less friction when so placed, as it then turns on a thin stud fixed on the front plate of the clock, instead of a very wide socket. It may either be driven by an intermediate wheel and pinion from the centre arbor, in order to make it go the right way round, or directly by the great wheel, which involves less friction and no inconvenience except the perfectly insignificant one of having to move the hour hand separately from the minute hand when you want to alter the clock much, which cannot happen once a year in any good clock.

Dial.—Two different ways of fixing a dial are shown in fig. 25, and both of them different from the common one, in which four separate pillars are screwed permanently into the dial and the other ends go through the front clock plate and are pinned behind it, the main pillars of the clock itself being only long enough to connect the two plates, and having nothing to do with the dial. Either of the two plans in the figure seems to me to be better than this, though it is a matter of very little consequence. The dials of the best clocks are made of brass plates polished and silvered: the common ones are of sheet iron: the American and Dutch of wood. The hands are always of black steel in regulators, but in common clocks they are of brass gilt, which is a very bad colour on a white face, though very good on a black face. The same remark applies to watch faces. I never could understand how such an absurd thing as gold hands on white faces—and still worse on gilt faces—came into existence, especially as gold hands of that small size have not even the vulgar merit of being dearer than steel ones.

Winding keys are generally made too short in the stalk or leverage, which makes the clock harder to wind and tends to strain the arbor besides, as you may see from considering that if the stalk was very short indeed, any force applied to it would be chiefly consumed in trying to bend the arbor. As there is absolutely no advantage in a short key, this is one of the many instances of doing wrong for the pleasure of it, when it would be quite as easy to do right and the effect much more pleasant. All my seconds pendulum clock keys are from 3 to 5 inches long, according to the weights. The French spring clocks without fusees, in which the winding is very hard towards the end, have keys like a very large watch key or a piano-tuner, to prevent the strain upon the arbor. But this makes the winding a much longer operation and a very unpleasant one, as you have to stop at every half turn. Sometimes such keys are given with small English fusee clocks, for which there is no excuse. You should take care that the wood or ivory on the handle of a key is quite loose, or it increases the resistance materially in winding.

Year clocks.—Clocks without striking parts are sometimes made to go a month, and occasionally even a year. For a month they only require one more wheel and pinion with a multiplier of 4, between the centre and great wheels. Year clocks require 3 wheels below the centre, and the pinions ought to be of 10, 10 and 12 at least, on account of the great weight required, which will have to be still greater if the pinions are of low numbers. Assuming the clock to go 380 days, and the barrel to have 16 turns as usual, the product of the 3 wheels must $= \frac{380.24.12.10.10}{16}$, for which 100, 90, and 76 will be the best numbers. This is far better than trying to do it with only two wheels of 192 and 190 (the lowest possible numbers), and pinions of 8, in which the friction will be very much greater. The best way is to have two barrels and great wheels acting on one long pinion of 12, with the weight hung by the same string from both barrels by a pulley which only turns while you are winding up; and there must be a winding stop to each barrel.

Clock cases are necessarily connected with the construction of the clock. The old tall case with a base as big as the top, standing on the floor instead of screwed to the wall, is sufficiently exploded to require no more to be said against it. The case need be no longer than is required for the pendulum, as the weights can have 3 lines, or smaller barrels, or larger great wheels, of which the second is the worst. The weights themselves of course have to be half as heavy again as with the fall one half longer. The best case for a superior clock is one of which the front and sides take off together. There need be no door to the face, but only small brass or white metal shutters over winding holes in the plate glass front, which enables you to lock up the clock completely and leave anybody to wind it up. Ornamental case making I have nothing to do with, and it is not much of an exaggeration to say that the value of the inside of a clock generally varies inversely as the decoration of the outside.

Moon dials.—There is an old calculation that if a pinion of 6 is put on the hour arbor of a clock—*i.e.* the centre one—and drives a wheel of 91 with a pinion of 9 on its arbor driving another wheel of 91 with a pinion of 37 driving a wheel of 171, that last wheel will turn in 29d. 12h. 44m. 3·4s., which is only about half a second more than an average lunation. A pinion of 6 does very well for a driving one, though it is much too low a number for a driven pinion, except of the fly of the striking part. But if the centre pinion is 12 and the great wheel 182, those two will do for the first pair of such a lunar train. In that case, as the arbor of the great wheel must turn for winding, it would be necessary to fix the 37 wheel on the back of the great wheel, and let that drive either the 91 or the 171 as may be convenient; for the order in which wheels and pinions are arranged is immaterial as to the velocity-ratio between the first and the last. The common moon dials of old-fashioned clocks are only driven by a single tooth on the hour arbor driving a wheel of 59 teeth by jumps twice a day, leaving an error of 44 m. in a month, to be set right by hand every now and then.

A simpler moon train, quite exact enough for this purpose, may be made by a pinion of 15 on a 24-hour wheel of the clock driving a 98 wheel with a pinion of 25 driving a wheel of 113, which will turn in 708·736 hours, which is only an excess of 7 seconds over one lunation of 708·73415 h., or 3 minutes in 2 years, which would be quite imperceptible in any dial. But nobody seems to remember that no flat moon dial can possibly imitate more than one phase of the moon, either just half-moon, or else the narrowest possible crescent; for the *terminator*, or boundary of light and shade on the moon, is always a semi-ellipse, varying continually from the nearest visible approach to a semicircle down to a straight line and then back again. The only way in which real phases can be shown is by a globe half white or gilt and the other half black, turning on an axis in the plane of the dial in a hole just fitting the globe. And that only will look like the moon when seen right in front and from a considerable distance, as the moon is. Practically a vertical axis will do, though I need not say that the diameter which joins the 'horns' of the terminator is seldom quite upright, and sometimes leans a good deal either way (see my 'Astronomy,' p. 140 of 7th ed.). The axis must be turned by a bevelled wheel, of which half will come through the dial—or above, if you want to hide it. But all such contrivances are mere playthings, and show the moon's phases much less distinctly than an almanac.

Day of the month dials have to be dealt with rather differently, because the move must take place only once in the 24 hours, at some time in the night. Consequently the wheel which carries either a small dial with figures showing through a hole, or a hand pointing to figures, must have ratchet teeth driven by a lever or click or tooth moved once a day. There must also be a slight spring or 'jumper' somewhere on the ratchet teeth to keep them exactly in the proper place for the click to catch next time. In the old-fashioned clocks the month wheel simply had 31 teeth, and you had to move it on by hand in February, April, June, September, and November.

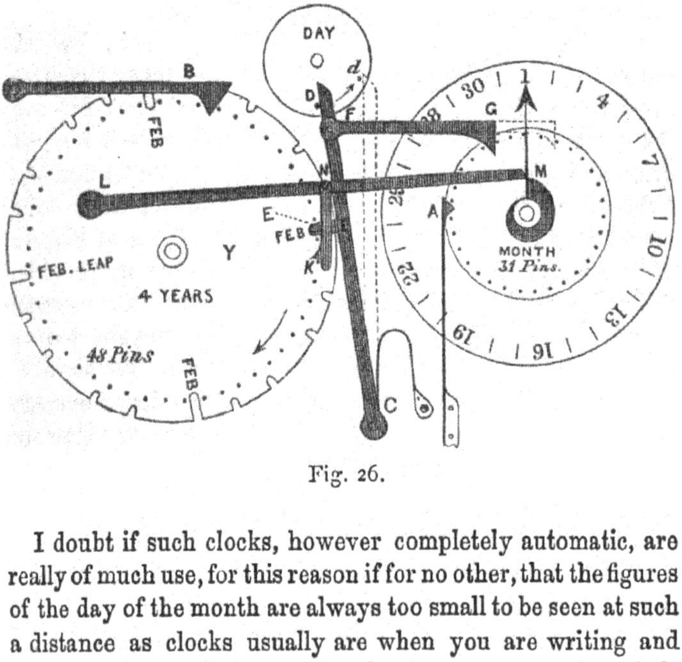

Fig. 26.

I doubt if such clocks, however completely automatic, are really of much use, for this reason if for no other, that the figures of the day of the month are always too small to be seen at such a distance as clocks usually are when you are writing and want the date. Large cards, changed every day, are infinitely more used. Nevertheless, as the machinery is, or may be, much simpler than you would suppose beforehand, I may as well indicate the nature of it, on what seems to me the simplest of the various ways of doing it. But for the present foolish arrangement of months, which the world seems impotent to set aside, we should only want six months of 30 days, and six of 31 in leap years, and in other years seven and five. But, as things are, we have four kinds of months to provide for. We must clearly have one wheel M with 31 teeth, carrying a hand pointing to a month dial; for that is much easier to see than numbers peeping through a hole, and it is enough to print every third number, except at last 28—30—1. Then we want some contrivance to push on two teeth in April, June, September, and November, three in February in leap year, and four in other years.

Let there be a 24-hour wheel Day, which, by the pin at D moving on to *d*, will move the long lever CEFD so much to the right as will carry the horizontal click FG over the space of 4 teeth or pins of the month wheel, *i.e.* will drive that wheel 4 days forward. That provides for common Februaries. Then if the lever CD is stopped from going more than the space of one tooth of M back again at the end of the long months, as of other days, they also are provided for. It must fall back the space of two teeth at the common short months, and three teeth at the end of leaping Februaries. Several ways have been invented for doing this. The simplest is to have a four-year wheel, or disc Y, with 4 deep February notches in it at quadrants, into which a long tooth E, on the lever CED, can drop, and short notches for the 30-day months, and none for the long months, but the outside of the wheel Y is the space of one tooth off the tooth E generally.

But we have to consider how the year wheel is to be moved at the end of every month, as it must be, and so as not to be held fast by the tooth E in any notch. I think the best way is to do it by another lever LNM on a stud in the clock frame at L, and a gathering click NK, which takes hold of one of the 48 pins in Y when that lever is lifted, and carries Y one step or month-space forward when the lever is dropped by the snail M just before the last move of the month is finished, or as the hand comes up to 1. This divides the work of moving the two wheels and levers more uniformly than any other plan, besides making sure of the tooth E being out of a notch before the year wheel wants to move: not that it would signify on this plan if it were not, for then the lever LNM would only wait to drop. Each of the wheels wants a safety click or jumper, A, B, of the usual kind for such motions. I only put pins instead of ratchet teeth to M, because the click FG will be safer not to slip out.

The work might be simplified a little and the lifts made less, and the year wheel need only be a quarter of the size, if you will be content to move the month hand at the end of February by hand. A week dial is quite superfluous, and so I have not crowded the picture with one. It only wants a seven-star wheel by the side of the day wheel, one ray of the star being moved by the pin D after it has come to *d* and done its monthly work. It is desirable to distribute the work as much as possible. The month dial ought to be at least twice as large as in this drawing, to be seen easily.

In any clock which has all this extra work to do, special care must be taken that the 'motion' wheels (or dial work) are held on the centre arbor by a squared spring, and a pretty strong one (see p. 136), or it is sure to slip and gradually drag, though the clock may be going right.

Several persons have taken patents for clocks and watches showing the time by figures appearing through a hole in the dial instead of by hands. Not that there is anything new in that, except as to the machinery for doing of it. Days of the month were generally so done 100 years ago at least; and my old regulator by Holmes, already mentioned at p. 62, shows the hours in that way. I do not profess to judge of public taste, but it seems to me that all such inventions proceed on the fundamental mistake of supposing that we *read* the figures of a dial. We do nothing of the kind, but judge at once from the mere position of the hands and the well-known marks for the hours and minutes. In fact dials are better and clearer with no numbers at all, but merely 12 large spots for the hours and five-minutes, and 48 small ones for the other minutes, as I have often convinced people by asking them if they want such a dial as they see in several of my clocks. They always answer 'Yes;' and then are much surprised to find no figures, but only the 12 strong marks and 48 small ones. And the hands upon a dial, whether large or small, are more conspicuous than any figures which can be conveniently made to appear through a hole, especially if the hands are of the proper colour—*i.e.* black on a white or gilt dial, or gilt on a black one.

Moreover, there is another objection to those 'chronoscopic' dials: the figure-discs must be driven discontinuously, or by jerks at the end of certain periods; and if that is done directly by the watch, as in Barlow's patent of 1866, the strain on the wheels is so much greater at the short time of action, and especially when the units and tens of minutes and the hour all have to be changed at once, that it is impossible for such a watch to go well. Besides the mere work of moving the 3 discs, all from the 5-minute wheel of the train with different intermediate wheels, there are jumper springs to be moved too. Whether such watches have been ever made I do not know; I have only read the specification.

It was patented in 1869 for clocks in a more practicable form by Siddons and Meese, who brought me one to examine. They have 3 cylinders, A B C, near together; A with the ten digits 0 1 . . . . 9 on its face for units of minutes, B with blank 1 2 3 4 5 (blank being better than 0 there) twice over for the tens of minutes (making 12 spaces), and C with 1 up to 12 for the hours. An 8-minute wheel in the clock has 8 pins in it, which are always lifting a weighted lever, except at the moment when it drops from one pin to another. The lever has a pall* or click which

* If this word has any etymology it clearly should be spelt *pall*, not from the *pallium* of an archbishop or a corpse, but from *παλλω*, to strike, the origin of pallet. 'Paul' or 'pawl' are mere nonsense, of the same order as the vulgar conversion of the expressive Yankee

first slips over and then brings down in falling one of 10 ratchet teeth on the side of A, and so changes it from one figure to another. Whenever A changes from 9 to 0 it also moves on B one figure by the usual contrivance of numbering machines for railway tickets, bank-notes, and pages of ledgers; and when B A change from 5 9 to blank 0, B in like manner changes C one figure, B having 2 levers on its side corresponding to the two blanks. There are other details of minor importance in Siddons and Meese's clocks, for which the patent has been transferred to Gillett and Bland, of the Steam Clock Factory at Croydon.

As descriptions of the numbering machine are not easy to find, I had better give one, at least sufficient to illustrate its principle. Take first the simple case of only two wheels or cylinders set near together on the same axis, but loose, with the 10 digits engraved on their faces, one A for units and the other B for tens. A carries on its side a small bent lever which generally hangs loose, but at one place in the revolution, where A is changing from 9 to 0, the lever has its tail caught by a fixed obstacle, and for that moment the other end of the lever cannot yield, but travels with its wheel. B has 10 pins on its side, and whenever the lever of A is in that condition it catches one of those pins and so drives on B one step, or changes it one figure. Then for the hundreds, B has a similar lever on its other side which changes a wheel C in like manner, and so on for as many figures as are wanted. In clocks the change of the 10-minute wheel has to be made after 5 so as to show 0 or a blank next, instead of after 9 as in numbering machines.

Equation of time clocks.—A still more obsolete con-

word *bunkum*, for bluster, into *buncombe*, for no other reason than because newspaper writers know there is a noble family of Duncombe, or the auctioneers' conversion of the site of a house into scite, which only shows themselves to be insciti.

trivance, but worth recording for the principle of its machinery, was that for making the hands of a clock show solar instead of mean time, at least in a rude and approximate way, which I suppose was used in the French public clocks which did so up to the year 1826. A*a* in this figure is a bevelled wheel on the centre wheel arbor, which in this case is made to turn the wrong way round, and another equal bevelled wheel B*b* rides upon it, with a hollow spindle *bc* to which the minute-hand is fixed. Between these two is another small bevelled wheel D of any size, which would merely reverse the motion, if it was set on a fixed spindle or arbor. But it is not, for it rides on the end of a bar or lever D E, which itself turns upon the centre arbor and has its end D beyond the wheel resting on a plate of the odd shape shown at Q*q*, which is fixed to the face of a wheel which turns in a year. Now if the lever D E, or the centre of the small wheel, is moved at any time in the same direction as the hand is going, it will evidently push it forward just twice as much as the lever itself is moved, and *vice versâ*. If then the equation

Fig. 27.

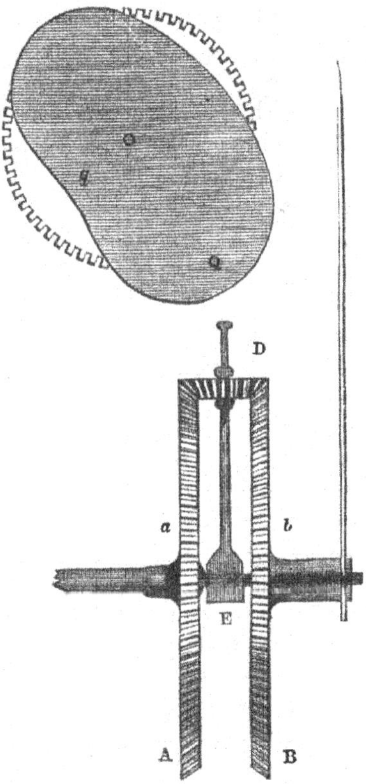

plate Qq is on the right side of the clock-frame, the hand will go ahead of its mean motion whenever D drops below the mean radius of the plate from O the centre of the year wheel, and will fall behind mean time when the protuberant parts of the plate are uppermost. The plate then may be so shaped as to make the advance and retardation of the hand agree with the 'sun before clock' and 'sun after clock' of the equation of time.

There are two other ways of giving a secondary motion of this kind; one, by substituting a common small wheel or pinion for the middle bevelled wheel, and putting it between Aa made as a common wheel, and Bb made as an *internal* wheel, *i.e.* with teeth inside its rim; but in that case you must remember that A and B will have different velocities, and therefore A must turn in less than the hour. The other method requires neither bevelled nor internal wheels, and is on the same principle as the one I shall describe more fully under *train remontoires.*

Clocks for showing other celestial motions are mere curiosities, and are always getting out of order from their complication; so I shall not waste time in describing them, but go on to something more practical.

MAINTAINING POWERS OR GOING BARRELS.

Winding up a clock evidently takes the action of the weight off the great wheel, and so the clock movement stops for the time, though the pendulum goes on swinging. This of course will not do in a clock of any accuracy, whether a large or a small one, and as the same methods of keeping the clock going (or some of them) are applied to both, I shall describe them all together here, except one.

The oldest of them all is *Huyghens's endless chain.* In fig. 28 P in the 'going wheel' is a pulley fixed to the great wheel of the going part, and having short spikes set in it, or roughened in some other way so as to prevent a rope or a chain hung over it from slipping. A similar pulley rides on another arbor *p*, which may be the arbor of the great wheel of the striking part, if the clock has one, and attached by a ratchet and click to that wheel, or to the clock-frame if there is no striking part. The weights are hung as you see, the little weight being only big enough to keep the string in the pulleys; but the string or chain is much longer, or one at least of the weights is always lower down than I have been obliged to draw it here. If you pull *b*, the left hand of all the strings, down, the ratchet pulley moves under the click, and the great weight is pulled up by *c*, without taking its pressure off the going wheel at all. This plan was generally used in the old 30 hour clocks, but went out with them, as the action of a chain or even a rope hung in that way is rough and uneven; and moreover the pulleys must be of only half the usual diameter for the same time of going.

Fig. 28.

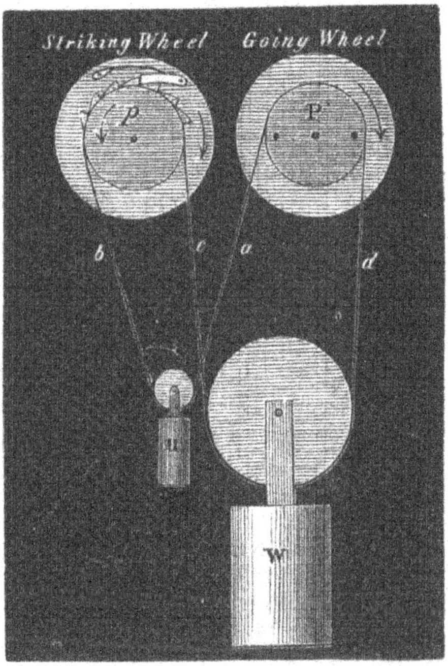

Harrison's going barrel is the maintaining power used in all regulators now. The larger ratchet wheel *r* is the one designated by the same letter in fig. 25 (p. 134); and the click R is fixed to that wheel, which is connected with the great wheel by a spring S S′. While the clock is going the weight acts on the great wheel G through the spring; but as soon as you take off the weight by winding, the click T*r*, whose pivots are set in the frame, prevents the great ratchet

Fig. 29.

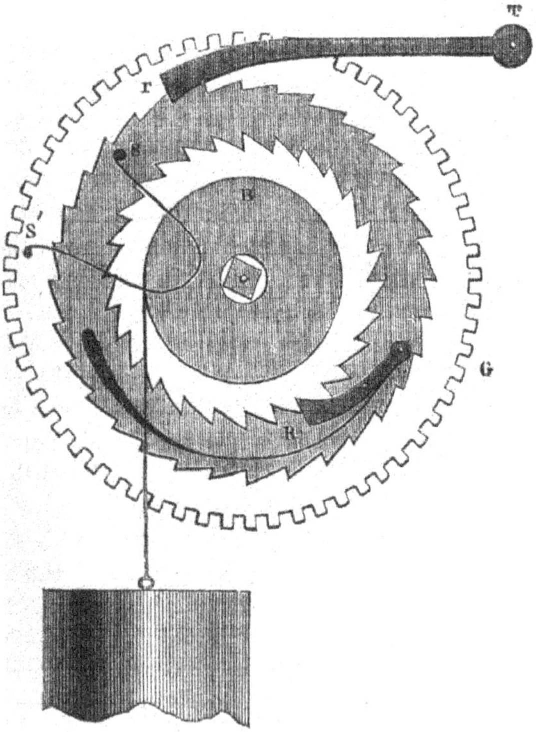

from falling back, and so the spring still drives the great wheel during the time the clock takes to wind, especially as it need only just keep the escapement going, for the pendulum will take care of itself for that short time. The drop of the great click over the teeth of its ratchet may be heard every 10 minutes or so while the clock is going. Watches have the same apparatus, with a spring click.

Bolt and shutter.—Another contrivance, which is now used only in large clocks, is an arbor with a weighted lever at one end of it, with a click in the form of a spring bolt on another lever; when the weighted arm is lifted up the click 'takes into' the teeth of some one of the train wheels, and the weight then keeps the clock going till it works itself out of gear in a few minutes and drops. The weighted lever is outside the clock and is made with a cap or shutter which shuts over the key-hole when it is down, to make sure of your lifting it before you begin winding. With the usual ingenuity for doing things wrong, this click is very often made not as a sliding bolt, but with a hinge, so that there is one position of the lever in which it jams against the teeth and stops the clock for good, unless the winding man finds it out and releases it, which he probably will not. Sometimes too the click sticks and sometimes it slips, even if made rightly. There is another defect besides in the common bolt and shutter, viz. that it may work itself down and rest upon the winder or key before the winding is done if the man is slow about it, and then it does no good and the clock stops for the time.

Improved bolt and shutter.—To prevent these evils, and to simplify the construction, I introduced the plan of substituting for the 'bolt' a segment D, in fig. 37, p. 183, of a small wheel suited to the teeth of the great wheel, and making the arbor C, which carries that and the shutter M, to pump in and out of gear, and the shutter not covering the key-hole, but made as a circular arc to the centre C, which all but touches the winder when it is on. The winder has a ring, shown by the circle at M, fixed round its end, which prevents it from being put on until you have lifted the shutter, and put it into gear with the great wheel, to hold it up. As you go on winding, the clock goes on and the shutter descends, now behind the ring, which secures your pulling it out of gear again when you take off the winder, and yet it will keep in action full 10 minutes if left to work itself out. This plan is now generally used in superior large clocks. The weighted arm should be long, so as not to require a heavy weight to be lifted.

I shall describe hereafter another totally different plan for keeping a very large clock going when the winding takes a long time; but as the spring going barrel does very well for small clocks, and the improved bolt and shutter for any but a clock of quite unusual size, I shall postpone that description till we come to the Westminster clock. For the same reason I need not repeat the description of Sir G. Airy's going barrel, which was applied to the Exchange clock, but is so expensive that it is certain never to be used again. A full description of it was given by him in vol. 7 of the Cambridge Phil. Trans. The object of it was not only to keep the power always on the clock, but exactly the same power; for the power of the spring falls below that of the weight, and that of the bolt and shutter doubles it just at the times of putting it on and taking it off; but that is of no consequence, except in a revolving pendulum clock, for which his apparatus was invented; he afterwards simplified the construction, but it is still an expensive one, and the thing can be done at a tenth of the cost by the Westminster method without either loss or duplication of force.

There is another maintaining power which has a tempting and scientific look, but is not so good as it looks. A rim at the back of the great wheel, M (fig. 29), has internal teeth at A—troublesome things to make—and in them works a wheel ABP on a stud B in the end of the barrel, and also in a pinion C fixed on the arbor which runs loose through the barrel ends. When you wind up you apply some force P to the intermediate wheel, and the same pressure P is communicated to the great wheel because B P = B A, P being whatever is necessary to lift the weight. Let W be the effective weight at A when the clock is going: at B it is

Fig. 30.

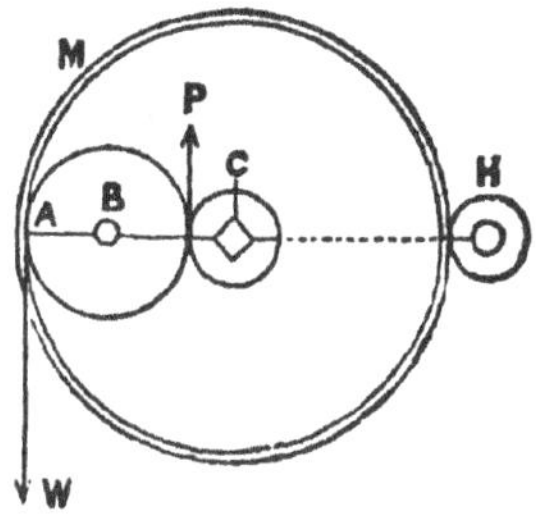

$W \frac{CA}{CB} = W'$; and $P = \frac{W'}{2}$, since $AP = 2 AB \therefore P = \frac{W.CA}{2\ CB}$; which cannot possibly $= W$, however small the winding pinion is, and if it is very small it takes a long time to wind. So this maintaining power must be deficient, though it may do for clocks with the common escapements, which will go for a few minutes with very little power on. Besides that, the winding arbor has to work under double friction of the pressure of twice W, both above and below it. For all these reasons I have never adopted it, either for large or small clocks.

Spring clocks.—Hitherto we have supposed all clocks to be kept going by a weight. But many clocks have no weight, but have a spring made of a long ribbon of steel coiled up in a barrel for their moving force. This construction however belongs so peculiarly to watches that I shall defer the description of it till we come to them. I will only mention here that the French clocks, like French and Swiss watches, generally have the great wheel fixed to the barrel, which of course makes the force on the train unequal. In that case the barrel arbor goes loose through the barrel ends and is fixed to the inner end of the spring, and has also a strong ratchet squared on to it, with a click on the clock frame, which holds it when you are not winding up. And a barrel of this kind is of itself a 'going-barrel,' for it keeps the power on as much while you are winding as at other times—in fact rather more. English spring clocks always have a fusee with a chain, made as in the figure of a chronometer movement, which will be given in the chapter on watches, to equalise the force on the train. I understand that this class of clocks is now sold more than any other English ones, with half-second pendulums, either in cases made to stand on a bracket or to hang up against a wall.

American clocks.—There have been great fluctuations in the quality of these clocks since they first came here above 40 years ago, and there is now a great variety of them, and they are certainly cheaper than most if not all others of the same general quality. Their principal defect is in the lightness of the pendulums. I have improved several of them by increasing its weight. The wheels and plates of the frame of these clocks are stamped out of sheet brass, and they contain several ingenious contrivances. They have advanced considerably in appearance at any rate since the original Sam Slick form; and, by the way, it seems that the original Samuel Slick was one Eli Terry, whose name ought to be preserved in a book on clock making. But even these new ones are equally defective in the weight of the pendulum, though they are longer; and yet I can hardly suppose that the makers of such clocks are as ignorant as a shopman here, who confidently assured me that the weight of pendulums is of no consequence! It would not cost a penny more to make them heavier. Some superior ornamental clocks and regulators appear to be made at the Howard watch and clock factory at Boston, U.S. I read lately in the *English Mechanic* that they make regulators as well as turret clocks with my gravity escapements, and that their own principal regulator has the four-legged escapement.

The old-fashioned 'Dutch,' *i.e.* German, clocks, with wheels of boxwood driving lantern pinions of wire stuck in wood, have been in a great degree driven out by the Americans with their superior appearance. But a more ornamental kind of German clocks is now made, and I see in the *Horological Journal* of October, 1873, that there are now nearly 2,000,000 clocks a year made in the Black Forest 'from the modest wooden clock up to the costly regulator,' and that there are 1429 clock manufacturers there, employing, one way or another, 13,500 people. Many of these are called trumpet and cuckoo clocks, in which the sound is made by a small pair of bellows, worked at every blow of what would ordinarily be the striking. The cuckoo is heard much farther over a house than striking. I need hardly say that the bird who appears at an open window is only a pretender, and does not contain the bellows in his inside.

Improved French clocks.—The old-fashioned French chimney-piece clock, of which the form is well known, was generally inferior to the commonest Dutch clock as a time-keeper. But a very superior kind has been introduced of late, far better than anything made in this country for two or three times the price. They have the pin pallets made of jewels, described at p. 100, and good pendulums of fair weight for their size. They have also a peculiar means of putting themselves in beat if they require it; for the pallets are set what is called 'spring tight' on their arbor, *i.e.* embrace it with a spring collar which will yield under sufficient pressure, though tight enough to give the impulse to the pendulum. If one pallet happens to work too deep, from the clock being out of level, it reaches the bottom of the teeth, or perhaps even the slight recoil of the pin pallets against straight teeth is enough to push it back a little into equal beat with the other pallet. These clocks have also adopted the English, or 'repeating,' striking part, which will be described afterwards. Every one knows that the old-fashioned French and German clocks oftener strike wrong than right, and are apt even to be set wrong in winding up the striking, which never happens with English or with the improved French ones. The ornamental English clock trade has ceased to exist, having been entirely driven out by these, and also by the

Austrian clocks, with cases about 2½ feet high, which

first appeared in our 1862 Exhibition. Their chief defect is the shortness of the arc of vibration, which makes them very sensitive to the least disturbance, not merely as to rate (as I showed at p. 92), but even as to going at all. They generally strike the quarters in a peculiar, and, I think, very puzzling way, *i.e.* 1, 2, 3, 4, on a single bell, and repeating the hour at every quarter; so that from 12 to 5 you have no idea what the clock means to say.

Self-winding Clocks have been made as long ago as the 1851 Exhibition. I have seen one by Mr. Horstmann, at Bath, with an endless chain made of a perforated wide watch-spring running over pins in the main wheel, and worked on the usual endless chain plan, but by a piston in a tube filled with naphtha, which expands and contracts with the temperature. I cannot say how far it is successful; and such clocks are evidently of no real use. Sometimes one sees an independent pendulum swinging in the hand of a bronze lady, apparently motionless, on the top of a small clock; but she really has a small twist at every beat in connection with the clock below, and that pendulum does nothing. Another curiosity sometimes exhibited is a balanced hand turning all alone on a glass dial. This is managed by a watch movement in the counterpoise, which tends to turn a weight inside it all round the watch in twelve hours. It is so poised that when the weight hangs at right angles to the hand it will lie horizontal, pointing to either III. or IX. When the weight is farthest from the centre, and therefore most effective, the hand must point upwards; and when the weight is nearest the centre the hand preponderates and points downwards, and similarly at intermediate positions. The weight in fact always hangs downwards absolutely in the watch, and the hand relatively revolves round it.

Water clocks.—It is evident that a flow of water may be made to drive a clock in various ways. A few years ago I was asked to see some water clocks in which the water came into a hollow spindle, and thence into four arms radiating from it. As each arm passed its lowest position it opened a cock in itself which discharged the water, and no more was allowed to enter from the hollow spindle until that arm came again to the top; so that the descending arms were always weighted with water, and the ascending ones empty and light. In some cases the weight was increased by a bulb or closed bucket at the end of the arm; and some of these were striking turret clocks, and they required very few wheels. Whether they ever came into use I do not know. I should think it would be difficult to keep the water from spoiling the rest of the machinery, and if such clocks require any attendance there would be no material saving over a common clock which requires winding.

Electrical clocks, properly speaking, are clocks going by electricity, instead of by a weight wound up periodically; but no such clocks have yet been invented capable of keeping as good time as the commonest weight clock. They may be divided into two classes. The first are those invented by Mr. Bain in 1840, modified more or less by other people, of which the principle was this: the pendulum bob is apparently a hollow brass cylinder, with the axis horizontal, and passing over two permanent magnets, without touching them, which are fixed to the clock case at one end, and nearly touch each other with opposite poles, which are marked N and S on the several magnets shown in fig. 32 (p. 160), omitting the upper set for the present. The cylinder really contains a long coil of insulated wire, of which the two ends run up to a pair of suspension springs, which make a complete galvanic circuit with wires somewhere connected with the poles of a battery, and capable of being 'broken.' The pendulum pushed a light sliding rod backwards and forwards, which made and broke contact. Whenever the circuit was complete there was attraction one way or the other between the magnets and the coil, and that attraction was enough to maintain the swing of the pendulum against friction, and the work of driving the train which it also had to do. For the escapement was rather a 'propelment,' like the action of the pendulum on a common recoil escapement if you take off the weight, for then the pendulum will drive the train the wrong way, which may of course be reversed. But these clocks never answered in any practical sense; nor would anything but the strongest evidence, independent of the inventor, convince me that any independent pendulum directly maintained by electricity can succeed in keeping good time for any considerable period.

Shepherd's clocks, by which it was announced that all the time of the 1851 Exhibition was to be kept, seemed more promising, but they soon failed totally there, and the time was kept by Dent's large clock, made from my design, now at King's Cross. In them the electricity was employed to lift a small gravity arm at every alternate beat, which gave the impulse to the pendulum by falling on a pallet like the down-pallet of a dead escapement, which had the advantage of giving a constant impulse when it gave any. But unfortunately it did not always lift. And any one who sets to work to invent electrical clocks must start with this axiom, that every now and then the electricity will fail to lift anything, however small: and if his clock does not provide for that, it will fail too. It is therefore unnecessary to describe Shepherd's plan further now, as I did in former editions.

The other class of so-called electrical clocks are those where a 'normal' or governing weight-clock drives or controls other subordinate ones; and these again may be either (1) a mere dial, or 'motion-work,' driven by electricity attracting a lever which works an escapement; this was Wheatstone's invention, about the same time as Bain's: or (2) when the normal clock keeps up the motion of a pendulum which drives an escapement—Ritchie's plan; or (3) Jones's, where the normal clock controls the pendulum of a common and inferior kind of clock wound up as usual, the pendulum being of the same length as the normal one. I have placed them in mechanical, not historical, order.

The first has never completely succeeded, and probably never will, from the cause already stated. Old Mr. Dent also tried at it for some time, but in vain. Mr. R. L. Jones, when manager of the Chester station about 30 years ago, hit upon the ingenious plan of controlling rather than attempting to drive a subordinate clock, by sending a current over its pendulum (in the same way as Bain's) at every beat of the normal or governing clock. It does not matter if the current sometimes fails for a few seconds, or even minutes, unless the subordinate pendulum has time to get more than a whole beat wrong, in which case the error would be soon augmented into two when the connection was resumed, which is rather a serious evil; and I understand this plan has been abandoned in some places which adopted it.

Mr. Ritchie, of 25 Leith Street, Edinburgh, has gone a step farther, and is able to dispense with the winding of the subordinate clocks by making the normal clock drive subordinate pendulums, which drive the escapements, and will maintain themselves for a few beats if the electricity fails for that short time. This will evidently bear less of such failure than Jones's plan; but still it appears from sufficient experience and the testimony of the Astronomer Royal of Scotland and others, that it answers very well. The construction of the escapement is a matter of some consequence. It is perfectly easy to drive a wheel and train by common recoil pallets, provided the wheel is good-natured enough always to stand still while the pallets are not moving it; but not if the hands are subject to disturbance by wind; and the very smallest motion of the hands would make a large one of the scapewheel. Mr. Ritchie accordingly uses several forms of pallets designed to hold the wheel as well as to drive it. The principle of them all is that of dividing them into two like the pallets of a gravity escapement, rising and falling with the pendulum.

In this figure (31) the pendulum is returning from the extreme right, and has just deposited the right pallet, B F *b*, with its end pressing on the tooth at *b*, and so trying to turn the wheel that way. But it cannot turn until the pendulum has gone farther and lifted the left pallet which is now locking the wheel at G. As soon as it does that, the weight of B moves the wheel, and itself falls a little more until another tooth locks at F. Then the left pallet falls and presses on a tooth at *a*, but cannot move the wheel until F is unlocked, and so on. In large clocks exposed to wind, a click is added to prevent the wheel from being blown backward by a force of wind greater than the weight of each pallet for the moment; it cannot be blown forward by reason of the stops F G.

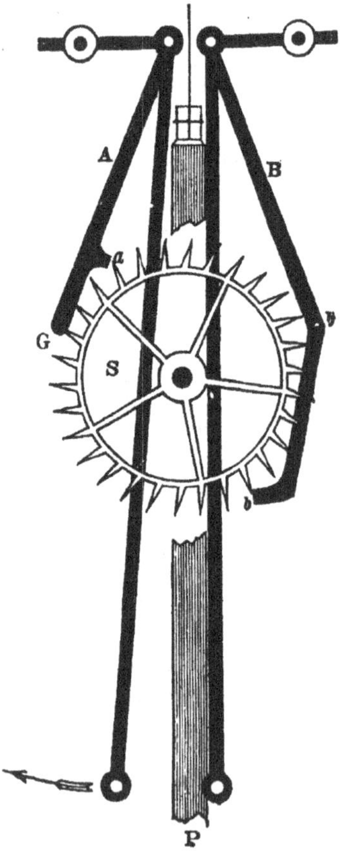

Fig. 31.

As the secondary clocks are kept in order by the normal one, they will bear a short or slow pendulum on the principle described at p. 41; and so a clock with a one or even a two seconds pendulum can be comprised in the length of an ordinary half-seconds 'dial,' as they call those very short

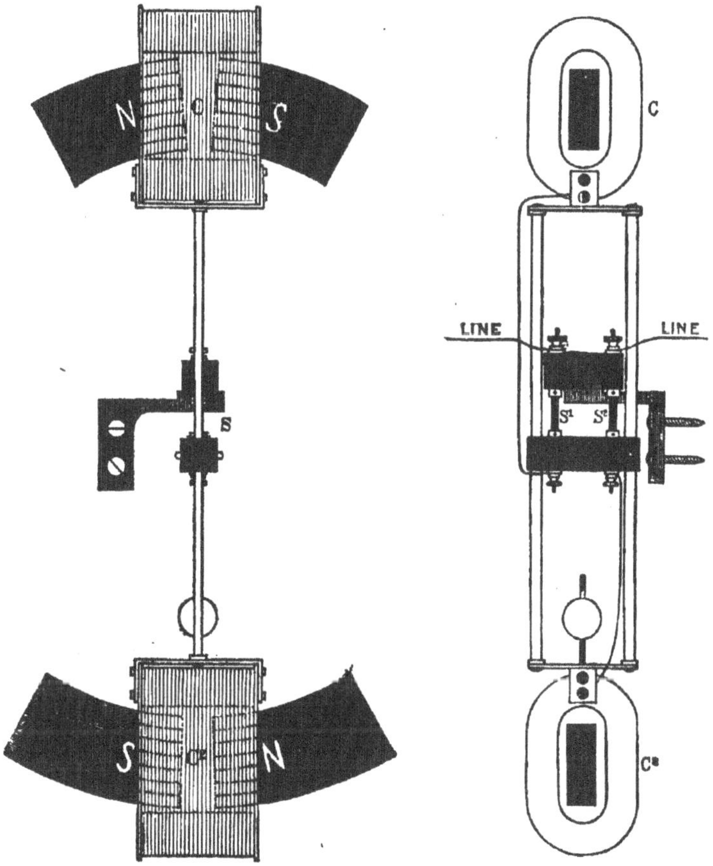

Fig. 32.

clocks, which look no bigger than their dials. Accordingly Mr. Ritchie makes these pendulums, as in this figure (32).

The lower half of the pendulum is rather longer than the upper, but both have bobs of wire coils in a cylinder, passing over a set of magnets as before described. The suspension is by two springs at S, as shown in the right hand part of this figure, which is the section across the plane of vibration; and one is connected with one battery wire, and the other with the other, the pendulum itself being made with two rods; and so the current goes down one and up the other, being reversed at each beat as before. In this way also a double attraction is got, as both bobs act in the same way. The small ball above the lower bob is for regulating the pendulum. The left hand figure is the front view.

This is taken in substance from a paper by Mr. Ritchie read to the Royal Scottish Society of Arts in April 1873, and published separately. His clocks are in use in connection with the Liverpool Observatory, besides Edinburgh itself, and a variety of other places. The great objection to having all the clocks of any large establishment worked by only one is the possibility of all time coming to an end by any temporary failure of the one which 'makes' the time of all. It would be prudent to have two normal clocks always ready to be connected with all the subordinate ones in case one has to be stopped; and the normal clocks should be good ones capable of going without any control.

When the last edition of this book was published, the Royal Institution, which is a sort of head-quarters of electrical science, had just been converting all its clocks into electrical ones, not on any of the controlling plans, but on a driving one, by currents from a sort of small turret clock with a hollow electrical coil for a bob, swinging a very large arc over two permanent magnets. As I expected, and intimated then, it turned out a failure, and was given up after a few years.

Another controlling plan is Lund and Blockley's, in which a Λ-shaped fork is brought down over the minute hand exactly at the hour by electrical connection with a standard clock, so as to bring it to the time if it is a little out either way. That requires a complete clock wound up and going as usual, only it need not go very well independently. There are also clocks driven by pneumatic connexion with a standard one; but I have no authentic information about their success.

Time balls and guns are let off by electrical connection with a normal clock which attracts an armature on a trigger which lets off the ball or fires the gun when the connection is completed by contact springs in the 12 or 24 hour wheel and the minute wheel simultaneously. In like manner the Westminster clock reports itself to Greenwich daily. A time ball is usually a large wicker globe covered with painted canvas, fixed to a piston which falls down into a bell-mouthed tube just air-tight enough for the air to act as an elastic cushion. It is hauled up by hand a few minutes before the time for falling. But these also sometimes fail like electrical clocks, and a better plan is to let a common strong clock electrically controlled discharge the ball or gun mechanically, just as it lets off a striking part.

STRIKING CLOCKS.

The simplest form of a striking clock is represented in fig. 33. It only strikes one at every hour, which is sometimes more agreeable than striking the full hour, especially where you can easily see at once what the hour is. It is very seldom that anybody has to count the striking of a clock except in the night. Striking one requires no striking part, or separate machinery to be wound up, the hammer being lifted by a snail (or sometimes, but worse, by a pin) on any wheel which turns in an hour, and dropped as the minute hand reaches the hour. It cannot be done however without affecting the friction of the clock, and therefore its rate, except with a gravity escapement or some equivalent contrivance to prevent the inequality of force from reaching the pendulum; and the remark I made at page 135 about the importance of fitting the dial spring on to the centre arbor with a square instead of a round hole, applies still more strongly here, as the friction of that spring has to over-

Fig. 33.

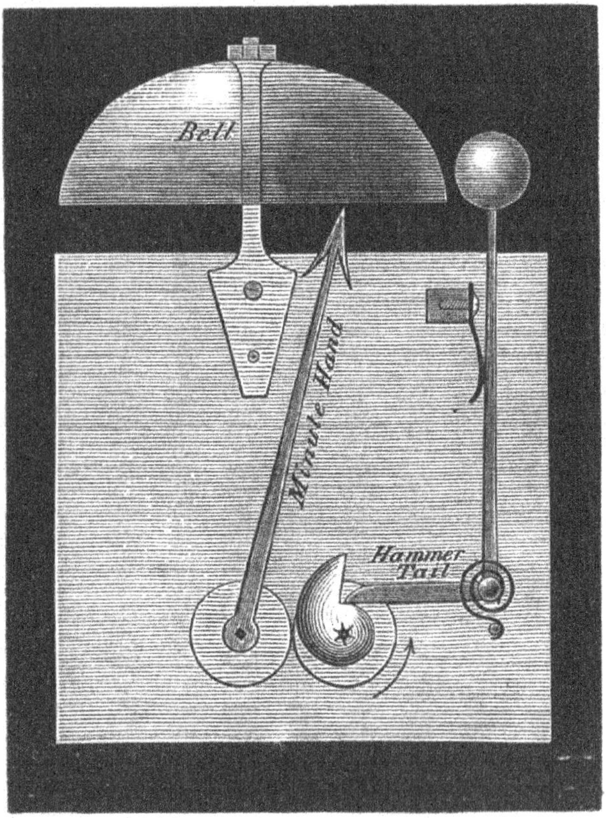

come the resistance of the hammer spring. The short spring against which the hammer shank falls is to prevent it from jarring on the bell, and this is necessary in every case both in large and small clocks, unless some other contrivance is resorted to for the same purpose, such as I shall speak of under turret clocks. In common clocks this check is given

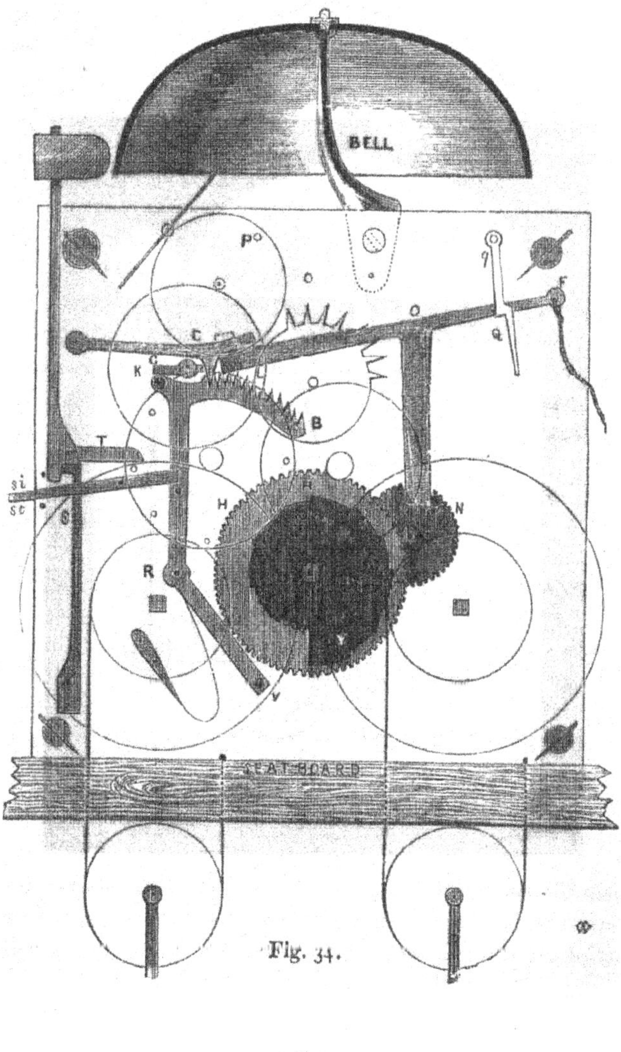

Fig. 34.

in a simpler way, as shown in fig. 34, where the square top of the stiff spring S butts against the square piece on the hammer shank, whose own elasticity lets the hammer strike the bell and then pulls it back again just out of contact. A piece of vulcanised India rubber tied round the pillar also answers very well.

This figure is a front view of the common construction of an English striking clock: the foreign ones are different, as I will explain presently. The wheels shown only by circles (with a few of the scapewheel teeth) are within the frame; only those with teeth are outside, and they are indicated by the same letters as in p. 134. The hammer tail is raised by 8 pins in the second wheel of the striking train, which corresponds to the centre wheel of the going train. The pinion of that wheel generally has 8 leaves, and is driven by the great wheel of 78, which therefore turns in 12 hours; not that that is at all material, and of course higher numbers would be better. The striking wheel drives the wheel above it once round for each blow, and that wheel drives a fourth (in which you see a pin P) six or any integral number of turns for one of the third wheel, and the fourth wheel drives a fly to moderate the velocity of the train and the time of striking.

The number of blows to be struck is regulated thus: the dial-wheel N has a pin on its face which raises the *lifting piece* L O N F a little before the hour, just far enough for it to lift the long click C out of the teeth of the *rack* B K R V, which then falls back (helped by a spring at its tail) as far as the tail V can go by reason of the position of the snail Y on the hour-hand wheel H; which has steps in it, one for each hour, so as to let the rack fall the distance of one of its own teeth for every hour the clock ought to strike. This fall of the rack makes the noise called *warning* a few minutes before the clock strikes. The reason why it cannot begin to strike yet is that the pin P cannot pass a stop which is turned inwards from the lifting piece, through a large hole in the frame, until that piece drops again, which it does exactly at the hour by the advance of the pin in wheel N. Then the striking train is free, and that little piece K G, called the *gathering pallet*, which is squared on to the prolonged arbor of the third wheel, gathers up the teeth of the rack, one for each blow of the hammer: the click is lifted as each tooth passes, and prevents the rack from falling again, and at last all the teeth are gathered up and the tail of the pallet is stopped by the large pin K in the top of the rack, and the train can go no further.

The great feature of this English striking work is that you may 'strike' the clock as often as you like within the hour, or stop it any number of hours, and yet it will always strike right, because the striking depends on the position of the snail attached to the going part, and not at all on the number last struck. These clocks are therefore sometimes furnished with a string to the outside, from the click, so that you can pull it in the night and hear the hour. But this is just the wrong way of doing it: for if you hold the string too long the click will miss some of the teeth and the clock strike too many, and if you drop it too suddenly the rack will not have fallen its full distance and it will strike too few. The right place therefore to put the string is to the lifting piece, as at F. (The piece Q*q* belongs to something else which I shall speak of presently.)

The improved French clocks, whose escapement I noticed at p. 154, have a better mode of stopping the striking than the usual one in English house clocks: the better one is used in our turret clocks when they have the rack movement. Instead of the gathering pallet G (fig. 32) on the third wheel of the train being stopped by a pin in the tail of the rack with a heavy blow, the striking is stopped by the pin P in the fourth wheel coming against a stop in the long click C, which drops into a deeper notch in the rack than the others at the last blow of the hour, and so falls low enough to catch that pin.

Strike and silent.—There are several ways of throwing the striking work out of gear, so as to keep the clock silent. I think the best, though not the usual one, is that shown in fig. 32, a small lever whose end *x* falls before and stops a pin in the rack when the other end of the lever is put up to *si* by an index or handle coming through the edge of the dial. I have seen methods used which are very likely to stop the going as well as the striking of the clock by leaving the rack to fall. Another way is to make the piece L O N F push forward so as to escape the pin at N, and be never lifted; and this is unobjectionable.

It may save people a little time if I tell them what hardly anybody seems to know, that you may move the hands of an *English* clock forward through all the hours without waiting for any of them to strike, except 12, where the rack-tail has to get over the great step in the snail; and even that is often provided for by sloping the front of that step and the face of the pin V, to let the snail push it aside, the tail being elastic enough to give way a little. In the same way the tail of the lifting piece at N is always twisted a little to let the lifting pin pass it backwards without lifting, so that you may turn the hands back and the clock will not strike at all, unless it has already given warning. The snail is sometimes set on a separate wheel below the hour-wheel and moved by jumps by a pin in the minute-wheel M; this is called the *star wheel* and *jumper;* but I can see no use in it, and therefore shall describe it no further.

Locking-plate striking work.—The principle of the striking work still used in most foreign clocks, French, German, and American, is shown in fig. 35 (p. 168), though the actual arrangement of the pieces may be different. It is generally used also in English turret clocks. You see the rack and its click are gone, and instead there is a wheel Y called the locking plate or count wheel, which turns in the

Fig. 35.

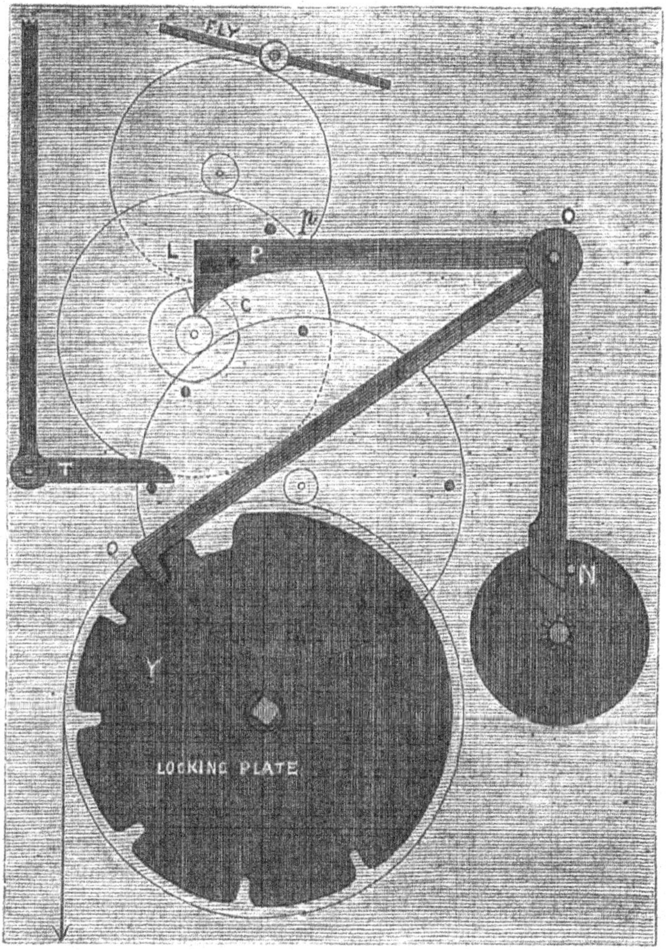

12 hours, and may therefore be put on the arbor of the great striking wheel, or driven by the pinion of the striking

wheel. It may be considered as marked out into 78 divisions, and notches made in it at the distances 1, 2, 3, &c., into which a lever O Q can drop, which is connected with the lifting piece. That is generally made with two stops upon it, one a little behind and below the other; or else there are two levers, one lifted by the other, with the stops upon them respectively; and the pin P in the pin-wheel, or in large clocks, on the fly arbor, stops against the first when the clock has done striking, and against the other when the lifting piece is lifted by the wheel N in the dial work. But I introduced an alteration in this respect in turret clocks, because one of the stops on the lever must be out of the right position for direct action of the pin. Instead of two stops on the lever, there are two pins on the wheel, P and *p*, one a little behind and nearer the centre of the wheel than the other, and *p* is caught by the stop when it has been lifted high enough to let P escape and 'give warning.'

There is, or ought to be, a disc C with a notch or cam in it on the arbor of the third wheel, which turns once for each blow, and so moves faster than the locking plate, and therefore is more certain to lift the lifting piece quite out of the way of both pins before the pin-wheel gets once round; otherwise the clock might be, and clocks without this cam sometimes are, prevented from striking at all, especially if the parts are not adjusted with great precision. The lifting piece evidently cannot fall again until another notch in the locking plate comes under the tooth Q. Sometimes the lifting piece L O N is made in two pieces, one lifting the other, but I see no advantage in it, except where the pivot C happens to be at an inconvenient distance from the locking plate. In turret clocks with three-wheeled trains it is generally more convenient to make the wheel C turn only half round for each blow, and then there must be two cams or two drops in the disc, as shown in the view of a large quarter clock at p. 196.

Half-hour striking.—For striking *one* no lift by the locking plate is required, but only a long notch reaching from 12 to 2; and for the same reason the clock can be made to strike one at the half hours by dividing the locking plate into 90 (= 78 + 12) and leaving a wide notch between every two hours, and putting a half-hour pin into the wheel N (fig. 32) besides the hour one. Most of the French clocks are so made; but they have the inconvenience of striking one three times between 12 and 2; so that between those hours the striking tells you nothing. I once saw a turret clock made to strike one feebly on a smaller bell, from the going part, as shown at page 163, which gravity escapement clocks will bear, though not others. It was not satisfactory, and led me to devise the following plan.

To make 12½ and 1½ silent, with the locking plate movement, we evidently want something which will stop the lifting piece, L O N in fig. 37 (p. 183) of a clock of this kind, from falling after it has been lifted to give warning, until the next hour. And the way to do that is to have a 12-hour wheel (the one with 24 ratchet teeth in fig. 37) with two steps in it, as you see, which come under the tongue of the lifting piece just over that wheel at those two half hours. The best way to drive it is by a gathering pin or single tooth, which is shown in the hour wheel marked 40, and which moves the ratchet wheel one tooth just after warning. There is also a spring click or jumper to keep it in its place, which wheels driven in that way always require, to make sure of the gathering tooth taking them up again. But another thing has to be attended to. The locking plate, instead of having 78 teeth or spaces, as when there are no half hours, or 90, as when all the half hours strike, must have 88, and each notch must again be wide enough for striking one, only it must be divided as if there were no half-past twelve or one, for one o'clock is the same as one half hour between twelve and two.

Half-hours with Rack Movement.—I have never seen this in any English clock. Indeed the English house clock-makers seem determined to lose every bit of the trade rather than allow any single improvement to be made here, and so they are losing more and more yearly. The modern French arrangement of the rack movement has the rack above the 'motion' or dial wheels, acting by gravity without a spring, and the striking is stopped by the long click dropping lower at the end of the rack than it does in the teeth, and low enough to catch a pin in the third wheel, instead of the gathering pallet being stopped by a pin in the rack and always tending to force it back. Half hours are struck by a separate lever being brought under a pin anywhere on the rack allowing it to fall one tooth only, the lifting piece being of course lifted by a second pin in the hour wheel at the half hours. In order to stop the striking at 12½ and 1½ there would have to be a 12-hour wheel with 2 steps or long teeth, as before described, to prevent either the rack or the lifting piece from falling, whichever might be most convenient. But this is not of so much use in small clocks standing in a room, when you can look up and see whether it means one o'clock or not, as in large ones, where it is a great convenience, and saves the much greater cost of quarters with the two more bells. It would never do to have a turret clock striking the hours on one of the quarter bells, as it would produce frequent confusion when they are heard at such a distance that people would not be sure whether they heard the smaller bell or not. It was done for some months at Westminster by the genii who managed and advised things there, as will be noticed afterwards. When Mears's Big Ben II. cracked enough to spoil the sound, but no more, they stopped the striking of the hours on him, and had them struck on the fourth quarter bell, until I got the confusion that it caused noticed in the newspapers, and then they resumed striking on the great bell with a much lighter hammer, which is there still. And for fear that should knock a piece out of it, like the one in the famous broken bell of Moscow, they put a huge wooden tea-tray under it to catch the piece, and there it is to this day, with a hole in the middle for the clapper 'tail to come through,' which reminds one of a celebrated poem of either Southey or Porson, for it is attributed to both.

The locking plate striking is specially objectionable for moveable clocks, and for any that are liable to run down for want of winding, like the American ones; because if the striking once gets wrong, or stopped from not winding up, or let off by accident, or the clock stopped by house-maids, it strikes wrong afterwards, until it is struck round to the right hour again. The American clocks have a wire specially provided for this purpose, but the French have not, and you have to put your finger in behind the left side of the clock and lift the lever you will feel there, to make it strike as often as is necessary to bring it right; which is a great nuisance, and in some cases almost or quite an impossibility. Striking on springs is a bad imitation of a large bell, and is very inferior to a common bell, except when heard very near.

Quarters.—If the clock is to strike the quarters, a third 'part' or train of wheels is added on the right side of the going part, with as many bells and hammers as may be required. There is indeed a method of making the same striking part do both for the hours and quarters, by sliding the hour hammer tail out of gear with the pins, and the quarter hammer tails into gear, and *vice versâ;* but it saves very little in cost, and is very seldom used; and it requires a much heavier weight or stronger spring, and stronger wheels, if it is to be done effectively. In that case there can be no quarters at the hour; but that is of no consequence, and perhaps rather an advantage with *ding dong* quarters. The construction of a quarter part is substantially the same as of the hour striking part. If there are only 2 bells, the 2 hammers are lifted by pins on each side of the striking wheel, or they may be the same pins if the arbor of one hammer is put above that of the other. They should be so placed that the interval between each pair of blows, or each chime, is twice that between the blows of each chime, whether there are 2 or 4 bells. When there are more, the interval between each chime requires to be as much as 3 spaces instead of 2. When there are more than 2 bells the hammers are worked by a chime barrel, because the chimes are not generally the same thing repeated, as they are with *ding dong* quarters. But this belongs more to turret clocks, under which I shall go more fully into it. The chime barrel is generally put on the third wheel, but it would require less force to turn it on the second, for the reason I gave before, that the more wheels there are between a slow power and a quick work, the more is lost in friction, in a proportion beyond what anybody would expect. As the barrel naturally turns in an hour, the proportion of the pinion and great wheel would be just the same as in the going part.

The quarters may be let off either by the English repeating method, or by the French locking plate. If they are merely the same chime repeated 2, 3, and 4 times, the repeating movement should be used, as it has the same advantage as in the hour striking part. But if each quarter is a different tune, it should not be used, because repeating the striking of the quarters in that case will throw the whole tune into confusion; though this plain distinction is often overlooked. The connection between the two striking parts, when the quarters have the locking plate, is made by that wheel performing the function of the wheel N in figs. 34, 35, in discharging the hour, as the 4th quarter finishes.

The repeating quarter movement is not so simple: the principle of it is this:—The quarters have a rack, snail, &c., just like the hours, the snail being fixed on the wheel N so as to turn in the hour, and with 4 steps instead of 12 (see p. 164). The rack is so placed that when it falls for the 4th quarter (its greatest drop) it falls against the hour lifting piece somewhere between O and N, so as to raise it and the click C. It is then held up by the lever marked *Qq* catching hold of the pin close by it, and as the last tooth of the quarter rack is gathered up it pushes *Qq* aside again, which lets the lifting piece drop and the hour begin to strike.

There is a very simple construction of a clock for striking repetition quarters only, when it is wanted in the night, by pulling a string which goes round a spring barrel, and so winds it up as far as it is allowed to go by the position of the quarter snail on the going part, which stops some pin or lever connected with the barrel. This may be easily made to indicate half quarters; for if there are 8 steps in the snail, then at about 50 min. past the hour the lever could go 7 depths and the clock would strike 3 *ding dongs* and one bell more; and it may either begin or end by letting off the hour. This construction is in substance that of *repeater* watches, of which the striking part is both wound up and let off by pushing in the handle.

Quarters—Tunes.—By this I mean chimes not merely repeated. And as these clocks for halls and rooms have come much more into fashion since the Westminster chimes were heard, I will give a picture of the construction I arranged for them in a clock of my own before I designed that clock, and which was followed in several others made by old Mr. Dent and afterwards by other makers. The chime tunes themselves I shall leave till we come to turret clocks. Y, the locking plate, is fixed on the arbor of the second wheel, which turns once in an hour, driving the chime barrel marked 50 for Westminster chimes, as will be explained afterwards. For the present it is enough to say that it turns twice in the hour, making 10 chimes, but the fourth and first quarters together are the same as the second and third together, and

Fig. 36.

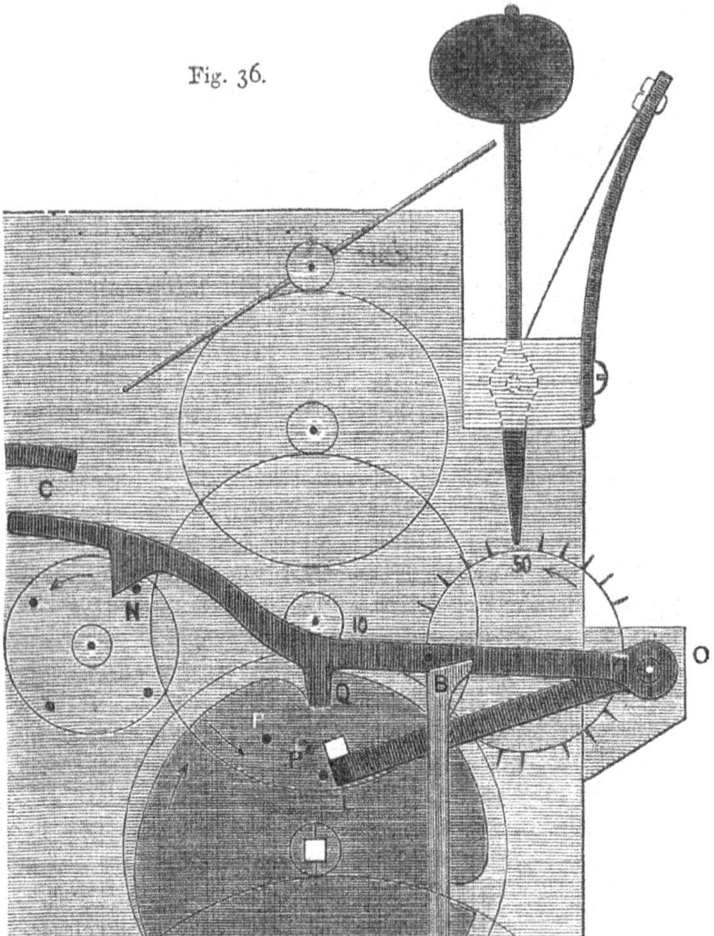

therefore it is not necessary to have a barrel with 10 different sets of pins on it, but only 5. The wheel on its end therefore has 50 teeth, and is driven by the second wheel of the train, of 100, which also drives a pinion of 10 on the third wheel, which consequently turns once for each chime of 4 bells. That wheel also has two pins in it, P and P′, P being rather nearer the centre, and they are alternately stopped by the piece at L projecting sideways from the lifting detent LON, of which the arbor is carried in cocks from the frame, as you see, its arm OQC being in front of the front plate of the clock and OL inside, near the back plate where the third wheel is. The chime barrel is also really set in cocks, not in the solid frame, in order that it may be adjusted as to its teeth after the clock is put together, which often saves a great deal of trouble, but it would confuse the figure to show them here. The fly should be longer than will go inside the frame, to clear the arbor below it, and that also is set in a long cock behind.

The action is extremely simple. The 4 pins in the hour wheel N lift the long detent a little before each quarter, and then the pin P′ slips under the stop L, which then catches P, the wheel moving on a very little. At the quarter the detent drops off N, and before the third wheel or its pins come round again the locking plate has lifted the detent by the inclined edges of its notches, such as Q, so high that both pins clear the stop L; the motion of the locking plate here being much greater than in an hour striking plate like fig. 35 and figs. 37, 40 afterwards, where consequently a cam wheel turning faster is required to make the lifting safe. The locking plate keeps up the detent for 1, 2, 3, or 4 chimes as required, and you see that by this method they can never get mixed, as they can with the repetition or rack movement. If they have got to strike wrong they only want 'striking round' till they are right again, and that is easily done thus. You see the right-angled lever BAD, with *Si* and *St*, for *silent* and *strike*, against two pins in the frame. When you lift the tail D to *Si* the bevelled end B lifts the detent by a pin in it, just enough to 'give warning,' and it cannot fall again. But when you bring it down again to *St* the detent does fall and the quarters strike whatever the locking plate is ready for, which lets you know whether they require more 'striking on' to set them right.

But we have still to let off the hour striking. In large clocks that is best done independently, as we shall see; but in small ones it is done by the end of the detent being raised by the snail kind of rise at Y in the locking plate as the fourth quarter finishes, which lifts C, the rack click of the striking part (see fig. 34). That is one way of doing it. Another, which I designed for a later clock of my own, allows the hours to strike even if the quarters are silenced, by retaining the usual lifting piece of the hours (which in the former plan is gone) and making the quarter detent stop it from falling again till the quarters are done—if they are in action; if not it simply does nothing. The quarters are silenced by pressing a spring against the side of the fourth wheel, by a slider at the edge of the dial.

Alarums.—If you suppose a short hammer instead of a pendulum fixed to the pallet arbor or the crutch of either kind of recoil escapement, it would swing backwards and forwards very quickly, and strike both sides of a bell of proper size placed so as to inclose the hammer. This is the way an alarum strikes, and not by the lifting of a hammer at distinct intervals. The hammer is driven by a wheel like a strong recoil crown escapement wheel (p. 28) with a spiked pulley or barrel attached to it by a ratchet and click, over which a rope goes, with a small weight at one end, and a smaller one at the other to keep the rope stretched and to wind it up by. The alarum can only go when a stop lever is lifted by a pin on a collar which is fixed on the hour hand wheel by a friction spring, so that it can be set to go off at any hour you like. You must not wind it up till within 12 hours of the time it is intended to go off, or it will go 12 hours too soon.

Tell-tale clock.—This is said in one of the Parliamentary papers about the Westminster clock to be an invention of Mr. Whitehurst of Derby, for watching watchmen and telling whether they are on the watch and in the proper place all the night. That unpleasant little clock which one hears striking the quarters 3 or even 4 times in some Westminster Abbey sermons, is of this kind, and there are some in the lobbies of the Houses of Parliament. There are a set of spikes sliding in holes in a 24 hour dial, one for every quarter of an hour, which can be pushed in by pulling a handle in the clock case during a few minutes of that quarter only. So if any pin is found sticking out in the morning it indicates that the watchman was either asleep or away at the time belonging to that pin. The plate carries the inner ends of the pins over an inclined plate or roller at some other period of the 24 hours, which pushes them all out again ready for work the next night.

Musical clocks.—Clocks that play tunes—not short quarter chimes, but tunes of several minutes, either on bells or organ pipes, are not clocks in respect of their music, but simply musical boxes or barrel organs turned by an independent spring or weight and lot off at the required time by a lever from the clock. I shall speak of chimes on church bells farther on. And nothing else occurs to me as belonging to small or house clocks of sufficient use or importance to require notice. So I pass on to the larger branch of the art, which has been the subject of greater improvements within the last 30 years than in the previous century, and has reached a degree of accuracy equal to that of the best astronomical clocks, and superior to that of chronometers, and that not only without an increase, but with a great reduction in the price.

CHURCH OR TURRET CLOCKS.

It may be supposed that as the work of these clocks only differs from that of house clocks in the size of the hands and the weight of the hammers they have to move, you have only to enlarge the machinery and the business is done. But there is a very important fact in the way of that conclusion: viz., that as you increase the strength of machinery you increase its weight in a ratio as much higher as the cube is higher than the square of any of its dimensions; and when you increase weight you increase friction, and friction is a word which ought never to be long out of the mind of a clockmaker, or at least of a clock-designer, inasmuch as the timekeeping part of a clock is the only machine whose sole business is to overcome its own friction, resistance of the air, and variations of heat, and to do that in a constant and uniform manner. And there is this further difference between large and small clocks: in small ones the force or weight required to work a hammer of an ounce or two is generally about the same as is required to keep the pendulum going, and so the two 'parts' or trains are about equal in strength; whereas in large clocks the lifting of the hammer generally requires a great deal more power than driving the hands and pendulum, and therefore ought to have much heavier and stronger machinery. Nevertheless the object o. some clockmakers seems to be to make the going train of large clocks as heavy and the striking train as light as they can.

Pendulum.—I have already treated of pendulums for large as well as small clocks at considerable length, and there is little to add with reference to large clocks only. I will only repeat that the construction and suspension of the pendulum are of primary importance.

The great majority of clockmakers, till lately, set their faces against compensated pendulums, and used nothing but wooden ones. And so long as the clocks themselves are no better than they are, it would undoubtedly be a waste of money to compensate the pendulums, as the escapement errors will far exceed the temperature one. But when you have got a first-rate clock in other respects, it is absurd to prevent it from going accurately by not giving it a pendulum without which it cannot keep the same rate in hot and cold weather. It is true that a 2 seconds, or even a 1½ second compensated heavy pendulum, is a rather expensive affair if well made; and with a common dead escapement probably the advantage is on the whole in favour of a 13 ft. wood pendulum of 3 cwt. over a 5 ft. compensated one of half the weight, which will enable a clock with such an escapement as I shall describe to keep within a second a week of Greenwich time. The fashion of extravagantly long pendulums has very properly gone out, as their inconvenience and liability to be affected by the wind overbalances any advantage from them in a moderately good clock. There were several in Yorkshire until lately as long as 56 feet, or 4 seconds: 20 feet = 2½ seconds, which old Doncaster church had, is the utmost length I should allow; and I gave the calculations for a cheap form of compensated pendulum of that length at p. 61.

Position of clock.—The worst of all positions for a large clock is the usual one, on a stool on the upper floor of a tower, for the reason I have already given at p. 93. The best is on stone corbels built deep into the wall. The Westminster clock lies on independent walls, which of course are stronger still. Where this cannot be done, cast iron brackets bolted through the wall will do, or iron beams across the room if it is not very wide. Wooden beams are not to be trusted. When the clock is fixed as firmly as this, the pendulum may be hung from the clock framo, if that is itself strong enough, and the pendulum cock properly fixed to it, or cast with it, though the wall is generally to be preferred for a long and heavy pendulum, if the clock stands near enough to it. But again it is inconvenient to have a very large clock so close to the wall that a man cannot get some access to it from behind. Therefore no general rule can be laid down for the fixing of turret clocks, except that firmness is the first consideration, to which everything else must give way according to the circumstances of the tower. I have already mentioned at p. 93 the increase of arc caused by hanging a heavy pendulum from the wall instead of a strong wooden frame from the ground; and as the escapement errors vary inversely as the cube of the arc, the clock should go more than twice as well with the firmer suspension; and in fact it does.

Frame.—The old established form of clock frame was a sort of cage of vertical and horizontal bars, some of which contain the bushes for the pivots of the wheels, and have to be unscrewed from the principal bars in order to get any of the wheels out. It was a great improvement on this to fix the bushes themselves with screws instead of riveting them into the bars, as it enabled the wheels to be taken out separately, instead of all dropping loose at once and perhaps bending their back pivots as soon as the front bar was taken off. Mr. Vulliamy introduced this plan, and old Mr. Dent used it in the Exchange clock, of which a perspective view is given in Tomlinson's Cyclopædia under *Horology*. But he soon afterwards adopted a still better arrangement, borrowed in principle from the French, who were strangely ahead of us in this branch of clockmaking, until shortly before the 1851 Exhibition.

The French clockmakers are entitled to the credit of having introduced the horizontal frame cast in one piece, with the great wheel set in bushes or cocks below it, and the smaller wheels above in separate frames of the A shape bolted to the great one, But much more has been done since that. Dent's clock, with a spring remontoire, which I designed for him for the 1851 Exhibition, where it kept better time than had ever been attained before, for nearly 6 months, will be described afterwards, though even that has been superseded by the simpler and stronger construction of the gravity escapement clocks. I will now describe a moderate-sized turret clock of that kind, for a bell up to 5 or 6 cwt., and therefore only needing the striking part winding once a week, which will not do for large ones, and striking one at the ten proper half-hours as before described. This was made for a clock-tower of my house by Mr. Joyce, of Whitchurch, Salop, the maker of the great Worcester Cathedral clock, striking on a 4½-ton bell, and a vast number of smaller clocks. I ought to mention at the same time that Messrs. Potts, of Leeds, in 1881, put up a still larger one in the finest clock-room in the world, about 40 ft. square, in the tower of Lincoln Minster, striking on Great Tom and four new quarter bells; and Gillett and Bland, of Croydon, who made the clock for some still larger bells in the Manchester Town Hall. But all these, and a multitude of others not quite so large, by these makers, and by Smith of Derby, belong to the class which will be described afterwards, winding every one or two days in the striking parts. Dead escapements have now become quite obsolete for all large clocks that are intended to keep time within the maximum that ought to be allowed, viz. 5 seconds a week; for I hear that some of these large clocks do not vary 5 seconds a month, except from some temporary cause. Therefore, although I retain the description and picture of one pattern of a dead escapement clock farther on, the time is come to treat the gravity escapement as the standard one for this purpose. It is very easy too to make this next picture serve for a dead escapement by putting the necessary A-shaped small frame on to carry the pallets, making the third wheel which drives the gravity three-legs into a dead pin wheel, as described at p. 99.

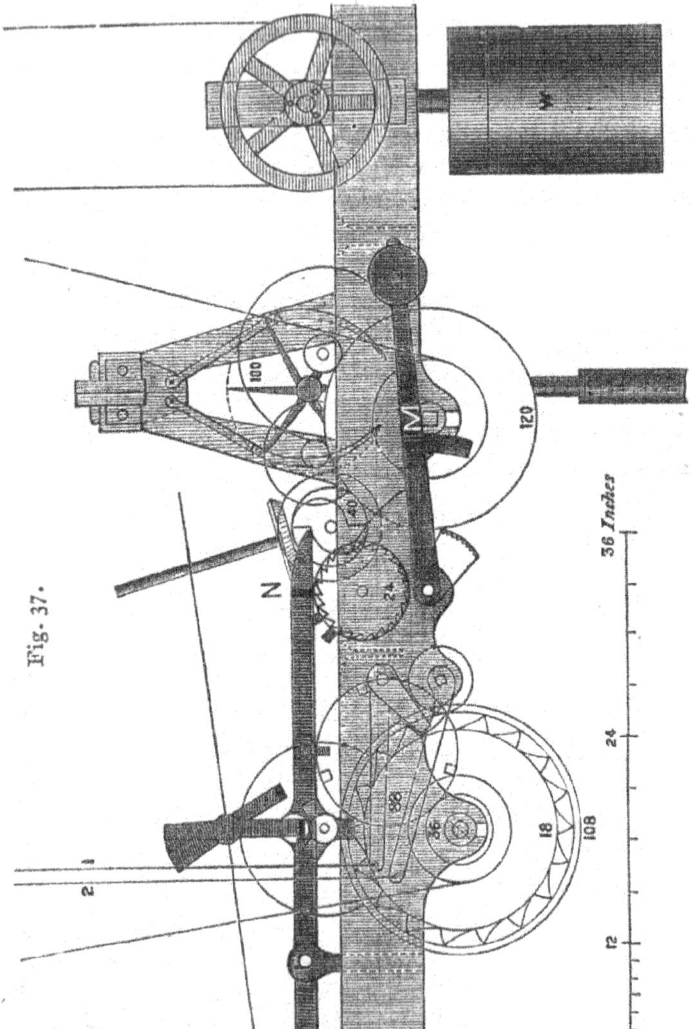

Fig. 37.

Very few architects have the least idea, or will condescend to learn, how large a clock-tower must be to hold a clock of moderate size properly, or bells either. Their notion seems to be that it is the duty of what they call their 'clients'—they being the 'patrons,' I suppose—to let them build what they think pretty, and then get other people to make it useful if they can. I must inform their 'unfortunate clients' then, that they cannot have a clock of the best construction suitable for a bell of only 4 or 5 cwt., and one or more dials of 4 ft. diameter, unless they provide a clock-room at least 6 ft. square, and 7 is better, and at least 30 ft. fall for the weights. The frame in such a tower as that is best built into the wall about half a brick deep at each end, as mine is, and of course made quite level and firm. It will then be firm enough to carry the pendulum, without resorting to an independent cock built into the wall. Mine is about 3 cwt., and just 99 in. long to the bottom, being 1½ seconds, which happened to be more convenient than a 1¼ seconds one, besides being rather better. As it is generally necessary to carry the ropes up to some place above the clock-room to get all the available fall, you have to provide space for them also, and evidently more than if they go straight down from the barrels, which is better if you have fall enough, as it saves a good deal of friction in several ways. This depends on the height of the clock-room from the ground, and the use you want to make of the lower part of the tower, which should all be considered beforehand, but never is, except by people who look after their own work, who alone get it done well.

Any one who is generally acquainted with clock-making will understand from this picture more than could be shown in it without confusion. The barrels, or great wheels of both parts, are set under the frame in bushes of the construction *a* and *d* together in p. 192, so that they can drive second wheels in bushes of the form *b* on the top of the frame, which top is about 1½ wide, and is now always planed in a machine, to carry all the cocks and bushes quite firm and level. There are three cross bars cast in it at convenient places, which are utilized also for carrying cocks for 'leading off,' for hammer-tails, winding-pinions, and anything else of that kind. The great going wheel generally has 120 teeth, and is 12 or 13 inches in diameter, and drives (first) the hour wheel with 40 teeth, on the arbor of which is the bevelled wheel driving other bevelled wheels up to the dials, both outside the tower, as usual, and in mine inside also to a dial at the end of a long passage in the house. That leading off goes downwards obliquely, and is omitted in the picture. The half-hour snails, with the main bevelled wheel, are clamped to the hour wheel by thumb-screws, to enable you to set the hands when necessary; and it drives no more wheels in the train, because it makes a better distribution of the teeth to leave it independent. The great wheel therefore drives the one marked 100 with a pinion of 10, which will turn in 15 minutes as the 40 wheel takes an hour, whatever may be the time of the great wheel, which is generally made 3 hours in these clocks. The 15 min. wheel of 100 drives another pinion of 10, ∴ in 1·5 min., and that having a wheel of 90 drives the scape-wheel pinion of 9, which, with the double three-legs and a 1½-sec. pendulum, turns in 9 seconds. But if the pendulum is 1¼ sec., the scape-wheel will turn in 6 × 1¼, or 7½ sec., and its pinion must either be 8 driven by 96 or 9 driven by 108.

In all clocks of this kind the pallet-arbors are set in small cocks, on the large one which carries the pendulum, and the scape-wheel itself has only a short arbor in 2 cocks behind the other wheels. The pendulum swings in a long slot in the flange at the top of the frame, reaching from one cross bar to the other; and the escapement fly also in another space, with the two three-legged wheels between the back bar and another which carries all the wheels. There is room enough for all this, and a sufficient length of barrel, in the width of frame that the striking part requires, which is always a good deal more than the going part, on account of the greater thickness of the rope, and the cams and levers and the winding wheel. The frame is generally 4½ in. deep, with a wide flange turned inwards all along the top, to set the various cocks on.

There is a little inconvenience in the third wheel turning in 1½ min. instead of either 1 or 2, as you want an index of some kind to mark seconds for regulating, unless you go entirely by the striking. But this train distributes the teeth so much the best that I adhere to it, and get over the other difficulty either by a pair of small wheels in the proportion of 3 to 2, the smaller carrying a seconds hand, or else depend on an index placed over the rim of the 90 seconds wheel, which is graduated up to 30, so as to give the seconds in every half-minute, leaving you to see by the other dial, on the hour arbor, which half-minute it is. With a 2-sec. pendulum there is no such difficulty; for the scape-wheel then turns in 12 sec., and its pinion of 10 driven by a wheel of 100 lets that wheel turn in 2 minutes; and that may have a pinion of 10 driven by 100, which will turn in 20 min. instead of 15, and consequently wants a pinion of 12 to an hour wheel of 36. Several other numbers also would do; but we shall see afterwards how the second wheel may be used to obtain greater precision in the time of beginning to strike than you can get from an hour snail, a good way off the escapement too, if the second wheel turns in 15 or 20 minutes. And if you want to apply it to the half-hour striking also, which only requires two pins (T in fig. 40, p. 196) instead of one, it must turn in 15 min. The great wheel need not turn in any particular time. Sometimes I have had them for 4 hours. In the Westminster clock it was convenient to have it turning in 3 h. 45 m. A minute-hand is, or should be, always set on the hour-wheel arbor, with a dial to set the clock by, which is best done by letting the gravity escapement run or stopping it for the necessary time, which is another advantage of these clocks.

Wire ropes.—The introduction of wire ropes, instead of the old hemp ones of 5 or 6 times the thickness, did a great deal towards enabling the barrels and clock frames to be made smaller than the clumsy things of old times, which it is no longer necessary to describe, as wire ropes and iron barrels have become universal. But I find it is still neces sary to warn people against using zinced wire ropes. I found long ago that for some reason or other zincing iron wire or sheet iron tends to make it brittle, and sometimes the zinc splits off, while tinning it has the contrary effect; only it does very little towards preventing rust, for galvanic reasons. But the best way of preventing rust on wire ropes is to keep them well greased with a mixture of tar and grease. Paint splits off with the bending of the ropes.

Striking from the great wheel was another important improvement which followed from the use of wire ropes with many more turns of the barrel for one winding up, and especially those of steel wire, which may be thinner still. The saving in friction, and consequently in weight, and the strength required, is more than any one would suppose; and that also has become universal in all clocks, except those of makers who have steadfastly set their faces against all improvements, and consequently never dare to accept contracts guaranteeing the rate of time-keeping prescribed above, or the raising of hammers that will bring the full sound out of bells, which the old clocks never did. I will explain afterwards the construction of the cams which are now generally cast on the great wheels, but in very large clocks are also faced with steel. For an eight-day striking part I have come to the conclusion that the best arrangement is to have about 18 cams working two hammers, so that each hammer-tail and cam has the same action as if there were only half the number of cams. In some of Dent's earlier clocks the hammer-tails were on opposite sides of the wheel, with two sets of cams, each having half the number. But this is unnecessary if they are placed as in fig. 37, keeping quite clear of each other, which is easily managed, taking care to place them so as to make the intervals between the blows exactly alike; *i.e.* their centres must be on prolonged radii of the wheel, 1½ cams apart. The reason for having 2 hammer-tails, instead of one shorter and working over half the space, is that the pressure of such a short lever sometimes cuts off the ends of the cams if the lever end was not blunted enough, though there are plenty of such clocks going without any such result. On the whole it is better to avoid the risk, except with small bells not above 2 cwt., which only want hammers of 4 or 5 lbs. If there are 22 or 24 cams, of course the teeth of the great wheel, and of the small one on it, which drives the locking-plate, must be increased; and remember that when there are two hammers that 36-wheel must still be twice the number of the cams, as each cam strikes two blows. I assume the locking-plate wheel to have one tooth for every blow struck, though you may vary the number if you keep the right proportion; *e.g.* you might have 30 and 66, instead of 40 and 88. Both these wheels are in front of the frame. The great wheel of 108, or 3 × 36, drives a pinion of 9, which therefore goes one third round for each blow, and accordingly has 3 cams to lift the detent. The winding pinion should pump into and out of gear, as there is no use in giving the clock the friction of turning it. Its back pivot is accordingly set in a cock bolted to the cross bar in the middle of the frame, with a key that drops into a nick round the arbor to keep it in its place. In the drawing it looks as if the second hammer lever pivot was in the winding-pinion one; it is only optically so, as they say of stars, which may either be really inside a nebula or billions of miles behind it.

The half-hour striking arrangement has been already

described with reference to this (see p. 170). The jumper spring is on the right side of the 24 ratchet wheel with the 12½ and 1½ o'clock steps on it.

Returning to the going part, the maintaining power which I always prescribe, except for very large clocks, has been already described (p. 151). Some clockmakers found, after much trouble (of which they were warned before), that the spring-going ratchet will not do for gravity escapement turret clocks, except very small ones. Where the dials are so large as to require more weight than can be wound up without an auxiliary pinion like the striking part, I think the Westminster maintaining power is the best (see also fig. 40); though this one in fig. 37 does perfectly well, and is generally used for larger clocks.

Stops for weights.—Where the weights do not go down to solid ground there ought to be a large box full of broken stones, not gravel, for them to fall on if a rope breaks. The reason why stones are the best is that they give the weight something to do in breaking them a little more, which uses up a good deal of its force. Sawdust or chips are too elastic, and sometimes throw the weight off on to the floor, and therefore through it, as happened at St. Albans lately, from over-winding. Gravel is too hard to break, but still gravel or even sand will take off a great deal of the force of a blow in displacing it; but I am convinced that stones are the best. They may be covered with a thin board, to look tidy. Their depth should never be less than 2 ft., and more according to the bigness of the weight and the possible drop. But another kind of stop also is requisite where the weights go out of sight, to prevent over-winding, *i.e.*, winding right up to the fixed pulley. A mere straight stop to catch the top of the weight will generally do, because it gives check enough to make any man who is winding feel it, though of course there is a little risk of the jerk and strain, for which extra strength of rope must be provided; but this will not do when a great multiplying power has to be used. Nor is it at all safe to rely on human stupidity attending to any mark on the rope, which there was at St. Albans, even if there is always light enough to see it and the real winding man is told about it, who is probably a mere blundering deputy of the one who ought to do it. At Westminster, you will see farther on, that I provided an absolute stop to the turning of the handles beyond the proper times, both when the clock is going to strike and when it is fully wound. But ropes on long ungrooved barrels do not travel uniformly enough for that method to be adopted with them; and besides that, the ropes often overlap, though they had better not. It would be easy enough to make each weight raise a lever with a sort of click to it, which would ring a bell when the weight has got to the top, in several ways. And now that electricity is coming into use for everything, the weights might make a contact to complete an electrical circuit when they reach the top, and so might make any kind of noise which the winder could not fail to hear, or drop a lever to stop him.

When the weights do not hang from the barrels, but the ropes have to be led off to a fixed pulley somewhere, it is necessary that it should be so far off and in such a position that each rope may feed straight off and on the barrel, without either separating or grinding against itself. It is so much better to let them hang down, that I would rather have them hang so by 3 lines, which requires no more pulleys, and only two-thirds of the height of 2 lines, than lead off for anything less than 15 or 20 times the length of the barrel. But more than 3 lines increase the friction enormously, and should never be used; nor should 3 together with a fixed leading-off pulley, which makes 4 lines or 3 pulleys. The weights in this clock go down within the frame, close to the walls, and are boxed all the way down. Even if they did not, the tower must not have been much smaller, on account of the fly, in which length is of great value always for steadiness of striking, and you must have a reasonable space for the handle to turn in winding, even if it goes the right way, as it does here, not requiring the man to stand beyond it.

Four-wheeled trains.—These horizontal frames evidently require rather more length than a well-arranged cage frame, in which the wheels stand over each other, and they would require still more to contain a four-wheeled train. Turrets are sometimes made so small that a horizontal frame clock even of 3 wheels cannot be got in. The clock can be got into less length by making the frame something like a pointed arch, which is also a strong form. Many years ago I designed some small quarter clocks for a confined space, to be sent to Mexico, and the arrangement on page 193 brings the work within the smallest possible limits. I believe they were the strongest clocks of the size that had ever been made. There are no loose bars whatever to the frame, and instead of cylindrical bushes (fig. 38 *a*) which can only be let in near the middle of the bar, the bushes are mostly of the form *c*, which admits of greater variety of position, and also enables the wheels and other pieces to be taken out singly with greater facility than the let in bushes. Another bush, which is used in clocks with horizontal and arched frames, is that at *b*, which is the best of all for convenience of fixing, and adjusting the place of the wheels, and taking them out separately.

There is yet another kind of bush which I used first at Westminster, and it is the best for the barrel arbors of clocks with a horizontal frame. A hole of the shape *d* is cut out of the piece, which is then cast as a projection downwards from the frame, large enough to hold a bush of the form *a*, and with a slot below just wide enough to let the arbor drop when the bushes are pulled out. Otherwise it is necessary to bolt large pieces, or cocks, on to the frame, and the back piece is sometimes difficult to get off. I shall call these 'drop-bushes' accordingly. At Westminster they are not used for the barrels, but for the third wheel arbors which drive the flies.

In connection with bushes it is necessary to warn people against making oil holes in them, which is sometimes done from overlooking the difference between the slow-going pivots of clocks which do not need oiling once a year, and the quick ones in other machines which require constant feeding with oil or they will heat. Such holes in clock bushes are very soon drained of oil and become receptacles and feeders of dirt and grit. What little oil is needed easily works in from the ends of the pivots, the old oil being first

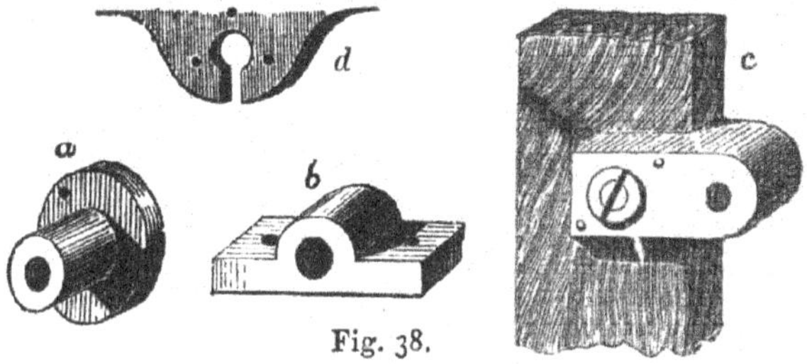

Fig. 38.

wiped off. I shall say a little more about oil at the end of this part of the book.

Fig. 39 is a front elevation of those Mexico clocks, showing as much as can be shown at one view without confusion, and representing the wheels only as circles for the same reason. They have 1 sec. compensated pendulums with bobs of nearly 1¼ cwt.; where there is room, of course it would be better to use 1½ or 1¾ sec. pendulums. They are calculated to strike the hour on a bell of 4 or 5 cwt. with quarters in proportion, and will drive four 6 ft. dials very well. The whole length of a quarter clock on this pattern is only 30 inches, and for one without the quarters 21 inches. The flies may require rather more, according to the size of the hammers. One fly may go behind the other if necessary. Those who build clock-towers should remember that there must be room for a man to stand and wind up the clock besides. The hole in the upright bar where the pallets are shown, is only in the back bar, which is widened out for

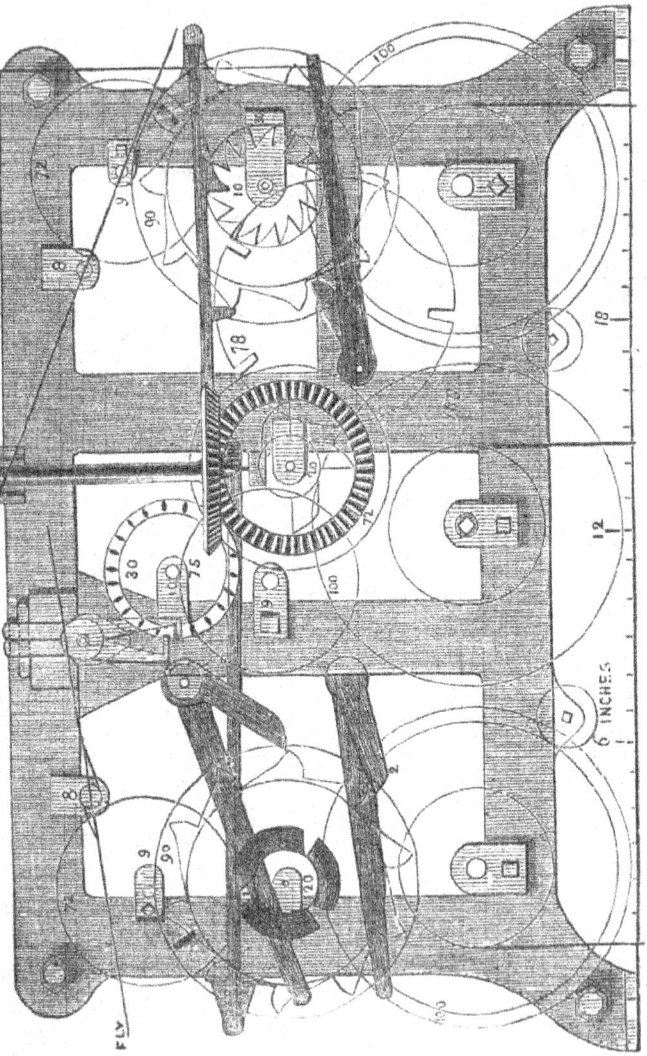

Fig. 39.

it, and the two lumps which form the pendulum cock are cast on the back. The cross bar which carries the locking plate 78 is only wanted on the front frame, and is omitted on the back one, as it would interfere with taking out the barrel. All the levers are so placed as to clear the ropes when they have to be taken upwards. The small cam wheel for lifting the locking lever of the striking part answers very well: that of the quarters is in fact the locking plate itself. If there are no quarters at the hour, the pinion of 20 should be 30, and drive a separate locking wheel of 36, which will turn in 2 hours, as 12 cams are used in that time.

Size of fly.—There is no more frequent mistake in turret clocks than that of making the fly too small to preserve uniformity of time in striking, and the defect is generally incurable for want of room. When the fly of the hour striking part is too small the velocity increases after the first few blows; and with quarters on four bells especially, some blows come quick and others slow, according as the heavy or the light hammers are being raised. You must have a considerable superfluity of force beyond what will just raise the hammers, and the regulation of time must be done by the fly. I do not however see my way to prescribing any rule for the size in proportion to the weight of hammers or bells beyond this, that each arm of the fly of any large church clock ought to be fully 2 ft. long; and for the very large bells which have lately come into fashion again, the flies must be still longer. Length is much more effective than width. There may be from 4 to 8 turns for each blow or each quarter, according to the size of the clock. At Westminster we could get no uniformity of striking in quarters with a fly going faster than 4 times to each chime; 6 or 7 is generally the best number.

When it is impossible to get room for flies of proper length and size to equalise the time, something may be done by using a three-armed fly, but that is by no means equal to one with two arms of sufficient length.

I find it necessary to add, that the flies should on no account be in front of the clock, for that involves the use of a winder with a very long pipe to clear them, which is harder to wind and strains the arbors. Fig. 47 will show how to get plenty of room for them behind in clocks of almost any size. In very large ones, such as Westminster, a vertical rod may be carried up and the flies put quite away at the top of the room. I have seen it done also in much smaller clocks where there was no room otherwise for the flies.

LARGE CLOCKS WITH QUARTERS.

Fig. 40 (*t.* .) is a front view of a larger clock than those previously described, substantially according to the plan which I settled for old Mr. Dent's factory many years ago, whereby we reduced the cost of large quarter clocks to little more than a quarter of what it used to be, and at the same time greatly increased both their accuracy and strength. This only shows two quarter hammers for simplicity. Indeed four could not be shown in an elevation, as the levers must then all be on one axis or pin, and the cams come irregularly, as will be explained under quarter chimes. As Fig. 37 was of an eight-day clock, to which a quarter part might be added much like the hour part, this is of larger clocks, with heavy bells, and winding up the striking parts every day or two days, according to the available fall. It is impossible to strike heavy bells properly with an eight-day clock unless it has a very unusual fall, and even then it would want very inconveniently long barrels, and the old-fashioned clocks never did strike properly.

The hour great wheel here has only 10 cams, which I consider the best number for the arcs that have to be described by the cams and the hammer-tails or levers, when we are free to use any number, which we cannot in eight-day clocks. Therefore this requires only 16 turns of the barrel for 24

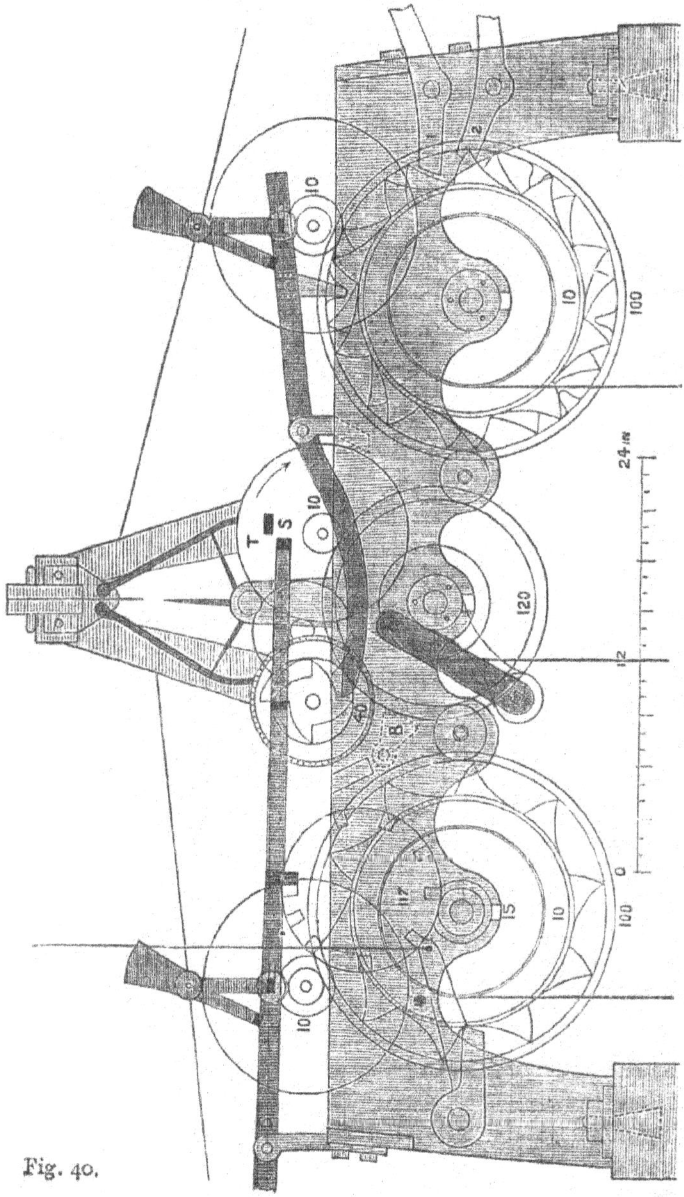

Fig. 40.

hours, or 156 blows, and 2 or 3 more for a few extra hours; or say 35 or 36 turns for winding every two days; and therefore quite short barrels will do, with wire ropes as much as ¼ in. thick, which are enough for all but very large clocks beyond the ordinary patterns.

The going part is always made to go a week, or rather 8 days and a fraction, to provide for the forgetting of a day.

It is impossible to give any rule for the size of the great wheels, as it clearly ought to depend on the work they have to do. For bells from about 30 cwt. to 53, a common range for the tenors of large peals, an 18-in. great striking wheel is the best pattern to keep, as it is inconvenient to use many. A 24-in. wheel with only 10 cams of proper size and strength will do very well for bells up to 5 or 6 tons, or even 10 tons, if the teeth and cams are wide enough. The great going wheel may be from 13 up to 16 inches for any four dials from 7 up to 12 feet, and there are very few larger. The other wheels may be in proportion. The scape-wheel legs should be from 5 to 6 inches long, and you must allow plenty of room for a long fly, not less than 9 or 10 inches, especially for large dials, which should have a considerable superfluity of force, to drive the hands in all weathers. For that reason the scape-wheel arbor is put on a higher cock than before. The pendulum is carried in the same way by the frame. If it is a very large one it is well to put one or two brackets or strutts from the wall to the back part of the frame near the pendulum, but that is not necessary generally, as this kind of frame for 18-in. great wheels is only 5 ft. 6 in. long, and forms a kind of arch when its feet are firmly bolted to the corbels, which I need not say should be a great deal deeper than there was room to show in this drawing. If they are iron brackets they should be built into the wall and wide at the top also, to prevent any risk of sideway motion under the swing of the pendulum.

For 10 cams the great wheel had better have 100 teeth,

and the pinion 10; or 120 and 12 of course run rather easier, which pinion will evidently go once round for each blow. But if you want the clock to go 4 days without winding and to have only one hammer, 16 cams will be better, and 192 teeth in the great wheel, driving a pinion of 12 two-thirds round for each blow. And then you may make the cam wheel on that pinion with 2 hollows in it one-third and two-thirds of the circumference apart, and they will always come right, because the number of the hours is odd and even alternately, if it is put in the right position; which you will find more easily by trial than by explanation here. But this would not do when half-hours are struck by the hour part, for two odds then often come together, such as 3 and 1, 5 and 1, &c. The general calculation for all numbers is this. Let p be the leaves of the pinion, which must be a multiple of 3 if you are to use that method; n the teeth of the great wheel, and m the cams. Then $\frac{2}{3}p$ must $= \frac{n}{m}$ or $n = \frac{2}{3}mp$: p must be taken as 12 for this purpose, and $\therefore n = 8\,m$. If p has 10 leaves and is only to go once round, n must evidently $= 10\,m$. It is not convenient to have two large hammers with a ringing peal, as it is difficult to get room for them in the frame, though it is easy enough for stationary clock bells. But it is unnecessary to consider all that if the clock winds up every day or two, which is always best for large ones. A few cocks and bushes are omitted in the drawing to avoid confusion.

In other respects this striking part is the same as in fig. 37. But I now show the plan for letting off the striking of the hours more exactly than can be done by the slow motion of the hour wheel and its snail, especially as it is a sort of outrider to the train. This was first done in the Westminster clock, but the plan now described for the first time was not expedient there on account of the great size and weight of the discharging levers. Here you see that the long lever or detent DS is carried over to the second wheel of the going train, which has a square pin T on one of its spokes; and there is another square pin S on the detent, now shown below the other, the clock having just struck. But at some minutes before striking S gets raised above the circle in which T moves, which can be done, because at that time T is out of the way. Just before the detent is going to drop off the snail T has come under J, and J cannot fall again till T slips from under it, which will take place with perfect accuracy at the last beat of the pendulum for the hour, as the motion is large enough to be visible, and a very little way off the scape-wheel. This is inconsistent with altering the hands less than 15 or 20 minutes, except by running the clock, as you easily can with a gravity escapement. In the Lincoln Minster clock Mr. Potts provided in this way for the quarters also, which however are not of so much consequence as the first blow of the hour. The second wheel must in that case turn in 15 minutes; but without the quarters it may be either 20 or 15. The pins S and T should be so placed and shaped that the pressure may not tend to stop the wheel.

As I have already shown the maintaining power which is best for clocks not requiring an auxiliary winding wheel, I show here the one which is best for those that do; though the other, of which a piece is shown at B, may be used for them. This is the one that I designed for the Westminster clock, and the principle of it will be described there. The winding pinion and the bars which carry it, the front one fixed and the back one hanging from the arbor, are out of the vertical, because the action of gravity is wanted to make the pinion and back bar which carries it fall on to a fixed stop, when the winding is done, or suspended for a short time. The back bar also carries the click which takes into ratchet teeth cast on the back of the great wheel. It might go into the other teeth, but for the risk of catching only on their tops and slipping, so that ratchet teeth are safer, and practically cost nothing when once the patterns are made to cast from.

The arrangement of the quarter part for 4 bells is practically the same as for 2, except that all the levers must be on one steel pin, which should be screwed tight at its ends to help to keep it stiff. Either for 2 hammers or for 4, the best way is to cast one set of cams on one side of the great wheel and another on the other, at the proper distances to make the interval between successive chimes at least twice as great as between the blows in each chime. For 4 bells two other sets of cam rings may be cast in one, and all bolted together. Some clockmakers prefer an independent cam barrel driven by the great wheel, to which there is no objection except the extra friction, which I once found to make the difference of requiring an auxiliary winding-wheel. When steel cams are used they are all bolted to a wide ring or barrel cast with or bolted to the great wheel, the nuts being inside it. I shall have more to say about cams afterwards. Four quarter hammer-tails cannot well be turned inwards as the hour one can, in the form which makes the pressure on the arbor only the difference instead of the sum of the two forces on the lever. But they can have their wires in the middle, instead of their fulcrum or arbor being there, as you see in the Westminster clock, and in that way the friction on their arbor is reduced to a minimum, the wires being put as near the cams as possible.

Two quarter hammers—or two alternate hour hammers—can also be placed as in this fig. 41, which I designed long ago with the same object of keeping the power and the work on the same side of the arbor, and many large clocks were so made at Dent's factory. It involves cams on both sides of the great wheel, which are as easily cast as on one side. The rope also should always pull on the same side of the barrel as the cams, in order that the pressure on the great arbor may be the difference instead of the sum of the weight and its work, and consequently the friction much less. You may fancy that these frictions of pivots, and of teeth when the cams are not on the great wheel, cannot come to much compared with a great striking weight, or a hammer of many

Fig. 41.

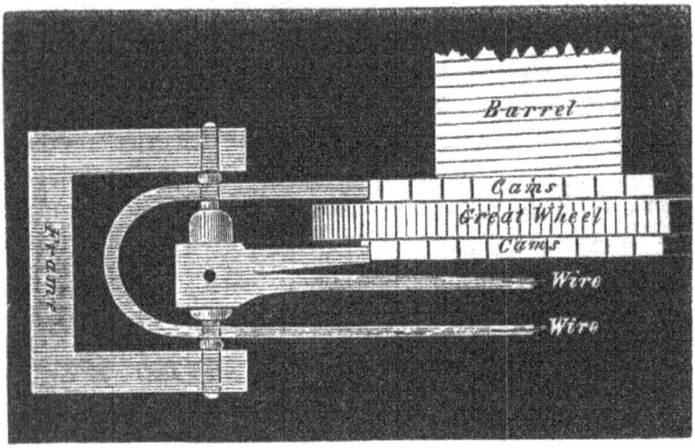

pounds weight raised 9 in., or about 6 vertically, 156 times a day; but you will find very few clocks in which the weight × its daily fall is not more than twice or even three times its theoretical 'duty,' or the hammer × 78 ft., which excess is all due to these various frictions.

If no quarters are struck at the hour it is better to have 12 cams, the locking-plate being on the great wheel arbor, so that they will turn in 2 hours. In that case the pinion may as well be of 12, driven by the great wheel of 96, and therefore turning ⅔ round for each quarter chime as before described for hours in some cases.

Lifting the hammer by pins on the striking wheel, as in house clocks, is totally wrong in large ones, and it is nearly discontinued. The pins necessarily begin to act a good way from the end of the lever, and therefore at the greatest disadvantage as to leverage, and that at the very time when the hammer is rising most vertically (in the usual way of hanging) and requires the most force to lift it. The lifting should be done by cams, which begin to act on the end of the lever, and are properly shaped to act with the least possible friction throughout, as I shall explain under the *Teeth of Wheels*. Pins with rollers on them are of very little use, and do not obviate these objections, and are weaker than fixed pins: they are nearly if not quite abandoned.

Cambridge and Westminster chimes.—Considering the thousands of men who had listened for 3 or 4 years together to the famous St. Mary's chimes at Cambridge, during three quarters of a century, it is odd that no one ever thought of copying them in any other public clock. The last but one Lord Lansdowne tried it, but his clock-maker Mr. Vulliamy made the mistake of ordering and hanging the bells of four successive notes before writing to ask me about them; whereas they ought to be such notes as E, D, C, G; or the fourth bell must be a musical 'fourth' below the third: bells are always counted from the highest note or smallest bell. At the Royal Exchange, in 1845, they did get so far as to adopt the Cambridge notes; but the Gresham professor of music thought he could improve upon the tunes, and spoilt them. I have been told by a Cambridge man of the last century, that they were invented by the well-known Dr. Crotch in 1780, from an air of Handel's, who has generally had the credit of inventing them, and perhaps deserves it.

Doncaster quarters.—For some time after I thought of introducing them at Westminster, it was assumed that the hour bell must be an octave below the third quarter, and that they were therefore impossible with a peal of only 8 bells, if the quarters were to be struck at the hour. Moreover playing 10 chimes in every hour, requires nearly twice as much power in the clock as playing 6, which is sometimes a consideration when there is not much fall for the weights, and it makes a little difference in the cost. Again, if you look at the Cambridge chimes (p. 204) you see that the blow on the lowest quarter bell (called 6), is repeated too quickly in one place for a heavy hammer to be relifted immediately; and accordingly in all very large clocks it is necessary, and it is always desirable, to have two of those hammers, lifted alternately. Avoiding it by a separate barrel, driven by the great wheel, is a much worse plan. For all these reasons, in a few church clocks, made before 1860, I adopted the plan of omitting the quarters at the hour; but I should not do so again with these quarters, because the hour chime is the best of them all, though *ding dong*, or any other mere repetition quarters, are neither useful nor pleasing at the hour. I also made a slight alteration in the third quarter in those clocks, to avoid the quick relifting of the largest hammer, and it sounds very nearly as well as the Cambridge one. The quarter bells are therefore, 2, 3, 4, 7, of a peal of 8 at Doncaster and Scarborough parish churches, and the cathedral at Fredericton, where the three-legged gravity escapement was also first used.

Worcester and Chester Cathedral quarters.—But after a time I came to the conclusion, and other people have gradually adopted it, that it is not at all necessary to have the third quarter bell an octave above the hour bell, and that the ear is quite satisfied if the fourth bell is two notes, or even one, above the hour, because the interval of time between the quarters and the hour ought to be from 6 to 10 seconds with large bells. Accordingly in the great peal at Worcester, the Rev. R. Cattley, the author of it, and I, as the designer of the clock and bells, agreed to take advantage of the tenor of the peal for the fourth quarter bell, though it is only 1½ notes above the great single hour bell of 4½ tons; and thereby we got far more powerful quarters than if we had kept them 2 notes higher. At

Chester Cathedral, and St. Chad's, Headingley, near Leeds, and some other places, the 4th quarter bell is only one note below the tenor, as at Doncaster, though they are the full Cambridge quarters; and that is the plan which I now always recommend when there are 8 bells, or even 6, with an extra one added above for the first quarter. At Ossington Church, having only a small peal of 6, I recommended to my namesake the late Speaker, a lower bell (out of the peal) for the hours; and the same at S. Shields lately.

The following are these several arrangements, which are suitable for any key, and therefore I have indicated them by the bells in a peal of 6, which these quarters require, leaving the hour to be any note below the lowest quarter bell.

Cambridge and Westminster.			Doncaster and Scarborough.		Royal Exchange.		
2nd	3126	4th	1st	1236	1st	3126	2nd, 4th
	3213		2nd	3126	3rd	6213	
3rd	1326			3213		1326	
	6213		3rd	2613		3213	
	1236..1st			6213			
				1326		But this is bad.	

I call the bells 1, 2, 3, 6, in all cases for more easy comparison, though they would be of various numbers according to the peal they belong to, on account of the numbers being always reckoned from the smallest, in the order in which they ring. Indeed a peal of 6 could not have these chimes at all without an odd bell, either above as at Headingley, or below as at Ossington.

You see from the table that the Cambridge chimes are repeated twice in the hour, and therefore the cams may be fixed on a barrel which turns in any multiple of the 5 chimes in that table. In some clocks the 5 sets of cams alone are put on a barrel driven by the great wheel, but with much more friction than in other clocks that I know with bells of about the same weight; one, with this second barrel, required a double multiplying power to wind it up, while another, with the cams for an hour on the great wheel itself, winds quite easily with a single pinion. At Westminster, as you may see in the frontispiece, there are 3 sets of cams, or what may be called an hour and a half's chimes, on the great wheel, *i.e.* on a wide barrel attached to it, because it happened to be convenient with the great fall we have there. The cams may be cast in separate rings of cast steel, but it is better to have smooth-faced steel cams screwed firmly to a plain barrel, as they are at Westminster, and in all the best very large clocks now.

The interval between each chime (*i.e.* between the chimes of each quarter) requires some attention. At Cambridge it is 3 times the interval between the blows of each chime. That appears to me decidedly too slow, and at Westminster, Doncaster, &c., I made the interval only two spaces instead of three. That again is perhaps rather too quick, and at Worcester, and all the subsequent large clocks for which I have given designs or specifications, there are 2½ spaces, which sound the best of all. The barrel, or each portion of it which contains the 5 chimes, must then be divided into 55 spaces, of which each blow occupies 2, and the intervals between the chimes 5 spaces.

When there are two hammers for the 4th bell, as there always should be, those cams may, and should be, made longer than the others, as there is then ample room for it; which diminishes the pressure on them, and makes it more continuous, and so tends to equalise the time. This is important, because the 4th quarter bell is much the heaviest, and there is the least interval between its blows.

In all quarter clocks, striking on a peal of ringing bells, provision should be made for lifting off the hammers, or they are almost certain to be broken when the bells are rung, and perhaps the bells cracked besides. This is easily done by a lever; or, when the hammers are very heavy, it is better done, as Mr. Cattley arranged at Worcester, by an eccentric brought down over all the levers just where they come out of the clock. The hammers must not only be lifted off the bells, but so high that the cams do not lift them at all, or all the wires will soon be broken by the jerks. And as the clock is thus relieved of its work, it is better also to make the turning of the eccentric, or pulling down the lever, let down a wooden brake on the fly arbor, to help the fly to stop the velocity and diminish the blow on the stops.

Quarters on two bells.—I have little more to say about turret clocks with these than I said about house clocks at p. 172. The bells must be at the musical interval called a fourth, and the higher of the two should be an octave higher than the hour bell if there are quarters at the hour; though that is not so material, and they may be the 1st and 4th of a peal of 6, or even 5, or the 4th and 7th of a peal of 8, because there should be a longish interval of time between the quarters and the hour, which saves the ear from being offended with the want of the proper musical interval. The old York Minster quarters, and a few others, were at the interval of a fifth (which in music, though not in arithmetic, is larger than a fourth), but they do not sound so well.

Time of striking.—The quarters are generally made to let off the hour, as described at p. 177. But where accuracy of time of striking is required, this plan is insufficient; for it makes the first blow of the hour depend on the time both of letting off, and of striking the quarters, both of which may easily vary several seconds. It is therefore essential to accuracy that the hour-striking should be let off by an independent snail of its own, exactly at the hour, with the quarters let off the requisite number of seconds before the hour; though the other three quarters may be allowed to strike at their proper times. Even this is not sufficient where extreme accuracy is required, as I have already explained.

The hammer should also be left 'on the lift,' or nearly ready to fall; which is in other respects also a good thing in a very large clock, because it relieves the wheels and the stopping pieces from a heavy pressure, and throws it all on the cams and hammer work, which must in any case be strong enough to bear it. In this case it is necessary to put a small click somewhere, to act on a pin in the fly arbor to prevent the train from running back when you wind up the clock: indeed it is as well to put one always, as the winding always tends to drive the train back. Where there is no special provision for accurate striking, as above, care must be taken to make the discharging snail large, and its corners sharp and hardened, and those of the discharging lever or lifting piece also. See also p. 199.

Chime tunes.—There has been a considerable revival of the old fashion of having tunes played on bells by machinery for a few minutes at certain hours of the day. The old machinery for this purpose was extremely simple, consisting of a large barrel 2 or 3 feet in diameter, generally made of wood, with strong iron pins screwed into it like a huge musical box barrel, and pulling down levers which lifted hammers on the bells like a common striking part. If there were several tunes on the barrel, either that or the bed of levers had an endway motion, shifting at the end of each tune, or shifted once a day by the clock. The Royal Exchange chimes are of this kind, only the barrel is of cast iron, with a vast number of holes in it, in which pins or short cams are screwed to play any tunes that the bells admit of.

Some new ones of the old kind have lately been put up in St. Albans Cathedral by Mr. Godman, a clockmaker there whom I have mentioned before, entirely from his own design. The barrel is 7 ft. long, of wood on iron rings, and the levers are worked by cams of 'phosphor bronze' screwed on, for 8 tunes on 8 bells, which I think are quite tunes enough, and more than were generally used in old times. The tunes shift themselves. The Westminster quarters are also put there on the same plan and in connection with the chimes in a room above the clock. Notwithstanding the superiority of those I am going to describe next when constantly looked after by a competent person, which they require, and which of course costs money, I have returned to the opinion that for ordinary churches, where there are seldom any funds to spare, the old-fashioned chimes are the best.

The chimes in some foreign churches are played by hand, I understand, without anything that can be called machinery. The hammers, being light, are easily lifted a little by a man playing on keys like a piano, only with his fist instead of his fingers. I have seen tunes played on church bells here with even less machinery than that. They just tie the bell-ropes to the clappers, and the lower ends to rings on a board screwed down to the belfry floor, bringing the clappers so near to the bells that a slight pull will make them strike. The man 'operates' them (in American grammar) by pulling some of the ropes with his hands and pushing others with his arms, and so manages to play a feeble sort of tune.

NEW CHIME MACHINERY.

The new kind just now referred to were introduced by Gillett and Bland of Croydon, and by Lund and Blockley of Pall Mall, both of whom bought some patent rights of an inventor named Imhoff, and have both made improvements of their own. The principle of the invention consists in this, that the hammers are raised by a long barrel full of cams which have nothing to do with dropping them, but are continually at work raising any levers that happen to be down. The levers are then caught and held by triggers, which are let off by quite small pins on a wooden barrel just like that of a common grind-organ. The barrel can also be taken out and changed, besides each barrel holding several tunes which change themselves by an endway motion. This figure shows Gillett and Bland's mode of lifting. Each lever has a jointed beak which is lifted by one of the 7 cams on the barrel belonging to that lever; and when it has been caught by the trigger at the other end the cam slips away, and the beak drops, so as to clear the cam next below, and gets caught again by the next to that, because it falls on to a block which stretches out the beak. This gives a greater lift for a given number of cams, and so allows more cams to be put in the barrel and more lifting to be done with a given velocity.

Fig. 42.

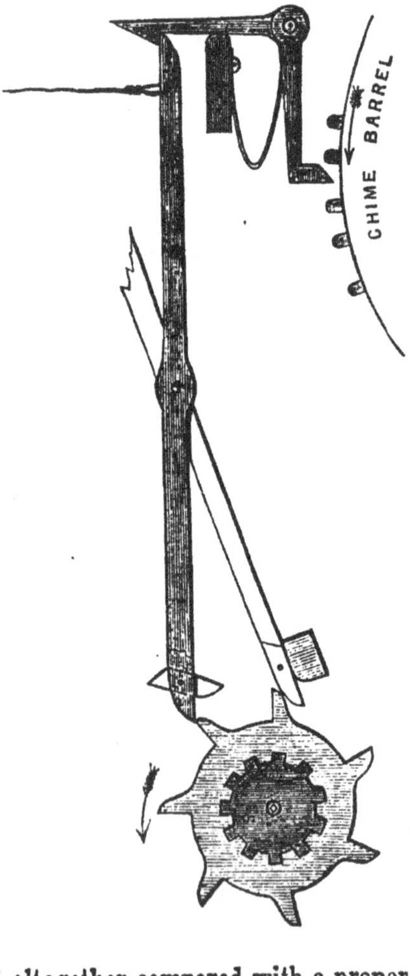

Lund and Blockley's lifting barrel has only 4 cams, and they lift from 90° before the vertical up to their highest position, by means of a hinged piece dropped from each lever. They also strike the Westminster quarters by means of the same set of pins as the tunes: which I must say I disapprove of altogether compared with a proper quarter-striking part. I have seen an otherwise very satisfactory large clock of theirs, with chimes on a peal of 16 large bells for the Bombay University; and some smaller ones. Gillett and Bland's principal chimes are at Worcester Cathedral, Boston and Croydon Churches, Bradford and Rochdale town-halls, and at Eaton Hall. The performance of both kinds is more accurate and satisfactory than in the old-fashioned machines; but, like most superior machines, they require a great deal more care and consequent expense than the old rough ones. No chiming machinery can bring the full tone out of bells, especially the large ones: but this is stronger in lifting as well as more exact in letting off than the old kind. Every bell requires two hammers, and at Worcester some of them have three, because of the quick repetition in some of the tunes.

It is necessary to give one caution most strongly to ambitious chime-cultivators. Avoid 'chords,' or two notes sounded (professedly) at once. At Croydon they thought they knew better, and a more horrible performance I never heard from the rudest old-fashioned chime barrel of 200 years ago. I believe the chords have since been abolished.

Another caution is, not to attempt chimes on large and small bells together. For some reason, which neither I nor the bell-founders know, it is a fact well known that bells below 4 or 5 cwt. cannot be made to sound homogeneous with large ones. I shall have more to say of that in the chapter on bells. They attempted it at Boston, and have got 42 bells (including the 8 of the peal), some of only a few pounds weight cast in Belgium, and chimes to play on them all together. I and other people who went specially to hear them considered them a failure. Even quarter chimes are never satisfactory when large and small bells are mixed. The same result is unpleasantly conspicuous at Eaton, though the bells there are not so numerous, and were all cast in Belgium together.

Clock hammers.—Turret clocks generally strike on bells of a different shape from the hemispherical house-clock bells, which do not answer beyond a very small size, as I shall explain more fully hereafter. Most people (except artists, who always draw them wrong) are aware that the general shape of church bells is that shown in figs. 43, 44. The clock hammer C S is always fixed at right angles to the swing of the bell, for two very obvious reasons: first, if the bell was free to swing under the blow of the hammer, the first blow of every hour would set it swinging a little, and at every blow after that the bell would either be out of the reach of the hammer altogether, or else jarring against it; and another reason (if it is worth while to talk of other

Fig. 43.

reasons after this) is, that if the hammer was put in front of the bell, all its machinery would have to be moved out of the way before the bell could be rung at all.

The hammers of large clocks also differ from small ones in acting by gravity, as you see. But they equally require a check spring, or some other contrivance to keep them from jarring the bell. When a church bell is rung *up*, *i.e.*, swinging once round for each blow, and *set* mouth upwards, the clapper lies on the bell, but not so heavily as a clock hammer would, because it stands at a higher angle. The usual kind of hammer spring is shown in fig. 43; and *s c* in fig. 44 shows as much of the spring as there was room for.

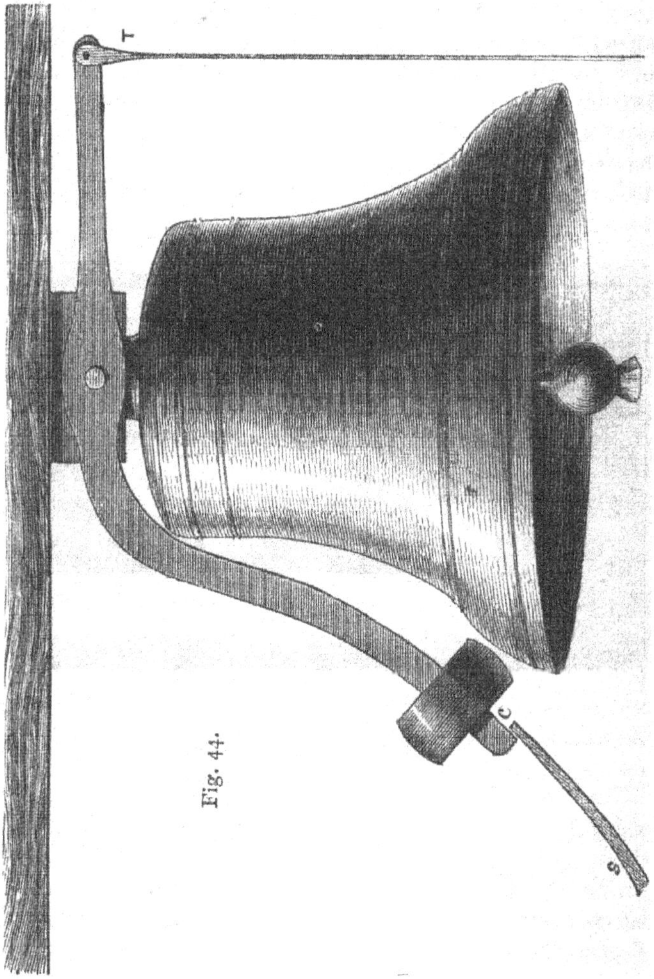

Fig. 44.

The spring is sometimes made adjustable, by having long holes for the screws to go through, so that you can bring it farther from or nearer to the bell as may be required in course of time. India-rubber buffers under the hammer shank are better in some positions, and have the advantage of never breaking, and being easily replaced or altered in thickness. I used them at Westminster.

I believe it is the fashion on the continent to fix the clock hammers with their pivots above the bell, when it has not to swing; and it has the advantage of securing a long hammer shank, and therefore less angular motion for a given lift, and moreover the effective weight of the hammer is not so much lost in the lift as it is in the common position. But on the other hand, with bells as tall as the foreign ones are, the hammer shank stands much more vertically than in the other way, and the rebound from the buffers or check spring is greater, and a greater lift (obliquely) from the bell is required to get the same momentum of the hammer; not that that imposes more work upon the clock. At Westminster we were obliged to adopt that plan, on account of the construction of the tower and bell-frame, and there it had this incidental advantage with regard to the great bell, that we were enabled to get the hammer tail T directly over the end of the lever in the clock, by setting the hammer frame, or the pivots which carry it, a little out of the cardinal position relatively to the tower, and so all cranking was avoided and a vast quantity of friction saved.

Cranks.—Nobody who has not tried it can have any idea how much of the force of a clock is wasted by having to lift the hammer through cranks. Where the hammers are not very heavy, you may sometimes use what I shall call *couple-levers* (for a reason which any mathematician will know), of the form in fig. 45 *a*, instead of a pair of cranks; but when the hammers are heavy, and the arbor of the lever has to be long, I find it is impossible to avoid some elastic torsion in it, which wastes quite as much force as the friction of a pair of cranks, and makes the hammers rise with a tremble, which checks the force of the clock and strains everything severely. For the same reason large cranks should be made with a light connecting bar as in fig. 45 *b*, which increases the strength enormously and helps to keep everything steady. The cranks and lever arms and hammer shanks and tails should all be long. I know that in modern towers, which are nearly always built too small

Fig. 45.

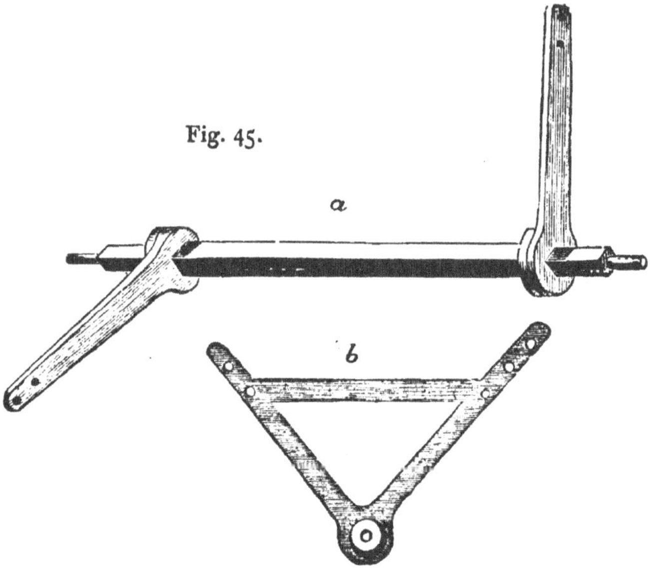

for properly hanging the bells, there is often great difficulty in getting room for clock hammers and cranks at all; but wherever there is room, the action will be easier and more effective if all these arms are made long instead of short.

When the clock is above the bells, as in the Leeds town-hall, and the Royal Exchange, it is a very common mistake to put a tail to the hammer to pull down first, which of course involves the necessity for another to pull up again. It ought to be done as I designed it at Leeds (if there is room) by pulling up at once from a lever set inwards or on the same side of the arbor as the hammer itself, unless it happens from some local peculiarities that this would involve as many cranks as the other way.

In several of the former editions I gave a design for catching the hammer at its rebound without any buffer-spring; but I have never heard of it being tried, and therefore I do not repeat it, and have modified fig. 44 accordingly to show a common buffer-spring.

The weight of hammer and its lift can only be determined by experiments. Different thicknesses and qualities of bells require different hammers. I have generally found that large bells whose diameter = 12 × the thickness of the *sound-bow* or thickest part (which is the best proportion) require hammers of about a 50th of their weight to bring out the full tone, and small bells require heavier ones. The lift has to be less for small bells than for large: the least that is effective in bells above 3 or 4 cwt. is 6 in. (measured obliquely in the direction of the motion), and beyond 13 in. we did not find any improvement in the sound of either of the great Westminster bells. Generally they are a great deal less either in weight or lift, and often in both, and therefore you hardly ever hear a church bell sound so loud under the clock striking as in ringing with the clapper. Thinner bells, as the larger ones of peals usually are, do not require such heavy hammers as thick ones; but it must be remembered that no hammer arrangement will get as good a sound out of a thin bell as a thick one, because it is radically inferior both in quantity and quality of sound. But I shall have more to say of that in the chapter on bells hereafter. I only add a word of warning against a piece of ignorance by which I remember a very fine old bell being cracked in a few months—viz. making the clock hammer to strike it with a sharp edge instead of a flat or slightly rounded face, as it ought to be.

DIALS.

The striking of church clocks, and perhaps of public clocks in general, is of more value than their dials, except in places where the public congregate. Indeed dials a good way above the eye are of no use for indicating the time very accurately, on account of the parallax which affects the minute hand, except when it is nearly vertical. Moreover many church towers would be utterly defaced by dials; and it is to be hoped that the splendid church of St. Mary's, Beverley, now that its fabric has been restored to a condition worthy of its architecture, will not long retain those abominable dials in the tower windows, which belong to the age when the inside of the church was divided into private boxes, in which people might eat their dinner or play at cards without their neighbours or the clergyman knowing anything about it.

But as dials must frequently be used, architects might as well condescend to learn something of their proper size, as they profess to provide places for them, as they do for bells, frequently in utter ignorance of what the provision ought to be. Luckily there is such a simple rule for determining it generally, which has now been long published, that they have no excuse for doing it wrong. That rule is that the diameter of the dial should not be less than a tenth of the height of its centre from the ground. If you want to verify this you have only to look at the dials of the Leeds Town Hall, planned by an architect with the usual knowledge of such things, 150 feet high and only 11 wide; and those of the Bradford Town Hall are no better, but I do not know the exact height; or that of St. Pancras Church, 6½ feet wide and about 100 feet above the ground. The neighbouring railway station however has inside it a large dial 15 feet wide, at the height of 55, looking down the largest space under one roof without pillars in the world, viz. 700 feet long and 240 wide (the length of the transept of St. Paul's Cathedral, and more than the length of the nave of any cathedral except Norwich, Winchester, and St. Alban's, the longest of them all). The external dial of that station is rather too small; but that, and some others which comply with the above rule are given in this list, which I have compiled from various sources.

	Dials.	Diameter. ft.	in.	Height. ft.
Mechlin	1	40		about 300
Westminster	4	22	6	180
St. Paul's Cathedral	2	17		126
Shandon Church, Cork	4	16		
Pancras Station	4	12	9	150
Scarborough old Church	1	12		on a hill
St. James's, Piccadilly	4	10		
King's Cross Station	4	9		90
Bow Church	2	9		70
Manchester Infirmary	4	9		80
Royal Exchange	4	9		90
St. George's Church, Leeds	3	8	4	57
St. Martin's in the Fields	4	8		
Horse Guards	2	7	5	
Marylebone Church	3	7		about 60
St. Luke's, Chelsea	4	6	10	72
The Queen's Stables	2	6	10	about 50

But if dials are sometimes too small, the opposite mistake is often made, of figures much too large, which is not a compensation but an aggravation of the evil, for they practically contract the size of the dial, in two ways; first by contracting the plain surface in the middle, over which the hands are most distinctly seen; and secondly, the larger the figures are, the more they run into each other and fill up the space of the figure ring itself, and make it still more difficult to distinguish the place of the minute hand. Ignorant people fancy that you see what o'clock it is by reading the figures; as if any single figure which you see in a clock dial indicated the figure which you read off; except for the hour hand, and the hour also is at once recognized by the position of the hand. You see the long hand pointing to VIII, and you say, '20 minutes to something.' Both for the hours and the minutes everybody really judges from the position of the hands, and 12 large spots would do as well or better than figures. I have several clocks without any figures at all round the principal dial, only 12 strong marks; and I never found anybody who even observed the fact that the figures were absent until it was pointed out to him, or complained of the want of them then. I came to the conclusion after various trials, that the figures and minutes together ought not to occupy above one third of the radius of the dial; the figures may be two thirds of that one third; and the minutes from half to two thirds of the remaining one ninth of the radius, with every fifth minute strongly marked by a larger spot than the others. The Westminster dials might have been clearer, considering their great size. They are not of my design, except that I gave the architect some suggestions for them, which are partly followed and partly not followed. They are unnecessarily confused with iron frame-work, and the clear space is unduly contracted by some broad rings supposed to be ornamental. The Midland Station dials are much better.

The only colours that seem to answer for dials and hands are black or dark blue with gilt figures and hands, or some very light coloured ground, such as white glass, with black hands and figures. Good gilding will last fifteen or sixteen years, as at the Royal Exchange, in the worst London atmosphere: bad gilding of course wants renewing much oftener, and is probably the dearest. Gilt hands on a light ground are a complete failure. The external counterpoises, if there are any, should be painted the same colour as the ground of the dial, except where they are very short in proportion to the hands, as at Westminster, in which case they cannot be mistaken at any distance for the hands themselves, and practically increase their visible length.

Dials may be made of almost anything—stone, slate, plaster, brick, iron, copper, and many old ones are of wood, which however is the worst of all, as it always shows the joints. In many cases the stone of the tower makes the best dial. Generally it requires painting; but if it is such a stone and in such a position that it will keep nearly white, it will do very well with only the black figures and minutes painted on it, and black hands, but by no means gilt ones. Dials on brick work must of course be painted. It is quite a mistake to suppose that a dial requires a very smooth surface. Some of the most distinct I have ever seen are painted on rather rough stone work; and brick will do as well, either all flat, or with the figure circle a raised ring of iron, or of any plaster that will stand. There are many dials of cast iron; but I should never make more than the figure ring of iron, unless there is a large hole in the wall which wants covering; and even then it is generally better to fill it with glass, which has all the effect of a dark ground outside and is often convenient within. Slate makes a good dial, but if it is not painted it becomes a pale grey colour. I believe 6 feet diameter is the largest size that can be got in one piece, but the joints are almost invisible if well done.

Copper dials are the commonest of all, and up to a moderate size, probably the cheapest, except of course when the dial is simply painted on the wall. But they are generally made in the very worst form that could be invented, viz., convex: the effect of which is that the point of the minute hand is thrown a long way off the dial, and the parallax is so great that you cannot tell what it is pointing at, except when it is nearly vertical, when seen from below, as a public dial always is. One way to avoid this is to countersink the middle, in which the short hand travels, leaving the long hand to lie close to the raised figure ring. I have lately seen some dials made of mosaic work, like pavements, by Messrs. Rust, of 16, Albert Embankment, which look very well, and the price was not more than of copper dials; and they can easily be made concave.

Concave dials.—It occurred to me some years ago that all the convenience of the light copper dials might be got, with even more closeness of pointing than in a flat one, and with as much stiffness as the convexity gives, and with less distortion of appearance, simply by making them concave instead of convex. If you draw a vertical section of a convex and a concave dial, and three lines of sight, from the top, the bottom, and the middle of each, to a spectator in the street, you will see at once that the convexity makes tho upper half appear much smaller than the lower, whereas in the concave one the two halves appear even more alike in size than in a flat dial; and the closeness of the hand pointing is evident. There are now a good many of such dials, and no one who has seen them can fail to perceive their superiority to convex ones.

Hands.—Large clock hands are so universally made of copper that it is hardly worth while to notice any other construction. There is indeed—or was, a notable exception, in Sir C. Barry's famous gun-metal hands at Westminster, which is not likely to be repeated. The hands of the clock at old Doncaster church, which perished in the fire of 1853, were of mahogany, and stood very well; but I should think copper ones are lighter, even including the stalk or centre piece. Where they are very large, say 5 or 6 feet long, the best form for them is that of the minute hands at Westminster, viz. a tube of thin copper, whose section is two segments of a circle, with a few diaphragms at intervals of about 2 feet to keep them stiff. The strength of this construction is enormous, and it is also good for throwing off snow, which sometimes accumulates on hands with broad edges heavily enough to stop the clock. Smaller hands may be made quite strong enough with a convex front and a flat back, the section being an arc and its chord, or even as a single flat piece of copper with the edges turned over square. A mere rib or hollow bead raised along the middle of a hand makes it strong enough for all ordinary sizes, but it does not look well. 'Galvanised' sheet-iron hands have been tried, but the zinc peels off, and they must be pronounced a failure. The minute hand should always be straight, and plain, with a bluntish point. At the broadest part, or near the dial centre, it should be about a 13th of its length, tapering to about half as much near the point. The hour hand should be the same breadth, ending just short of the figures in a broad piece called a heart, of any shape you like.

There should always be some external counterpoises to large hands, both for wind and weight. They should not be above 1/3 the length of the long hand, and should be broad, but of a shape not to be confounded with the heart of the hour hand. The advantage of counterpoising the hands to some extent for the action of the wind is evident; and the other use of an external counterpoise is to diminish the tendency of the hand either to twist the arbor, or what is more likely, to work itself loose and shake over from one side to the other every time it passes the vertical, as Reid says the old hand of St. Paul's cathedral used to do, and as Sir C. Barry's heavy hands did at Westminster to such an extent as to stop the clock. The only way to prevent this shake is to fit the hands on a tapered square or hexagon at the end of the arbor, and not a prismatic one. The latter may be called engineers' fitting, and is perfectly right for many purposes, but perfectly wrong for this, for which the old clockmaker's taper fitting alone will answer. It is found better not to put the whole counterpoise for very long hands outside: from one third to one half is quite enough, leaving the remainder to be done by adjustable counterpoises inside, which should be long rather than short, as they then do the same work with less weight and friction on the arbor.

Illuminated dials.—Occasionally it is possible, as at the Horse-Guards east dial, to illuminate a common white dial by reflection from a lamp on a roof projecting below it. This answers well enough for dials to be seen a short distance only, in the few cases where it can be done. Where it cannot, the common way is to make the dial of glass, or all of it except the figures and the rings to connect them, which form a solid framework of cast iron. The glass is ground behind, or painted, or covered with muslin stuck to it, and gas lamps are put behind it. But all these things have such a bad appearance by day that the advantage of illumination is dearly purchased at that cost. Now however a white glass is made by Messrs. Chance of Birmingham, and perhaps by other makers, which forms a very good and always clean white dial by day (if left open for the rain to wash its face) and a bright one by night: the hands and figures must be black as with other white dials. It should be 3-16ths of an inch thick, or 22 oz. to the square foot; the middle of large dials has to be in two or three pieces, which must be divided by bars not radial, or they will look like hands at night; and all but the figures and minutes should be gilt.

The gas lamps of illuminated dials are generally kept alight all day, turned down as low as they can be without going out. They are usually turned down in the morning and up at night by a 24 hour wheel in the clock, which has pins screwed into its rim and taken out again from time to time by the man who takes care of the clock. So long as any of the pins are in the position to hold up a weighted lever connected with the gas cock, it is turned down, and when the lever drops off it is turned up. In the Westminster clock three fan-shaped pieces were prepared on a 24 hour arbor, which can be opened out to 18 hours or contracted to 6, according to the length of illumination required; but it was afterwards determined by the authorities to turn the gas completely off and on by hand. There was a clock in the Exhibition of 1851 with completely automatic or self-adjusting machinery for turning gas off and on at the proper time throughout the year; and Gillett and Bland did the same in the Bradford Town Hall clock; but the tower is so small and crowded with bell beams, that they were actually hot by day, and I advised putting out the gas by day at least. The machinery puts it quite out now. It cannot be done at all accurately by any uniform automatic motion, because the clock times of sunrise and sunset vary irregularly from the equation of time (p. 6), and very slowly near the solstices, but at other times from 12 to 15 minutes a week.

It is desirable to have a wall, if possible, behind illuminated dials, instead of having them practically in the clock room; partly because the wall may be made useful as a reflector, and so save gas, and also because it protects the clock itself both from the variations of heat and from the watery vapour caused by burning the gas. Reflectors are a great saving of light and of cost. It must be remembered also that the counterpoises of the hands on glass dials must neither be long ones outside, nor immediately behind the glass inside, or they will cast a shadow and be confounded with the hands at night. There should always be ventilation over illuminated dials.

Dial wheels.—The construction of the dial-work of large clocks differs very little from that of small ones. The principal difference is that the numbers of the wheel teeth are differently distributed. Instead of two equal 60 min. wheels, there is a pinion on the minute-hand arbor which drives a wheel corresponding to the wheel N in pp. 134, 164, 168, only moving slower; and that wheel has a pinion on its arbor which drives the hour-hand wheel as in house clocks. If $t\ t_1$ are the numbers of the wheels and $p\ p_1$ of the pinions, they have only to satisfy this condition, $\frac{t\ t_1}{p\ p_1} = 12$, bearing in mind also that, from their position, the radius of one wheel must be as much less than of the other as that of its pinion is greater. The larger wheel is generally put on the hour-hand arbor. The most convenient numbers are 90 and 100 for the wheels, and 25, 30, for the pinions, or in smaller clocks 72, 80, and 20, 24.

The bevelled wheels leading from the clock to the dials ought to be of a good size, not less than 5 inches wide in small clocks, and 7 to 9 inches in large ones. They need not be very strong, as they have only to move the hands; but the advantage of their being large is that any given amount of shake in the teeth allows less angular motion of the hands. In the old way of fixing clocks on a stool in the middle of the room, which I have already shown to be the worst, there was generally a vertical rod from the clock running up the middle of the room, with 2 horizontal bevelled wheels, one on the bottom worked by the clock, and the other at the top working 4 others leading to each dial; and in that case it is necessary that the bevelled wheels on the vertical rod should be larger than all the others, both the first one in the clock, and the others leading off to the dials: otherwise the 4 leading-off wheels will take into each other as well as into the horizontal wheel. Where the vertical rod does not lead into the middle of the room this does not occur, but there must then be two nests of 3 wheels each, if there are 4 dials, besides the two wheels in the clock. I have seen several more wheels added, through a singular piece of ignorance that it is not the least necessary that the rod which leads upwards should be vertical. I was rather glad that it was necessary to put it very oblique in the Westminster clock as an example of such treatment (see frontispiece). With bevelled wheels of the common shape, intended to lie at right angles to each other, the rod must be in a vertical plane parallel to the clock wheels, but there can seldom be any difficulty in that: if there should be, the bevelled wheels have only to be made for the proper angle.

Even in such a clock as that of the Royal Exchange, undoubtedly the best that had been made up to that time, the odd mistake was committed of supposing that nests of bevelled wheels could not be connected obliquely. When I remodelled the clock in 1854, I removed all the intermediate bevelled wheels, and connected the central nest of them with the one in the clock by an oblique rod, like that of Westminster. When there is only a small parallel displacement it is managed by a loose cross riding with sufficient play endways in the sockets of two half universal joints on the rods. Where the rods are too much out of the straight line for that, a short oblique rod is interposed with common universal joints at each junction, and these will do well enough where the obliquity is not very great, provided always the oblique rod lies between two parallel ones; otherwise the velocity is not uniform. But where the obliquity is great, rigid rods with the bevelled wheels on them are better, for the wheels may be at any angle. Nevertheless universal, or half universal joints, are properly used wherever a rod is either too long to do without several supports, or where there is any risk of there being an unequal strain upon it in one direction, or where bevelled wheels can be saved by a jointed rod with very slight deviation from straightness. Turret clocks are generally made with the minute-hand on the internal dial turning the wrong way round, to provide for the case of the external dial arbor being able to go straight through the wall from the back of the clock, as it does when the clock can be placed immediately behind a single external dial.

I have lately seen a new kind of universal joint, almost ridiculously simple, in an universal or any-way-pointing drill, called Morrison's, from America. If you take a common spiral wire bell-spring and twirl one end of it while you hold it bent at any angle, the other end will twirl uniformly; which is not the case with very oblique universal joints. This kind of connection might often be used with advantage in leading off, and with less loss of force than from the friction of bevelled wheels, though of course there is some slight loss of force in continually bending the spring. The spring must be strong enough not to let the wind move the hands.

Sometimes the clock has to be placed a long way below the dials, 30 or 40 feet or more, and then it is necessary to provide both for the weight and the want of stiffness of such a long leading-off rod; and this is best done by a pair of friction plates and rollers at the top. The lower plate is set on the beams which carry the nests of bevelled wheels or *motion work:* on that lie three small flat cheese-shaped rollers on a horizontal tripod, with a hole in it for the long rod to go through quite loosely; and the other plate is fixed to the top of the rod, which is in fact hung by it, the rollers carrying the weight, and with no sensible friction on their own centres, for the three-legged pivots have no weight upon them. This kind of suspension is also used for heavy

Fig. 46.

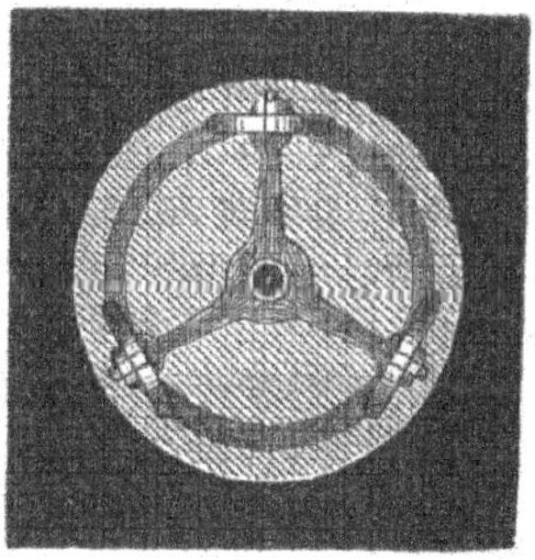

weathercocks which work a wind-dial inside the house, and the ease with which a very heavy weight can be turned in that way is surprising.

Weathercocks.—As these are generally fixed by clock-makers in such cases as those last mentioned, I may as well mention that a weathercock which is intended to answer steadily to the wind, ought not only to be long in the vane and thin in the tail, but equipoised; and so far from the vane being perforated for ornament, it should be double, with the two flat sides or vanes spreading out at a small angle from the axis. When the cock works a dial it must be fixed to a rod working loosely in a tube, and the top of the tubo covered with an inverted funnel on the rod; as also the rods or wires which work the clock-hammers should be funnelled, over a short pipe soldered to the leads, wherever they are exposed to rain: otherwise the wires lead down the wet into the clock. And the weights and ropes should be enclosed in a case to keep them from the rain and wind, if they are in an exposed place.

Ventilation of clock-room.—The clock-room at the Exchange was at first made with the object of keeping out the dust and damp in every possible way: even the slits in the floor for the ropes had sliders to them; the clock was enclosed in a glass case, the plate-glass cover originally placed over the escapement being found not enough to keep it from the damp. When the clock was repaired, and some of the brass-work replaced with iron in 1854 (for a reason which I shall mention hereafter), I suggested the removal of all this glass, and encouraging instead of preventing a draught through the room. This was done; and although the wet used to stand in drops upon the clock before in damp weather, it has been perfectly dry ever since. The same thing has been found in small clock-cases: they may easily be too air-tight. I do not mean that there is any objection to enclosing a clock in a case, and of course it is absolutely necessary where the clock-room cannot be kept locked against everybody but the man who has the care of it; only there should be a draught through the room, and the case itself not too close to let air through it. If the room can be kept warm enough to prevent the damp from condensing on the clock it is better still.

TRAIN REMONTOIRES.

I have postponed this branch of the subject till I had gone through all the more ordinary work of turret clocks. A train remontoire differs in principle from a remontoire escapement (of which I have already treated) only to this extent: the small weight or spring which gives the impulse to the pendulum is not wound up at every beat, but at some longer intervals, seldom more than half a minute; or the remontoire work, you may say, is put one step farther back, acting on the scapewheel instead of on the pendulum. So that if a train remontoire of constant force and friction is made to act on a dead scapewheel, the only variation of force to which the pendulum is subject will be that arising from the pallet friction. In small clocks the variations of the pallet friction are generally much greater than of the train friction, and therefore a remontoire would be of little or no use; but in large clocks with heavy wheels and large hands to drive, the contrary is the case; and there accordingly, either a train remontoire or a remontoire escapement is of great use, provided they really do what they profess—which many of them do not.

The simplest form of train remontoire acting by a weight is that described in Reid's book, on the endless chain principle, which I have already described for a going barrel at p. 148. The scapewheel is not driven by the clock train, but it has a spiked pulley on it which carries one loop of the endless chain, and the other is carried by a similar pulley on an arbor driven by the train and turning in the same time as the scapewheel. This remontoire arbor has a few long spikes sticking out from it, at different distances along the arbor, and they are just long enough to reach the middle of the scapewheel arbor, and can slip past it through a nick cut for the purpose, whenever that nick comes into the right position, which it does once in every turn of the scapewheel; then the remontoire arbor turns and winds up the endless chain a little until the next spike falls against and is stopped by the scapewheel arbor, till its nick also presents itself and lets that spike slip through, as you see in this drawing.

Nevertheless, that construction is far from satisfying all the conditions of a remontoire. The action of a chain cannot be made smooth and uniform, and a rope or string passing only half round a pulley is sure to slip in time, and so the remontoire would fail. Moreover, Reid says that although the Edinburgh clock went very well for a time, yet it became necessary to remove the remontoire in consequence of the banging of the spikes against the scapewheel arbor. That however would be easily cured by a fly; which may now be considered a necessary element of every remontoire: for there are several other kinds.

One is the bevelled wheel remontoire, which was used in many of the French turret clocks in the 1851 Exhibition. I have already explained the principle of it, so far as the bevelled wheels are concerned, at p. 146; but something more is now required. In fig. 47 (*t. o.*), consider the dark disc F G and the pulleys omitted for the present (we shall want them presently for something else). X is part of that wheel of the clock which would naturally drive the scapewheel *s* if there were no remontoire, and if it were fixed to the arbor of the pinion *t* which X does drive. It also drives another pinion P on an arbor P Q which I call the remontoire arbor, because it carries two long spikes not quite in the same straight line, or not at right angles to P Q, which can pass

Fig. 47.

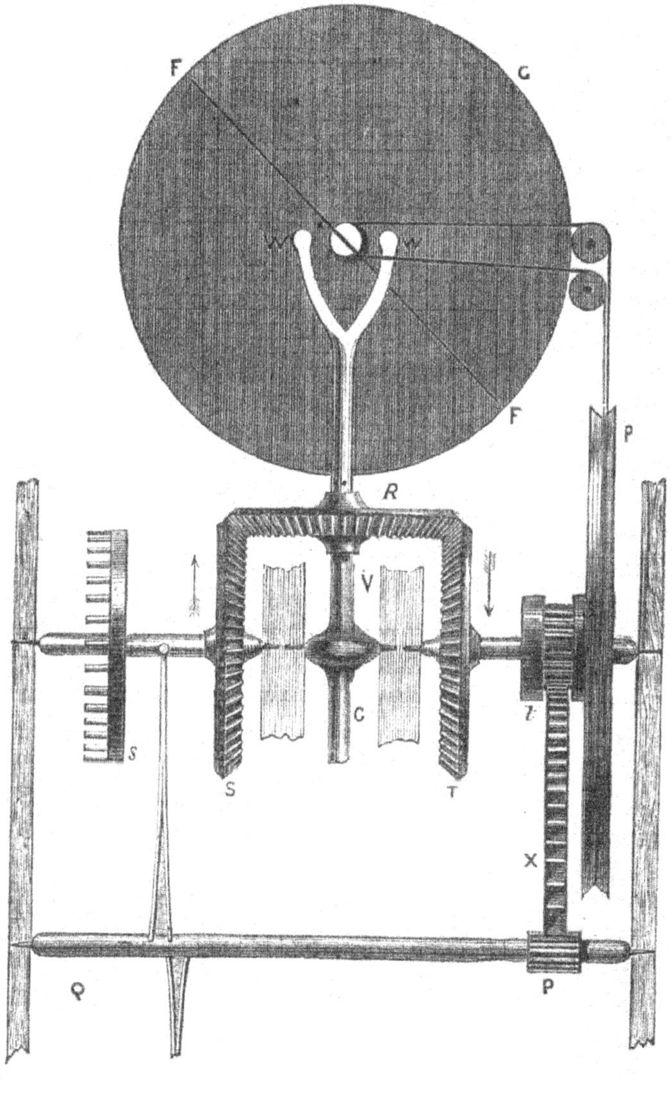

alternately through two nicks in the scapewheel arbor. When the arbor is stopped the train and the wheel T do not move; but the weight of the middle bevelled wheel R, with any weight (+ or —) attached to it, always presses down the far side of wheel S, and of the scapewheel *s* fixed on the same arbor, and that force is practically constant. When the scapewheel arbor lets a remontoire spike pass, the pinion P turns half round and lets the train move a little, and lift up the right side of wheel R, and therefore lift up its centre half as much, and then the other spike is held by the scapewheel arbor. The fly may be driven by the wheel D anywhere, either on the remontoire arbor or any other. We shall see another way of driving it presently.

Continuous motion remontoire.—In connection with this I will describe a remontoire exhibited in 1851, by Messrs. Wagner, the great turret clock makers of Paris, for the purpose of getting a continuous motion for telescope-driving clocks, or clocks to drive a barrel on which times of observation may be recorded at less intervals than a second, with all the advantage of a vibrating instead of a revolving pendulum. The action for the vibrating pendulum which is driven by the scapewheel *s* is exactly what I just now described, except that there is no spike wheel and no sudden letting off. Instead of that there is a large pulley *p* on the arbor of the wheel T which lifts the remontoire arm V R W. This pulley drives a much smaller one on the arbor of a fly F F, which runs inside a tin drum without a bottom, which is hung over it by two wires W W from the end of the remontoire arm. The weight of the drum is counterpoised so as not to let it preponderate too much. The fly is the thing which regulates the velocity of the clock train, which is always moving; the farther the drum falls over it and cuts off the air within from the air outside, the faster the fly will turn, and vice versâ; and things are so adjusted that the continuous motion of the clock train driving the fly will just keep pace with the average motion of the scapewheel driving the pendulum by beats as usual. If the clock falls behind the proper speed the remontoire wheel and its drum falls a little and lets the fly go quicker, and if too fast it rises and the velocity of the fly is checked. The one in the Exhibition seemed to go very steadily; and as there is nothing in all this at all difficult to make, I am surprised that more complicated contrivances should be used for such purposes as I have mentioned.

Mr. De la Rue and Mr. Cooke appear to have hit simultaneously on the following plan for a telescope-driving clock, which retains the advantage of a vibrating pendulum, and may have a gravity escapement also. In the figure at p. 230 suppose the two opposite bevelled wheels to be respectively fixed on the 'centre arbors' of two clock trains, the left hand one ending in a vibrating pendulum, and the right in a conical pendulum or a fly. And let the arbor of the intermediate wheel (omitting all the wind apparatus F G F) be held by a spring instead of the constant weight C. At every beat of the pendulum the left hand train will stop for a moment, but the fly-wheel train, of which the barrel is to drive the telescope, will go on by its own momentum, and because the spring allows the intermediate wheel, and therefore the right hand wheel and its train, to move a little. If that train is inclined to go too fast, the spring will diminish the force upon it; and if too slow, will increase it. Of course the weight and fly are to be adjusted so as to make that train go naturally as nearly as possible with the other.

A still simpler telescope-driver was invented by Mr. R. F. Bond, of Boston, U.S., brother of the well-known Professor of Astronomy there. As I understand the description, one arbor of the clock has two wheels on it, one fixed, and the other connected with it by a spiral spring. The second of these wheels is either the scapewheel or the one below it, and the other ultimately drives a fly, which allows the train to go continuously, the spring equalising the force sufficiently at every beat. Mr. Cooke said it did not answer, but Mr. Bond replied with a testimonial from the late Rev. W. R. Dawes, F.R.A.S., who had the reputation of being the best observer of his time, that the 'performance was exquisite; that it kept the thread of the micrometer dividing a star for nearly an hour together with a magnifying power of 800 or 1000 on, and that no jerk or interruption was perceptible.' Lord Lindsay, P.R.A.S., invented another, too complicated to describe here, on the ground that neither Cooke's nor Bond's clocks would drive a large telescope fitted with spectroscopic apparatus steadily enough for that purpose.*

The Royal Exchange clock was originally made with a gravity remontoire, though it was afterwards altered. Instead of bevelled wheels, Mr. Dent used an *internal* wheel, *i.e.* one with teeth on the inside of the rim, instead of the outside. That wheel D (fig. 48, *t. o.*) had the letting off spikes on its outside; at least a wheel on the same arbor had, which is the same thing. It was driven by the centre wheel of the clock, and whenever it moved it lifted the remontoire arm and weight by means of the small wheel B lying between the internal teeth and the wheel C on the arbor of the wheel F which drove the scapewheel: that arbor being of course independent of, though in the same line with the arbor of the wheel D and its pinion. The remontoire was let off at every 20 seconds; which however is not so good an interval as 30, because it is not easy to distinguish whether the hands are pointing to 10 seconds before or 10 seconds after the half minute; whereas it is perfectly easy to see whether they are pointing to a minute

* All those contrivances are described in the R.A.S. Notices, vols. 27 and 33, and *Horological Journal* of February and April 1868.

or a half-minute, if the dial is properly made, as I have already described. This facility for taking the exact time

Fig. 48.

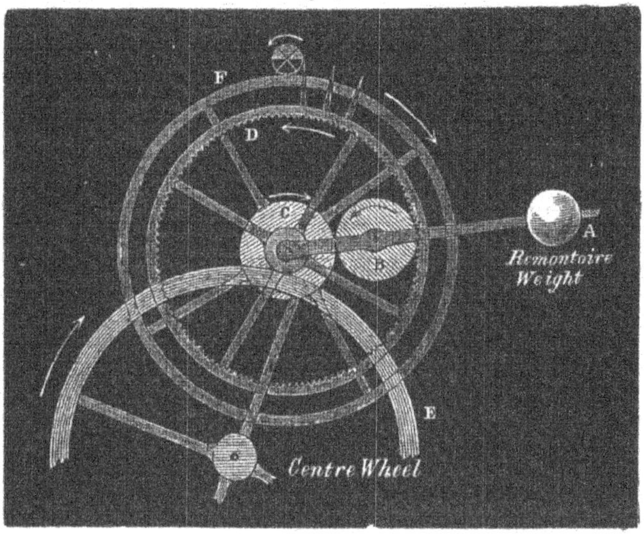

from the dial by the jump of the hands is one of the advantages of a train remontoire, where the momentum of the hands is not too great.

There is yet another way of making a train remontoire without either bevelled or internal wheels. In fig. 49 (next page) E is the scapewheel, and *e* its pinion driven by the remontoire wheel D which rides with its pinion *d* fixed to it on a stud in the remontoire lever A P. The centre wheel C drives that pinion and a smaller one *g* on a wheel which drives another pinion *f* on the fly arbor, which has also the remontoire spikes A B attached to it. The numbers of the teeth are so arranged that the fly will turn once for each turn of the scapewheel, and the scapewheel arbor has only two notches in it, so as to let off one remontoire spike every 30 seconds. It is evident that the scapewheel is always

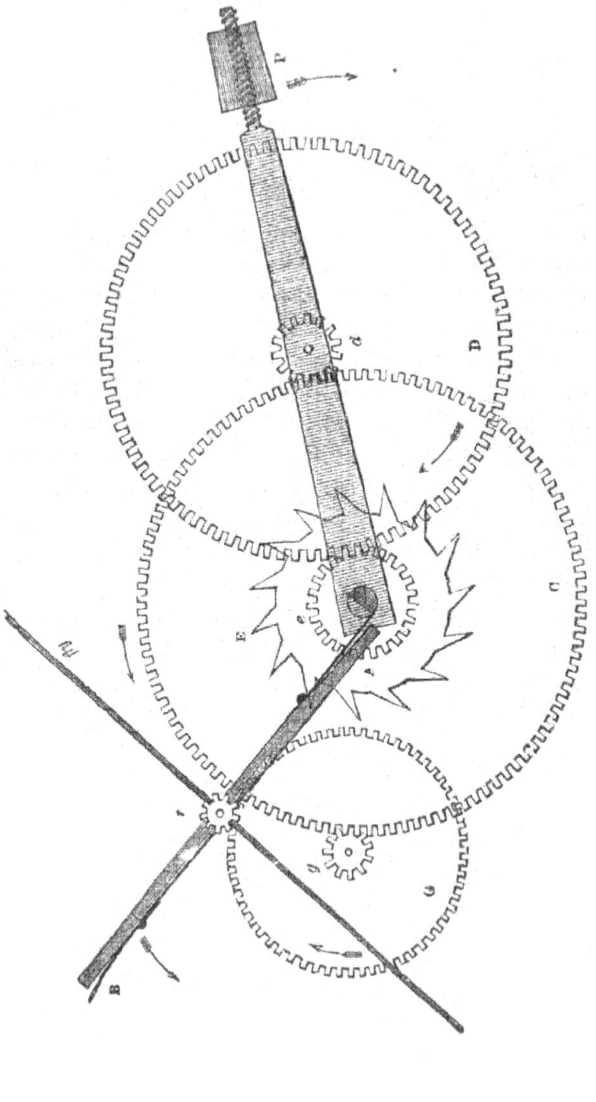

Fig. 49.

driven by the remontoire wheels and weight, and not affected by the clock train. There were several French clocks of this construction in the Exhibition of 1851, but they all had the fly driven by an endless screw, which is objectionable because it involves an immense amount of friction, and the motion of the train was so slow that you could hardly distinguish the jump of the hands.

Spring Remontoire.—But it must be observed that all these gravity remontoires are still subject to the friction of the remontoire wheels themselves, which is not inconsiderable, although it is much less, and less variable, than that of the clock train and hands. To avoid this, it was long ago attempted to contrive a spiral spring remontoire which would drive the scapewheel without any sensible friction. One of these is described in Reid's article in the seventh edition of the *Encyclopædia Britannica;* and another was invented by Sir G. Airy,* and two or three specimens of it were made by old Mr. Dent. They all went on the plan of connecting two wheels, or a wheel and pinion, on the same arbor by a spiral spring, one being fixed to the arbor and the other riding upon it; and the consequence was that the scapewheel was always subject to the friction of the other wheel set upon its arbor and pressed tight upon it by the action of the spring, which was probably worse than the ordinary friction of the train; and they were all failures.

This difficulty however may be got over by a very simple contrivance which I invented in 1849, and which was used in several large clocks, and would have been in many more but for its having been superseded by the cheaper and simpler gravity escapement already described (p. 121 to 132). In fig. 50 (next page) E is the scapewheel and *e* its pinion, not fixed to the arbor of E nor riding upon it, but upon an independent stud *k* screwed to the clock frame. In front of the pinion a small bush is pinned on to the same

* See *Horological Journal*, xvii. 22.

stud, which carries the pivot of the scapewheel arbor, on which is set a large watch spring *s*, of which the outer end is held by a pin *h* screwed into a small plate fixed to the

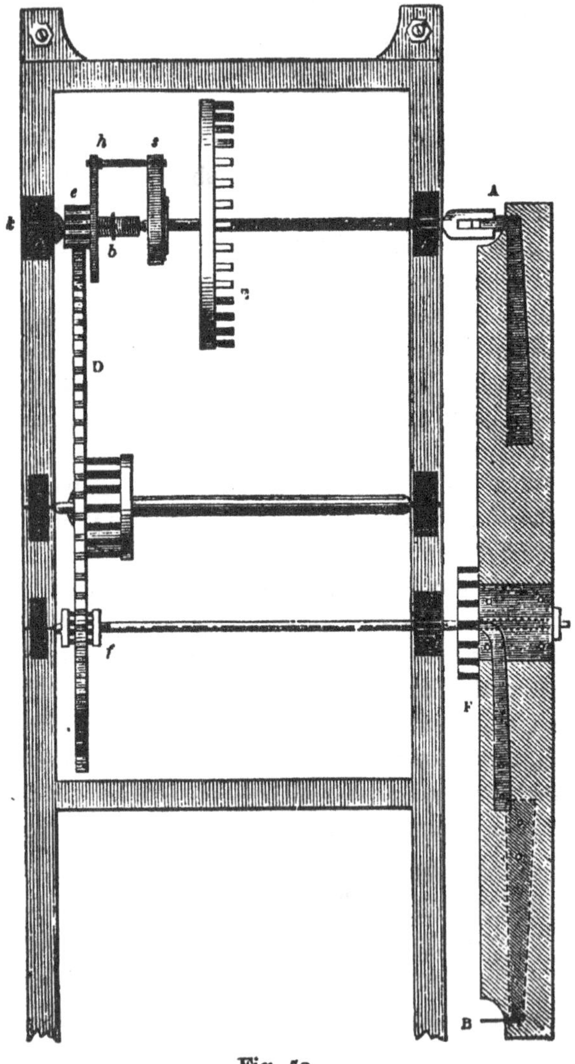

Fig. 50.

front of the pinion, so that the pinion acts on the scapewheel by the intervention of this spring without any friction except that of the coils of the spring upon each other, if they touch at all. The wheel D which drives that pinion also drives another on the fly arbor *f*. If the scapewheel turns in a minute, there will be two nicks across its arbor, as in fig. 47, for the remontoire spikes or stopping springs to act upon. But if it turns in two minutes, as the pin scapewheels generally do, a different plan is adopted: the scapewheel arbor is then brought through to the front of the clock frame and has a small steel cylinder set upon it, with two nicks across its face, not its side, one broad and shallow and the other deep and narrow; the remontoire springs set on the two arms of the fly have corresponding shapes, so that one will pass through the broad nick only and be stopped by the narrow one, and the other will pass through the deep nick but be stopped by the shallow one. The fly in this case makes half a turn for every quarter turn of the scapewheel, and therefore its pinion must be only half the size of the scapewheel pinion. In very large clocks such as the Exchange, in which the gravity remontoire was replaced by a spring one in 1854, the fly is made separate from the remontoire arms, with a ratchet and click as usual to let it run on a little by its own momentum; but in smaller ones it does very well to put the remontoire spring stops on the fly itself.

The wheel shown at F with a kind of detent in it, is for setting up or letting down the spring with reference to the remontoire arms, and so increasing or diminishing the force on the scapewheel.

This was the construction of old Mr. Dent's large clock in the 1851 Exhibition, now at King's Cross station, which was found only 3 seconds wrong at the closing of the Exhibition, though it had never been altered during the 24 weeks it was there, after its preliminary regulation. I never knew any clock with such a rate as that, and yet it has only a common unjewelled pin-wheel escapement, and a $1\frac{1}{2}$ sec. zinc and iron pendulum, and most of the wheels are of cast iron. It has sometimes gone since for several months together without any noticeable variation.

The Exchange clock was immensely improved by that alteration, and made as good as that of the King's Cross clock in general, but I have no exact record of its rate. I must confess however, that this kind of remontoire is not safe to put into clocks left in any common hands, as I have found by experience that the first time the clock wants cleaning or repairing—or even without that opportunity—a stupid workman, or master, makes a point of destroying the remontoire. I knew one case where it was done as soon as an ordinary clockmaker got the care of it—and that at Manchester, the city of machinery. This cannot happen with a gravity escapement, which is also much cheaper. The Exchange clock has been altered by Gillett and Bland to the double three-legged gravity escapement, which is now the standard one for all large clocks. Therefore it is no longer necessary to explain further details of the spring remontoire, as in former editions.

Cast-iron wheels.—The success of the contrivances for cutting off the variations of force from the pendulum led to another alteration which helped to reduce the price of large clocks considerably; and that was the making all the wheels below the escapement, and all the dial wheels, of cast iron instead of brass or gunmetal. Mr. Vulliamy had before recommended that as a good construction for cheap clocks, but it had always been thought that they could not be also good ones on account of the greater friction of the train. I believe that apprehension was very much exaggerated, even for clocks of the common construction, provided of course the escapement is light and well made; but as soon as you cut off the friction of the train from affecting the escapement it is obvious that cast-iron wheels are just as good as brass or gun-metal. The clockmakers in general violently denounced it, probably for no better reason than that it lowered their prices enormously, the price being now only £200 for clocks of greater power and far greater accuracy than those for which £500 used to be charged not many years ago.

The cast iron wheel controversy came to a head in some of the Lancashire papers soon after the making of the clock, from my design, for the Manchester Infirmary; and the advocates of brass wheels had clearly no case whatever. Their three points against cast iron were friction, rust, and liability to break. The friction of the train is absolutely immaterial with a remontoire, or a gravity escapement, and no large clock can go with great accuracy without some such contrivance. The next objection is obvious nonsense, because all except the acting surfaces are painted, and they are of course oiled as in all other iron wheel machinery. The liability to break is a mere question of experience. Those who condemn them for clocks must be very ignorant of the extent to which cast iron wheels finer than are ever used in church clocks are used in every factory in Yorkshire and Lancashire. I made particular inquiries once as to the sizes down to which the teeth are cast in iron wheels for spinning machinery (for that is what I mean by cast iron wheels), and I found that they are cast with quite sufficient accuracy with teeth as small as $^1/_{10}$ inch thick; which is smaller than any I have seen used in clocks, because there is very little saving in cost in using iron wheels so small as that. The great saving of course is in the large wheels of the train, and the dial and bevelled wheels, of which a good many of the same pattern and no very fine pitch are required.

Before I leave the cast iron wheels I should observe that they work better with cast iron pinions than with steel ones: indeed cast iron and steel seem never to work well together, at least in no clock-work that I am acquainted with, if there is much pressure between them. I have seen cast iron fly ratchets used with steel clicks, by clockmakers who would not listen to the proposal of iron wheels and pinions for any but the commonest clocks, and they had to be removed, and replaced either by wrought iron or brass ones. I have seen and heard of brass teeth worn out in an almost incredibly short time, long before iron teeth in the same clock showed any signs of wearing. Moreover few people have any idea how rapidly brass is corroded and in fact destroyed by such an atmosphere as that of London and other large towns. I have several times seen the brass tubes which had been used in dial work, and thin pieces of brass elsewhere, brought back to be replaced with iron because they had become completely rotten. It was so at the Royal Exchange in eight years.

In this respect gun-metal is better, which is copper and tin instead of copper and zinc, but for large wheels it is no way superior to iron; and it is generally made too soft. It is equally absurd to polish iron work, except the acting surfaces; that rusts even sooner than brass begins to corrode; in fact very often in a week after the clock is put up. Then somebody who has the care of it floods it all with oil, and it is filthy ever after. The only proper rule to lay down is that *all non-acting surfaces should be painted.* In the Westminster clock even the small brass wheels in the escapement are painted like the iron ones.

The truth is that all this 'finishing' of non-acting surfaces is what old Dent used to call 'working for fools.' It has literally no other object (as plenty of clockmakers have confessed to me) than to make an impression on the 'fools' (in and out of the trade) who go to see a new clock in the first month after it is put up, and never see or want to see it again. Such people as these think it a much finer thing to have a clock look bright for a month and never keep time within a minute a week—or a day, for what they know—

than 'to look like a patent mangle,' but keep time within a second a week.

I must warn people however against some altogether cast-iron clocks which got into vogue some years ago by their extreme cheapness, if they could be called cheap at any price. I heard constant complaints of them, and had to subscribe once to help a friend to get rid of one and substitute another of a proper kind. I shall give presently a form of specification showing how much of a large clock may be of cast-iron.

A few years ago I learnt when it was too late that a deputation from the corporation of a considerable town had come to London to negotiate for a first-rate town-hall clock, and after wandering about for some time they fell into the hands of 'an eminent firm' who boast of sending clocks all over the world, and engaged to pay them more than would have bought the best possible clock for one which was warranted to keep time within 5 *minutes* (not seconds) a week; whether it actually does perform that feat I do not know, but I heard casually of some other of its successes, which gave hopes of it breaking down altogether.

Another common cause of bad church clock making is the inveterate habit of jobbing for the benefit of a townsman by giving the order to a watchmaker who never made a turret clock in his life, and who immediately goes or writes to one of the few real makers for the cheapest clock that will serve his purpose, and probably charges for it as much as the people could have got a first-rate one for, if they had had a competition of the best real makers of large clocks, or even gone to one of them alone. The local watchmaker, by way of carrying out the fraud completely, generally insists on the real maker putting, not his own, but the pretender's name on the clock. If this were merely a question between the makers who consent to do so, and those who will not, I should leave them to find their own remedy; but as it affects the general credit of our public clocks, I think it right to give this warning, though it will probably not be read by one in a hundred, or attended to by one in a thousand, of those for whom it is intended. It would be far better to do the 'native talent' job in a straight-forward way by letting the local watchmakers draw lots for a *bonus* of £20, and then advertise for tenders under proper conditions, such as I suggest below.

The practice of insisting on clockmakers tendering for bells, except quite small and common ones, is almost equally objectionable. Every now and then, one party or the other has paid dearly for it themselves; for which also I do not care; but the more material thing is, first that two profits have to be got out of the transaction without any real necessity for a middle man, and that there is practically much less responsibility and control than when you deal directly with the bell-founder. You might as well insist on the clockmaker fitting up an observatory with telescopes,* or trust a builder to put in painted windows and the organ in a church.

SPECIFICATIONS FOR PUBLIC CLOCKS.

Every now and then I happen to see specifications for large clocks, either prepared by somebody belonging to the office which has to order it, or in the form of a tender sent in by some advertising clockmaker, and I hardly know which are generally the worst, except that the author of the tender does know what he is about, and takes care to put in as many adjectives and as few substantives as he can, and the author of the specification never does know what he is about. An elaborate specimen of this kind was shown to me some years ago, from a Government office, which stipulated among

* There was indeed one clockmaker quite competent to do both, the late Mr. Cooke, of York; but then he made the telescopes himself, and was one of the best makers of his time.

other remarkable things, that the pendulum was 'to vibrate 2 seconds, but to be as long as the room admitted of.' It was also to have *convex* copper dials a *quarter* of an inch thick, and a *dead* escapement; which were three specimens of the author's practical knowledge of the present state of science. The specification, though of many pages, did not contain a single thing, except the probably impossible pendulum, of the slightest value towards securing what (I suppose) was wanted, either in the way of time-keeping or striking. Sometimes the providing of a clock and bells is entrusted to the architect, by people who are not aware that modern architects rather boast than are ashamed of being totally ignorant of every thing of the kind. I have heard people who ought to know better say, that 'the only way is to throw all the responsibility of everything connected with the building on the architect,' and utterly puzzled when I asked them what those fine words meant, or what would they do if everything turned out wrong, as it generally does in that case. I will therefore give a specimen of the specification suitable for a large church clock, though I cannot adapt it to all possible circumstances; and a much stricter one is requisite for a general competition than for one limited to a few makers whose work I know to be of the best kind. Indeed no specification can secure a good clock from a bad clockmaker; and it ought not to be necessary to warn people (but I find it is) that those who blow their own trumpet loudest in advertisements are very often the least to be trusted, especially in an article of which most purchasers are quite incapable of judging, until it is too late. All that a stringent specification can do is to frighten away bad makers from tendering at all; and for that purpose the first clause of the following form is useful in a general competition, provided it is certain to be enforced.

1. To make and fix a clock with *m* dials of *n* feet diameter, and striking the hours and Westminster quarters,

on bells which are or would be the 2nd, 3rd, 4th, 7th, and tenor of a peal of 8, the tenor weighing ——.

2. The dials to be concave and made of copper, or painted on the wall, if smooth enough, or with the figures and minutes of cast iron, in rings. If the dials are to be illuminated they *must* be of cast iron, and should have opal glass of 22 oz. to the foot, except behind the minutes, which may be 16 oz. to the foot. There must be no straight radiating pieces of iron in the middle. [Architects should be made to understand (if possible) that dials intended to be illuminated must have a clear opening in the wall of the full diameter of the dial.]

3. The long hands to have a short external counterpoise, painted the same colour as the dial, if not illuminated, and the hands, figures, and minutes to be gilt. If illuminated, the hands, figures, and minutes to be black, and all the rest gilt. The long hands to be straight and plain.

4. The escapement to be the double three-legged gravity, and care must be taken to leave room for the fly of sufficient length, and to make the angles such as to run no risk of tripping, and generally according to this book.

5. The pendulum to have zinc compensation with iron rod and tube, and bob not less than 2 cwt. To beat 1½ seconds for a clock with several large dials, but may be 1¼ for a smaller clock, or if there is no room for a 1½ second pendulum. It is to swing $2\frac{1}{2}^\circ$ from zero, and to have a block under it in case the spring breaks, and a degree plate.

6. The clock must lie on stone corbels or iron beams, or brackets bolted through the wall, and the pendulum cock either bolted to the wall, or rising from the clock-frame, which is to be generally on the plan at p. 196 of this book; but the great wheel arbors are to be in drop bushes of the form *d*, at p. 192.

7. There is to be a minute dial, and either one for seconds or a wheel marked for that purpose, with a fixed index.

8. To have the improved bolt and shutter maintaining power (p. 151) or the Wesminster one for a very large clock The going part to go a little over eight days.

9. The striking parts to be wound up every one or two days, except in small clocks [or they will never strike efficiently]. All the striking to be done by steel-faced cams on the great wheels. The 4th quarter bell to have two hammers (see p. 188). In smaller clocks cast-iron cams will do.

10. The hour striking to be let off independently of the quarters, and the first blow to be struck exactly at the hour, the hammer being left on the lift: the other quarters to begin exactly at 15, 30, and 45 minutes.

11. The hour hammer to be not less than a 60th of the weight of the bell, and to be raised not less than 9 inches; the quarter hammers to increase in weight upwards from a 60th to a 40th of the weight of their bells, and all the hammers to be raised enough to bring the full tone out of the bells: small ones at least 6in., and the hour one at least 9.

12. There must be either levers or eccentrics to lift the hammer levers completely off the cams, when the bells are ringing, if there is a peal.

13. The small wheels of the going part to be of brass or hard gun-metal, driving lantern pinions. All the larger wheels and the pinions driven by them may be of cast iron, and their being of brass or gun-metal will be considered no reason for a higher price. All bushes to be brass or gun-metal. The large pivots to be case-hardened, the smaller ones and their arbors to be of steel. The pulleys to be large and pivotted in brass bushes.

14. The barrels of sheet iron brazed, with iron or steel wire ropes *not* coated with zinc [which makes them crack], and the pulleys to be so placed that the ropes do not grind, or go twice over the barrel.

15. The flies not to be in front of the clock, and to be long enough to make the time of striking quite uniform, and slow

enough. The fly ratchets to be either squared or keyed on, and not merely pinned. The ratchets *not* to be of cast iron, and each to have two clicks. [Many smashes have occurred for want of attention to these two conditions.]

16. All the iron-work except acting surfaces to be painted blue [black is more difficult to see if anything is wrong, and if not painted iron only gets rusty].

17. If the weights go out of sight there must be something to stop or warn against winding them too far. And also a large box with 3 feet deep of small stones, to catch the weights if they fall; unless they go down to the ground of the tower, when they can only break flags.

18. All wheels to take out separately by unscrewing the bushes; and pulleys to be placed where they can be oiled.

19. On all points not specified, the clock is to be made according to the directions of this book, applicable to clocks of the kind now required.

20. The clockmaker to provide (or design and superintend) a strong wooden case for the clock, with glass showing the minute dial if it is in the belfry, and so arranged that the striking parts can be wound without opening the case; unless the clock is in a room always locked up, and protected from weather. In many places it is necessary to have a case for the weights, and pulleys at the top, if exposed in a bell-chamber.

21. The estimate to state the cost of the clock and dials, and of the fixing, case, &c., separately.

22. Half the price to be paid on the clock being completed, to the satisfaction of such person, not objected to for good reasons assigned by the clockmaker, as the committee may appoint, and *the other half when it has gone for three calendar months continuously, without varying more than five seconds in any week*—tested by the striking of the first blow of any hour. [This is an ample margin to allow now that we know from the Westminster and other clocks that they may

be made to go without much more than a second a week variation. Every town now has the means of testing its clocks by electric telegraph time at the post office, but not so exactly as it ought to be. And if there is any neighbouring observatory, or meridian instrument, properly attended to, that is much better.]

CONSTRUCTION OF THE WESTMINSTER CLOCK.

The frontispiece of this book is a front elevation of this great clock, or so much of it as can be shown in one view without confusion, for which purpose I have drawn the wheels and pinions only as circles. As I have inserted the numbers of the teeth, and the size of all the parts may be taken from the scale, I shall say no more than that the frame is 15½ feet long and 4 feet 7 inches wide. The going part does not occupy above 2 feet of this width, the front bushes of the wheels being carried on a separate bar lying on the two cross pieces of which the ends are shown in the drawing bolted to the great back and front girders.

Double-barrelled Crab.—The space between the front of the going part and of the great frame happens to be convenient for the fall of the ropes from a double-barrelled crab (fig. 51 next page), which is fixed upon the iron beams B B which go across the room from east to west to carry the 'motion work' or nests of bevelled wheels above the clock, of which the first pair are shown in the drawing. This crab is the only means of getting anything into the clock room which is too heavy to be carried by a man up the stairs. It consists of two small equal barrels, with four pulley grooves in each, each having an equal wheel at the end turned by equal pinions on the winding arbor. The rope passes over the outer halves of the two barrels 4 times, one end falling down the clock shaft or well for the weights while the other rises; and it can either be used

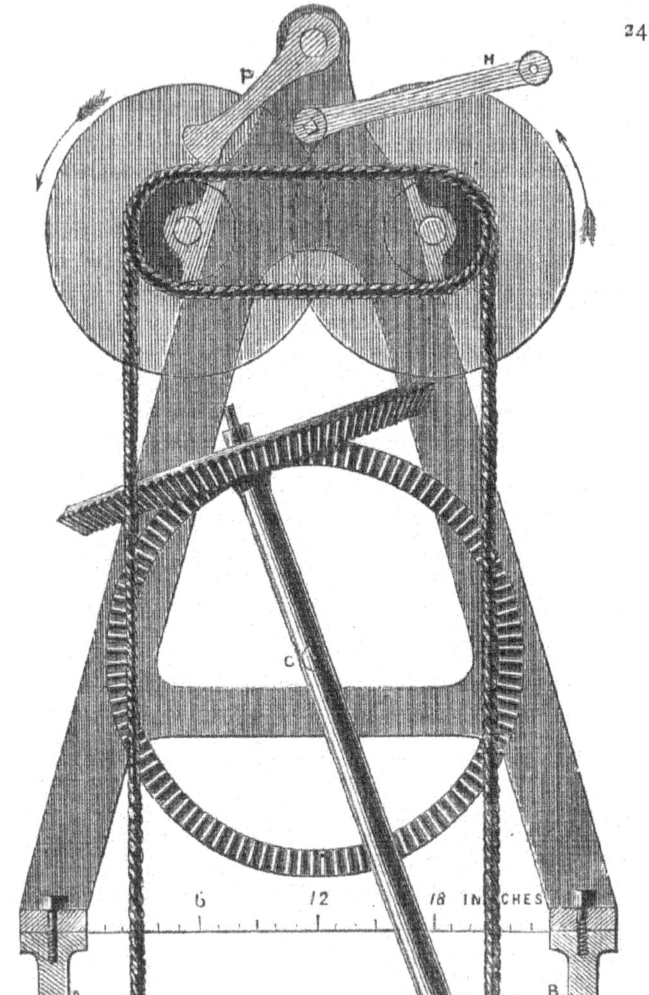

Fig. 51.

with a single rope, or with pulleys. The *pall* P is in effect a silent click to hold the rope when the handle H is let go; and it will turn over on the horizontal pillar at the top of the crab, to act the opposite way when the barrels are worked the other way.

The great advantage of it is that it avoids the crowding and overlaying of a long rope on the barrel, which makes the work harder at every fresh 'overlay,' as it practically increases the diameter of the barrel; and that becomes so intolerable that the plan of pulling off the rope by hand after three or four coils on the barrel has to be resorted to in long lifts; and that again involves the difficulty of having to slide the coils back to the beginning of the barrel, holding the rope by the hands of several men, or by some supplementary machinery, for very great weights, as was done with Big Ben in 1859. 'Great Paul' was raised on 3 May, 1882, by a double-barrelled crab, or rather two of them, and pulley blocks besides; with only this difference, that the ropes were crossed between the barrels to give greater frictional hold; but that also causes a considerable friction of the ropes against each other, and is really unnecessary. A crab like this, not with the ropes crossed, driven by steam, was used for taking up the ribs of the great domes of the 1862 Exhibition. I had recommended it in the 1860 edition for taking up heavy bells,* and in the later editions I suggested a further improvement, shown in fig. 52 (next page), by putting the pinions right in the middle between the wheels, and making them all roll on blank 'pitch circles' (see chapter on teeth of wheels), besides having small teeth cast by the side of the circles. This will take a great deal of pressure off the barrel pivots, and also off the driving pivots where it does mischief, and transfer it to the rolling circles, where it is useful in driving and relieves the teeth,

* The differential pulley with an endless chain in notched grooves is now largely used for moderately heavy weights, the chain being pulled by hand as in the old-fashioned 30-hour clocks; but it will not do for great weights.

and would almost drive without any. They may consequently be made much smaller, and the wheels also, and yet with the same power as a much larger crab of the common kind; especially if 4 wheels are used: which also has the

Fig. 52.

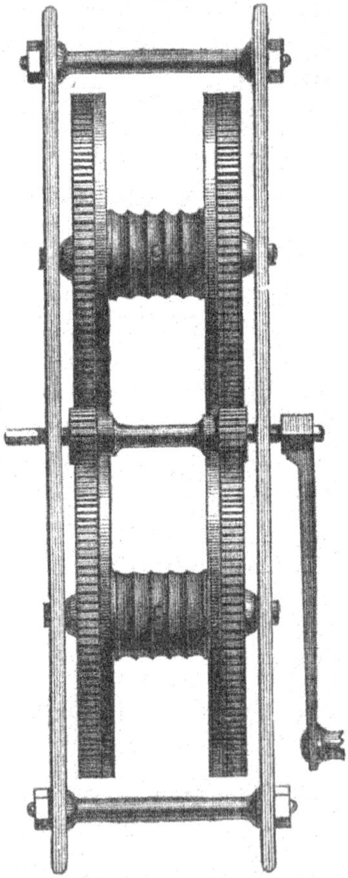

advantage of relieving all the pivots equally. 18 inch wheels with 1½ inch pinions would be strong enough for all ordinary work of single multiplying crabs. And it is easy to add another wheel and pinion for larger ones.

The teeth of the pinions and wheels respectively should be made as will be explained under 'Teeth of Wheels,' the pinion teeth alone projecting beyond the pitch circle. The rolling circles must be hard, or they will bruise each other. The arrangement of the frame also will depend on the work for which the crab is intended; I only show what affects its principle. The horrible noise of the common *pall* may be avoided by putting a wedge-shaped one between any two of the rolling circles, with a slight spring acting on it, according to the direction of driving for the time. This will act as a silent pall as soon as you let go the handle.

Returning to the clock, the back of the frame is 2 feet 5 inches from the west wall of the room, which is the east wall of the ventilating chimney or air-shaft of the Houses of Parliament running all the way up the tower. The room is 28 feet by 18, and the clock lies, as shown in the drawing, on the north and south walls of the shaft or well for the weights, which is 174 feet high, and the floor of the room is 2½ feet below the top of two iron plates which cover the walls and are spread out behind and built in quite through the wall of the airshaft, so as to prevent any possibility of endway motion of the clock frame, which is bolted to the plates. The pendulum cock is a large piece of iron frame work cast in one piece and also built in through the wall, quite independent of the clock frame. The pendulum chamber is made of sheet iron within the weight shaft, in order to protect the pendulum from the wind: you can descend into it by a ladder and a trap-door; which can seldom be wanted, as there is a degree plate set on the floor under the clock, which is a little lower than the general floor of the room, and is covered with a grating lying on the beams which carry the returned end of the ropes.

Regulation.—The mode of regulating the pendulum by small weights has been sufficiently described at p. 51. Whenever a larger alteration is required, in consequence of thunder-storms or an accidental disturbance of the pendulum, either stopping the scapewheel or letting it trip one beat by lifting a pallet alters it 4 seconds, whereas you cannot put a dead escapement forward at all, nor stop it without considerable risk of spoiling a tooth.

The pendulum weighs altogether very nearly 700 lbs., of which the bob is about 4 cwt. In other respects it has been already described at p. 44. The spring is a 60th of an inch thick and 3 inches wide, and the free part of it 5 inches long. The great pin through the upper chops has nuts on its ends to adjust it to the exact centre between the pallet arbors. Those are not shown at page 44, because they are not generally used. The pendulum cock, and the position of the floor behind the clock, are so arranged that a tall man can stand with his head inside the cheeks of the cock so as to look square at the escapement; which is

The double three-legged gravity escapement, as shown at p. 129. The scapewheel teeth are equidistant and 6 inches long; and though the clock has to drive about a ton and a half of hands and dial-work through all weathers, the pressure and friction on the stops are so little that I could find no difference in the weight required to lift the pallets (by a thread over a small pulley) whether the teeth were bearing on them or not,—*i.e.*, it was less than the friction of the pivots and the pulley. The weight required to lift each pallet through the angle of impulse was an ounce falling ·9 of an inch, which is therefore the amount of impulse received by the pendulum at each beat. In other words, the daily momentum of clockweight required to drive the pendulum, or *Wh* in all the calculations about escapements, is only about 200 pound-feet, while *Ml* is 9100, a much smaller proportion than in the finest astronomical clocks, but yet larger than it was in the three-legged dead escapement which we had at first (p. 102), on account of the loss of force by the impact of the pendulum on the pallets.

The fly is 11 inches long in each vane and nearly 2 inches wide: it is set on the arbor by what may be called a silent ratchet, or a steel-faced roller with stiff springs bearing endways against it, but obliquely, so that the fly can run forwards, but not backwards. It is almost impossible to make the escapement trip by any force you can apply to it by hand; and it once went for several days without tripping though the fly had been accidentally left loose.

Maintaining power.—The going part of this clock takes about 20 minutes to wind up, and therefore none of the common maintaining powers would do. A bolt and shutter (see p. 151) might indeed have been made to lift higher than usual and so keep in action longer; but it would have had to be very heavy, and moreover there was the risk of the man being interrupted while winding, or stopping to rest, and so letting it run out of gear or stick fast. Sir G. Airy had proposed a modification of his Northumberland telescope apparatus, which I have already mentioned as having been used in the Exchange clock; but that is also liable to run out of gear, and is open to other objections; and so the following much simpler plan was adopted. The barrel in any case would require an auxiliary pinion to wind it, taking into a wheel on the end of the barrel itself, close to the great wheel; and the only addition is, that the back end of the winding arbor runs in a loose bar, which hangs obliquely from the back pivot of the barrel (as shown in the drawing), and has a click on it which acts upwards in a set of ratchet teeth cast on the back of the great wheel. When the clock is going and not winding, these ratchet teeth pass under the click, just as in Harrison's going ratchet (see p. 149); but as soon as you begin to wind they stop the click and the bar from rising as it tries to do, and the great wheel itself thus becomes the fulcrum for the winding up of the barrel, and so the clock weight is for the time transferred to the great wheel directly, instead of through the barrel.

The winding arbor fits loosely in both the bushes, because the back pivot and its bush in the bar gradually move a little upwards as the great wheel turns, while the front one of course remains fixed in the clock-frame. When it has moved as far as it was thought prudent to let it go, a long tooth on the winding arbor catches against a stop in the back frame, and the man cannot wind any farther without turning the handle back a little to allow the bar to drop and the click to take up another mouthful of the ratchet teeth. The unusual length of the winding arbor, 4 feet 2 inches, makes this sideway motion insignificant for 10 minutes motion of the great wheel: if the frame were narrower it could still be used, taking care to put the stop so as to prevent too much oblique action. Very few clocks take as much as 5 minutes to wind up the going part. If you take the trouble to calculate the pressure, you will find that there is rather more force on the clock in winding than usual; which however is of no consequence. If the winding pinion were larger in proportion to the wheel the difference would be greater; but it might always be equalised by hanging a weight on the loose bar, just enough to counterpoise the difference. The winding pinion pulls out of gear with the wheel in the usual way.

The dials are 22½ feet in diameter, or very nearly 400 feet in area, and are made of cast iron frame work filled with a very expensive kind of opal glass, which appears to me no better than some much cheaper glass of the same colour by Chance of Birmingham, which is used in other clocks. The dials and the hands together cost no less than £5334, which is more than the whole cost of the clock and all the striking work up in the bell chamber. I shall have more to say of this in the history of the clock. The minute spaces are a foot square, and the figures 2 feet long. The dials would have been clearer, and the hands more visible upon them, if the framework rings, beyond the minutes and the figures, had been omitted, as they diminish the clear space in the middle of the dial by about one third of its area. The dials stand 5 feet from a whited wall which is the main wall of the clock room and clock-tower, and in front of which are the gas lights for illumination. I had provided by request that the clock should be able to light up and turn down the gas if required, in the way described at p. 223, but it was finally thought better to light them by hand, and I think so too. There is nothing peculiar in the dial-work except its size, which may be judged of from the drawing of the clock and fig. 51 (p. 249), and the fact that the minute wheel arbor is 8 feet long and 3½ inches thick; it is of course tubular, except at the ends. The dial centres are exactly 6 feet above the top of the walls, or 180 feet from the ground.

Hands.—The minute hands, as made from my design, after the architect's two successive sets had failed, are thin copper tubes of a section formed by two segments of circles, with a few diaphragms soldered in, set on a gun-metal stalk or central piece, which also forms a partial counterpoise both for wind and weight outside, there being another of cast iron inside the clock room, to divide the pressure between the two ends of the arbor. This copper tube of each minute hand only weighs about 28 lbs., though they are 9½ in. wide near the centre, running off to 5½ in. at the end; the gun-metal stalk of each hand weighs very nearly 1 cwt.; but the whole of that weight is near the centre, and so its moment of inertia at each beat of the pendulum affects the clock very much less than if the same weight were distributed all along the hand. It is remarkable that from the momentum of the hands, and the general elasticity of the leading off rods, they move continuously, and with no visible jump as usual. The length of each hand and its external counterpoise is 14 feet; and the total weight of each hand with its external and internal counterpoises is now within 2 cwt., whereas Sir C. Barry's

4 minute hands and counterpoises (of which I shall have to speak in the history of the clock) weighed a ton and a quarter. His hour hands are still there; for though very bad in construction and three times as heavy as they need be, their motion is so slow that they do not sensibly affect the clock as the minute hands did, and so they may as well stay until they become unsafe. One of them cracked and had to be taken off. The hour hand arbor is a tube 5½ inches wide, and lies on large friction rollers both behind the dial and within the clock room. It was easy enough to put the clock room end of the minute hand arbor on friction rollers too, as it of course projects beyond the other; but at the other end it is managed by setting the pivots of 4 smaller rollers in a pair of rings screwed outside the hour hand tube, and cutting holes in that for the rollers to go through and reach the minute arbor, so that those rollers move in a 12 hour orbit of their own, besides the pair in action for the time turning on their own pivots.

Quarters.—There is nothing very peculiar in the striking of the quarters. The position of the levers and cams is evident from the picture. The wires go up alternately on opposite sides of the winding-wheel arbor to prevent their fouling each other; each bell has two hammers because in that way we get longer cams with less pressure on them, and on the levers. The hammers are all nearly a 40th of the weight of their bells, which I am satisfied is the proper proportion for bells on this scale of thickness, which is that of the smaller and thicker bells of church peals, as I shall explain more fully afterwards. The levers are all 19⅛ in. long, and their centre is nearly 36 in. from that of the wheel. The cams are of wrought iron with hard steel faces, screwed onto a cast iron barrel which is bolted to the great wheel, and they are constructed on the circular section which I shall describe for cams under the *Teeth of wheels*. Between the clock and the bell chamber there is another low room in which the cranks are, for leading off from the vertical wires to the 4 quarter bells. These wires are in fact wire ropes: I had them substituted for the iron rods which were at first used both in the quarters and the hour striking parts, and thereby got rid of an amount of concussion and noise in striking, which sounded as if the clock was shaking to pieces. Now the action in the clock room is so silent that you hear nothing except the bells and the passing of the click which stops the train from running back in winding.

Hour-striking part.—The great striking-wheel has 10 cams 2½ in. wide cast upon it, and these have steel faces screwed onto them. The lever has a thick part in the plane of the cams and a longer and thinner piece lying behind the wheel and having the hammer rope attached to it, or rather a short rod with a swivel and nut to take up the rope. The rope is half an inch thick, the same as is used for the striking weights, and is about 25 feet long and reaches up to the horizontal arm of the great hammer lever, which projects 5 ft. 4 in. from the pivots, which are forged as part of the collar in which the great bell hangs (see page 212). The hammer shank is also double, going through 2 holes in the cast iron head, which weighs 4 cwt.* and is lifted 13 in. from the bell, or about 9 in. vertically. There is an iron plate screwed to the hammer shank, which falls on india-rubber buffers 5 in. thick, carried by a kind of long stirrup hung from the bell frame. The other hammers are prevented in the same way from jarring on the bells. The construction of the bell frame, which is entirely different from the usual form of frame for swinging bells, made it impossible to fix the hammers or the buffers in any other way: not that there is any objection to this, except that it is more expensive, and the hammer shank is farther from being horizontal.

* It did weigh 7 cwt., but was reduced when the bell became partially cracked, as related afterwards.

The second wheel turns 2/3 round for each blow, as I explained that it might at page 198, and the third wheel and fly make 4 turns for each blow. The fly arbors are placed vertically in order to get room for the flies, which have to be put near the top of the room. The vanes of the hour fly are each 2 ft. 4 in. square and extend 3 ft. from the arbor: the quarter ones are rather larger. To prevent all risk of accident, the ratchets are not pinned but 'squared' on the arbors under the flies themselves with an octagonal fitting, and each fly has two clicks. The stops are set on springs to diminish the blow at stopping, and the striking work is stopped on the lift, both for accuracy of discharging and for diminishing the constant strain on the wheels and arbors. Each striking weight is, or was before the hour hammer was reduced, nearly a ton and a half; and you may observe that each of the three parts is so arranged that the weight hangs between the arbor of the great wheel and the teeth or cams which have the heavy work to do, so as to reduce the pressure and friction on the arbors as much as possible. If this striking part had been made as they often are, and lifting by pins instead of cams, the weight would very likely have had to be 3 tons, or more than a man could wind up in a day.

The mode of letting off the hour-striking is peculiar. It was one of the original conditions, that the first blow of the hour should always be struck within a second of the real time; and in order to do this it was not sufficient that the clock should go far more accurately than usual, but there must also be the means of making the hammer fall exactly when the clock reaches the 60th second of the last minute of the hour. First then it was necessary to leave it on the lift and nearly ready to fall as soon as the striking part is discharged; and secondly, to have some more sudden and precise means of discharging it than the slow motion of a snail turning in an hour. It was at first intended to do it by the train remontoire described at p. 228; but when it became expedient to abandon this and rely on the improved gravity escapement which I invented after the clock was partly made, it became necessary to contrive some other plan for discharging the striking part with equal accuracy: and this is it. P Q is the ordinary discharging lever lifted by the snail on the hour arbor, P being the first stop against which the arm E F on the third wheel of the striking train

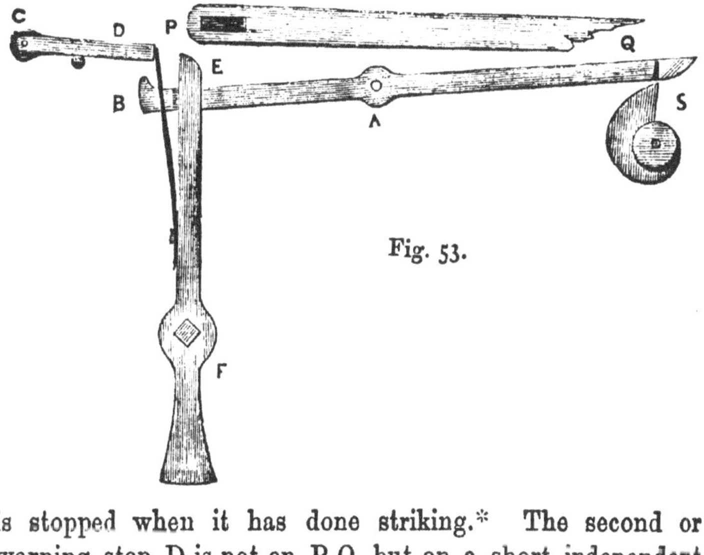

Fig. 53.

is stopped when it has done striking.* The second or warning stop D is not on P Q, but on a short independent lever C D. B A S is another lever set on pivots A on the cross bar of the great frame, and its heavy end S is lifted and dropped by a snail on the 15-minute wheel of the escapement; when it drops it tips up the second stop lever C D, not merely by dead weight, but with a blow, which is

* This lever, and several others, are omitted in the elevation of the clock (frontispiece), as they could not be shown without confusion; and as the snails are all shown, any intelligent reader will understand where the discharging levers must be: there is an intermediate one from the quarter snails to the quarter locking lever.

certain to overcome any friction there can be between the two stopping pieces. The lever B A S drops every quarter of an hour; but it does nothing except when the stopping piece is resting against the second stop, *i.e.* when the clock has given warning for the hour. It is let off at the 58th second, or the last beat but one of the pendulum in the hour, and the train just gets far enough into motion to let the hammer fall as the seconds hand makes its last jump for the hour, and it acts with the greatest precision.

This precision in letting off the hour-striking made the want of it disagreeably apparent in the quarters, which were at first discharged in the usual way by a snail on the hour arbor which leads off to the dial work; and we had to put a larger snail for them on a separate arbor under the escapement wheels, free from the shake of the hands and dial work; this answers very well and there is never more than 2 seconds difference now in the time of discharging the quarters. The 1st, 2nd, and 3rd quarters all begin to strike at those times respectively; but the 4th quarter is let off about 20 seconds before the hour so that it may have done striking before the hour begins at the real time.

That, by the bye, is an objection to a train remontoire let off at the usual half-minute intervals, viz., that it makes the interval between the quarters and the hour disagreeably long, if the hour is to begin striking exactly at the hour as it should do. Even 20 seconds is too long an interval for any but very large bells. However since the invention of these gravity escapements train remontoires may be considered extinct, as they cost much more, and are no better, and are almost certain to be destroyed, or at least put out of action.

Winding of striking parts.—Each of these takes about 5 hours to wind, and requires double multiplying wheels, which are as to size repetitions of the large wheels and pinions of the train, with a modification of the shape of teeth, which will be explained afterwards, because different shapes are proper for driving and for driven teeth, though that is very seldom attended to. The striking parts wind twice a week; for although they might have been made to go a week, the weight would have had to be twice as much —and more, on account of the increased friction; and a man could not have wound up each of them in an ordinary working day. Moreover the whole work must have been heavier and larger and had a greater strain upon it. Even in much smaller clocks than this I always design the striking parts to go less than a week. I do not believe that there is a clock in the world that strikes a large bell, say above a ton weight, *properly*, with winding once a week; unless it happens to have an unusually large fall for the weights.

There were various suggestions made both publicly and privately for winding the clock by water, by a steam engine, and even by a kind of weighbridge or sinking platform to be worked involuntarily by people walking over Westminster bridge, which would have to rise between every two walkers, to do any good: a turnstile would have been a much more practicable thing. I forget whether anybody proposed to turn the tide to account for the purpose, but I daresay they did. It never occurred to these inventors that the interest on the cost of construction and the expense of looking after any such machine would far exceed the wages of two winding men for five or six hours twice a week.

In most large clocks the winding pinion simply 'pumps' out of gear with the winding wheel on the front end of the barrel, and it does not much signify if it is left in, as the friction caused by it is not much. But this would not do for a pair of large wheels, and there was not room to make the great winding pinion pump out of gear. Moreover it was absolutely necessary to have some contrivance for automatically stopping the winder a little before every striking of the part he is winding; which occurs in no other clock, as none takes anything like a quarter of an hour to wind either part, and if it did you need only slip off the handle. That is managed thus. In each striking part a long lever (omitted in the frontispiece to avoid confusion) stands in the way of a tooth or arm on the winding arbor: when the man begins to wind he must lift this lever up on to a certain hook which will hold it up so long as the weight of the lever rests upon it; but when the weight is relieved the hook falls back: this is done by the snail a few minutes before the clock is going to strike, and just a minute before it strikes the snail lets the lever drop again, and the hook being then out of the way, it drops completely and stops the winder; and he then throws the winding-wheels out of gear, and in again when the striking is done.

This throwing out of gear is also done by a new method: the first pivot of the second (150) winding-wheel arbor is set in an eccentric bush, which can be turned in its own holes by a lever with a handle to it, as you see in the drawing, and the eccentricity is just enough to take the teeth of that pinion out of gear with the great winding wheel, leaving the 2nd wheel in gear with the winding-pinion (14). Besides this the ropes themselves stop the winder when the weights are wound up to the top, by throwing another lever off a hook on which it has to be set before the man can begin winding. In all these winding and maintaining power contrivances there are some further provisions for enabling the man to turn the winding handle back a little, to let the barrel ratchets down softly on to their clicks, but it is hardly worth while to describe them.

Rate.—Provision is made in the clock for reporting its own rate of going to the Greenwich observatory at any convenient hour or hours every day, by electric telegraph, as I have already described under 'electrical clocks' at p. 156; and it has been arranged by the Astronomer Royal that it should give two signals a day, at an hour before and after noon, for greater accuracy. The average performance of the clock is reported by him annually to the Board of Visitors of the Observatory, and for years the report has been practically the same. Last year, 1881, its error was under 2 sec. on 84 per cent. of the days of observation; between 2 and 3 sec. on 14; between 3 and 4 on 2. They were almost exactly the same in 1872.

Sir G. Airy sent me, for the purpose of the inquiry into the barometrical error (see p. 71), the complete rating of the clock for 1872, except Sundays and a few odd days. Rejecting three abnormal disturbances, one of which reached 8 sec. in a week of thunderstorms, and two others of 5 sec., which came suddenly, and evidently from something done in the clock-room, because the daily variations never reached 2 sec. at any other time, the average daily variation was only 0·2 sec. on each side of zero, or 0·4 sec. altogether; and taking it on all the Saturdays in the year, the average weekly variation was just 1 sec., or exactly half that of one of old Mr. Dent's best regulators which I happen to have had for nearly two years while we were trying the Exchange clock. Therefore the first blow of the hour may be presumed to differ considerably less than a second from G M T, and is practically always right within 2 seconds. In hearing it at a distance you must allow nearly 5 seconds a mile for the velocity of sound.

Some enemy set afloat the statement that the clock is 'controlled' by electrical connection with Greenwich, in some such way as is described at p. 158, instead of merely reporting its own performance; and the story has been repeated by people who ought to have known better, or might easily have learnt, by writing either to the Astronomer Royal or me, or from any edition of this book after the first. A signal is sent back from Greenwich at some hour daily, for the information of the winding men, but they never touch the pendulum unless the error is 2 seconds, which happens very seldom.

History of the clock.—It is no longer necessary to relate as fully as in 1860 the history of the 16 years of disputes relating to this clock; for it was the subject of violent disputes before I had anything to do with it. I shall only say as much as may possibly be useful with reference to any similar transaction hereafter. Not that I expect such warnings to be attended to, except perhaps in exactly the wrong way; for I know very well that the tendency of the official mind to get things done with as little trouble as possible is infinitely stronger than to get them done as well as possible. I was only brought into this business originally with the view of saving somebody else trouble; and as soon as it was found that by the legal effect of the contract I had real power to direct the work, every possible effort was made to get rid both of it and me. No official who joined in those attempts cared three half-pence how the clock was made. Luckily I did care, and knew what would become of it if I gave it up. And as the biographers of Sir Charles Barry have told their story on his behalf, I shall also, without the least resentment against them, repeat mine, of which not a word has been ever contradicted.

No one who has read in the Parliamentary papers the early correspondence, beginning in 1844, can have the smallest doubt that the architect had made up his mind that the late Mr. Vulliamy should make the clock, and no one else. Mr. Vulliamy himself was so confident about it, that even after the Astronomer Royal had utterly condemned his plan for it, as 'only suitable for a large village clock,' and recommended old Mr. Dent, he said to me, 'Dent will never make that clock.' After that, the matter went to sleep till November 1851, when the Duke of Somerset, then Lord Seymour and First Commissioner of Works, asked me to act as referee with the Astronomer Royal, on his recommendation. I examined the plans at the Office of Works, which had been sent in in 1846, and found that Vulliamy's was quite as deficient in strength (for which Sir G. Airy had commended it) as it was in all provisions for accuracy: indeed more so, because accuracy is a question of degree; but such a clock as that, striking on a bell of half the weight that had been fixed upon, would have destroyed itself in a week. By some strange omission, Dent had never been told that the bell was to weigh 14 tons, and he had assumed it to be much less, and consequently his plan was defective too, though not so defective, and that only in size, not in principle. The plans of Whitehurst, of Derby, which Sir G. Airy had reported on more favourably than Vulliamy's, could not be found in the office; and it was evident that Dent's had been somewhere else, where they had been very roughly handled, dirtied, torn, and mended, which was certainly not done at Greenwich.

Whitehurst was dead, and Vulliamy had refused to accept the Astronomer Royal's conditions as to the performance of the clock, which the Clockmakers' Company also pronounced impossible, though they have been far exceeded in accuracy now. So we agreed that it was useless to apply to any one but Dent, and that the only practicable way was to ask him if he would make for any definite sum such a clock as we should require, from his own knowledge and our information as to what was intended, and such general plan as we could give him, in the unfinished state of the tower. He agreed to do so, for a sum which was soon found to be inadequate, and the arrangement was altered into his keeping a separate account of all outlay on the clock, and receiving 10 per cent. profit on it, which ultimately came to £4080, only an eighth more than Vulliamy's estimate for a clock totally unfit for the work.

The Duke of Somerset went out of office soon after giving Dent the order in 1852; and then began, and lasted through the reigns of his successors, Lord John Manners and Sir W. Molesworth, a series of intrigues and attempts to get rid of or alter old Mr. Dent's contract, and then to repudiate his successor, F. Dent, and then—or rather, all along—to get rid of me; for they all knew very well that for various reasons the Dents alone could not contend against them, nor go on with the work at all. Sir C. Barry demanded working drawings of the clock from Dent, who told him very truly that there were none, and would be none, except such sketches as I might give his men from time to time. Then he applied to the Astronomer Royal for them, who desired me to make and send them. I answered that I should do nothing of the kind: the Office of Works might show him the signed plans if they liked (as in fact they did), which were all that could possibly concern him; at any rate I should make no others. Thereupon Sir G. Airy resigned; and as his knowledge of large clocks was purely theoretical, and not one of the suggestions he had made could be adopted, his resignation saved a good deal more of unprofitable correspondence. The Office of Works made that another excuse for trying to get rid of me, and went so far as to consult the Attorney and Solicitor General about it.

Then they tried the obstructive policy of refusing any information about the tower, in spite of my warning that the only consequence would be that they would probably have to pay for altering the clock when it got there; for which they cared nothing, as the nation and not they would have to pay for it. Still the work went on, and as the inscription on it records, the clock was made in 1854, and after going in the factory for 5 years was fixed in the tower in 1859, and was set going permanently in 1860.

Sir C. Barry was particularly anxious to design the hands, and I rather unwillingly and with some misgiving consented, having already reason to put very little faith in his mechanics. He first produced some cast iron ones, of such frightful weight that I would not even let them be put on. He then tried some of gun-metal, which were lighter, but still far too heavy for the clock, besides being so fixed on by his engineer that they fell over a minute or two every time they passed the vertical. And so we removed them, at least the minute hands, and made new ones, as described at p. 256, which have always gone quite easily, except once when loaded with snow which froze upon a partial thaw, and broke down hundreds of miles of telegraph wires, at the same time that it stopped the clock. Before Dent's final bill was paid, the Astronomer Royal inspected the work, though his legal control over it was gone, both from his resignation, and from other causes, and he candidly expressed his entire approval of it, and especially of the escapement, on which its performance chiefly depends. It is due to Lord Llanover, though he is dead, to say that in his time I had no trouble with the Office of Works.

As soon as it was finished, Mr. W. Cowper Temple, who had then been made First Commissioner of Works by Lord Palmerston his step-father, told the House of Commons, with evident satisfaction, and the gratitude natural to some people, that my connection with the clock had ceased, and that it was handed over to the Astronomer Royal. The next I heard of it was that he had replaced the great check spring of the hour-striking part by a contrivance of his own, which soon produced a smash, and then the old spring was quietly restored, and is there now, and there have been no more such experiments. I must however do him the justice to say that he had previously written to me 'that the clock would be far better in my hands than in any other person's;' not that there was really any more for me to do, beyond nominal superintendence of the clockmakers who have the care and winding of it by an independent contract, viz., the successors of the Dents in the Strand. By an odd coincidence, I had the pleasure of giving a sort of clinical lecture on the clock to as many members of the Horological Institute as the room would hold, on the 21st anniversary of our signing the plans for it, 29 January 1873.

THE WESTMINSTER BELLS.

I had better give the history of them also here. In 1852 the Astronomer Royal declined to have anything to do with the bells, as he did not profess to understand them; and nothing was done towards getting them, beyond some abortive correspondence with me by Sir W. Molesworth, and his giving a commission to Sir C. Barry and Professor Wheatstone to learn what they could about bells at the Paris Exhibition of 1855, which proved to be nothing. In 1856, Sir B. Hall (Lord Llanover) asked me to take them in hand, and it was then arranged that I, with Sir C. Wheatstone, and the late Rev. W. Taylor, who had paid some attention to the subject in a theoretical way (and must be distinguished from his namesake the bell-founder), should be the referees. Sir C. Wheatstone never acted, beyond telling us the result —or rather, no result—of his inquiries at Paris, and Mr. Taylor would take no responsibility beyond giving the final certificates. I therefore prepared a specification, which was sent to the three English bell-founders.

Mr. Mears refused to accept the referees because they had among them spoken ill of his two condemned Royal Exchange peals, of his great York Minster bell, and a rather larger one which he had sent to Montreal. He also declared that nobody else could make the bells; and his tender was not the lowest. Mr. Taylor's (of Loughborough) was; but he wanted some terms which could not be acceded to. Messrs. Warner required the referees to take the responsibility of giving the patterns for the bells; *i.e.* they confessed that they did not know how to make such large bells of the proper notes: they had previously copied all their bells from existing ones. However I was able to do that for them, and so their tender was accepted, though they demanded ten guineas a cwt., while the usual price was seven, and they were to recast any of them (unless condemned for bad casting, in which case they were to recast for nothing) for £2 a cwt., and also to cast any small experimental bells for the same price.

They made the great bell first; and from some mismanagement it came out thicker than the pattern, and two tons heavier than was intended, and required a clapper twice as heavy as we had reckoned on by analogy to other bells. Undoubtedly we had a right to reject it; but it appeared a sound casting, except some holes at the top, and was generally praised by the public who heard it, though there was always something unsatisfactory in its tone. And no wonder; for after being rung occasionally for some weeks, it one day cracked, no doubt from the weight of the clapper which it required to bring out its tone, and when it was broken up there was found a great flaw in it, where the two streams of metal meeting round it had never joined. So we were in every way well rid of Big Ben the first.

The founders however had cast then the fourth quarter bell of four tons successfully, and there was no intention of taking the job out of their hands. But they demanded a price for recasting enormously beyond the £2 per cwt., which they had agreed to before, evidently presuming that neither of the other founders would be employed. Mr. Mears had learnt something by experience, and no longer objected to the referees, and offered to recast the bell at a more reasonable price, and so this time his tender was accepted. He however was still more unlucky; for he produced a bell which partially cracked also, after a few months striking; and Dr. Percy pronounced it, on cutting a hole down to the bottom of the crack, and analysing the metal, 'a defective casting, porous, unhomogeneous,' and at the place where it is cracked, not of the composition I had prescribed, and therefore much more brittle.

Mr. Mears also determined to conceal this porosity from the referees by filling up the holes with cement, before he let us know that the bell was ready to be seen. And when I publicly charged him with having done so, he put a bold face on the matter, and brought an action for libel, and had, no doubt, found half-a-dozen engineers and brass-founders ready to swear that porous castings are as good as sound ones. But he also found that I had got a piece of the bell analyzed, and knew that the composition was wrong, besides the porosity and its concealment. So his counsel accepted his costs without a verdict, after making a speech in which he confessed and declared that the composition had miscarried, and become unhomogeneous; that he had filled up the holes, *because he thought them immaterial*—as if he was to be the judge of that; and that it was impossible—*i.e.* that he did not know how—to cast large bells without holes in them!

His successor, who had bought Mears's declining business, twenty years afterwards thought he would try again, evidently with the object of advertising himself, on my once more publishing the fact that Big Ben II. was a disgrace to its founders. His workmen who had been with Mears had to confess the concealment of the holes, and again they and all his party asserted that it is impossible to cast large bells without holes, and that it was their regular practice to fill them up with cement coloured with bell dust: which both Taylor and Warner's people utterly contradicted, and I have every reason to believe, quite truly.

Though Dr. Percy had been asked to examine the bell as soon as the cracks were discovered, he was not allowed by Mr. Cowper Temple, then First Commissioner of Works, to cut into it for nine months after. If that had been done at once, as I desired, Mears's action would never have been heard of.

He had also got a report from Sir G. Airy, who, although he declined meddling with the bells before because he did not understand them, felt no difficulty in giving judgment on them afterwards, with the help of two musical assistants and a fiddle 'tuned to the pitch of the Italian opera'—as if the pitch of the Italian opera had anything to do with the question whether five bells are good or bad, or in tune with each other. Mr. Turle, the organist of Westminster Abbey, had given his opinion before we certified them, that the four quarter bells were 'all that can be desired, both in tune and tone,' which are very different things (for a set of iron pots may be in tune), except that 'the fourth is a shade too flat.' But Sir G. Airy came to a very different conclusion, viz., that the fourth bell was *a note and a half too sharp*, and the first about the same, the third 'rather sharp,' and the second alone right. On the whole, he was 'much dissatisfied with those bells,' and thought two at any rate must be recast. Afterwards he began to have a little misgiving about it, and thought it might be advisable to try again before recasting them; but he never did try again. I suppose it had then occurred to him, as a mathematical fact of which he certainly was not ignorant, that the two bells could not be a note and a half too sharp without being a sixth less in diameter than they are, or else a great deal thicker. He also pronounced the great bell, which Dr. Percy had not then ascertained to be cracked three inches deep, and to have almost every possible unsoundness, 'perfectly sound for all practical purposes, and the cracks probably superficial,' and that it only wanted a lighter hammer *faced with tin!*

This seems to have been too much even for Mr. Cowper Temple to swallow; for although he sent me one of the polite and intelligent epistles which I used to receive from him and Mr. Alfred Austin, then the secretary of that office, saying that he 'disagreed with every word of the strictures I had sent him on Sir G. Airy's report' (as if his adoption of it made it less ridiculous), he afterwards consulted Mr. Turle again, and no quarter bells were recast, nor was the great hammer faced with tin; and after getting Dr. Percy's report he told the House of Commons that the bell was, not 'perfectly sound,' but 'irretrievably cracked,' as it is, though the defect is much less perceptible than one would expect. The hours were struck on the fourth quarter bell for three years, which made them difficult to distinguish at a distance, and after a paragraph in the *Globe* to that effect, the striking was resumed on the great bell with a lighter hammer striking in a different place, which was easily done by my arrangement for turning the bell any degree round by giving it a mushroom or button-shaped top (see p. 212). When the new great bell of St. Paul's has taught people the difference between a cracked bell and a sound one, perhaps 'this great country' and 'the model legislature of the world,' as it is the fashion to call them, may screw up their courage to the vast undertaking of recasting their own great clock bell at the cost of six or seven hundred pounds.

The cost of the bells, including £750 for recasting 'Big Ben,' was under £6000, while the cost of the iron frame provided by the architect was about £6600, partly in consequence of its being made too weak at first, so that it shook under every blow of the hammers, as I had told him that it would. And as the clock, with my hands, cost only £4080, while his hands and dials alone cost £5334, you see that the actual clock and bells cost much less than the architect's appendages to them.

TEETH OF WHEELS.

There are various treaties on this subject, and I only intend to say as much on it as it is necessary that a clockmaker should understand, if he means his wheels and pinions to run together smoothly instead of wearing themselves out by jerking and scraping, which I have known to happen in a very few years. The most comprehensive view of the whole theory of tooth-drawing (at least of this branch of the art) is in a paper by Sir G. Airy in the 2nd vol. of the *Cambridge Transactions*, and it was further expanded by Professor Willis in his *Principles of Mechanism*.

Some persons have a mistaken impression that the object to aim at in constructing wheel-teeth is to make them roll on one another without any rubbing friction. This can indeed be done by what are called *involute* teeth, of the shape described by a point in a string unwound off the circumference of a wheel: but they are really useless, because they are so oblique that they produce a squeezing pressure between the two wheels which is more than equivalent to any saving in friction. The great thing to aim at in describing teeth is to make the relative velocity of the wheels uniform from the beginning to the end of the contact of each pair of teeth, which of course involves also the absence of all concussion or drop of the teeth. Another point is to have the action entirely or chiefly between the teeth which are separating from each other, and not between those which are approaching, which is commonly expressed by saying that the action should be after the line of centres of the wheels and not before it. The reason of this will be evident at once if you draw some teeth with a very rough outline so as to give an exaggerated view of the effect of friction, for you will see that there is a degree of roughness which will make the teeth jam against each other and not let them slide at all as they approach the line of centres, but that no degree of roughness will do this when they are leaving contact or are past the line of centres. The most perfect thing is when the contact takes place for a very short distance only close to the line of centres; and this can only be with very small teeth, and therefore very high numbers (except with involute teeth, which I have already said will not do for another reason). There is indeed a contrivance, which I have somewhere seen called White's gearing, for getting this kind of action with large teeth in heavy machinery, by putting several large-toothed wheels close together, with the teeth of each a little behind the other, as in figure 54 *a* (next page): but this is never used in clockwork.

Helix-teeth.—A modification of this plan, though very unlike it in appearance, has been occasionally used in clocks under the erroneous name of the *helix lever*. The teeth certainly do at first sight suggest the idea of an endless screw, but they are essentially different, the arbors being parallel, and not at right angles as in an endless screw. If you suppose a good many very thin toothed wheels put together side by side, each with its teeth a little behind its neighbour, they will present a surface like a rough tooth with a sloped face, as in the upper part of this figure *a*. Then go a step farther and suppose the rough edge to be smoothed off, and the result will be a smooth-faced oblique tooth, like the lower one in the figure, which will drive teeth of corresponding obliquity on another wheel, and the contact will be solely at the line of centres, where there is no friction. But when the teeth thus become smooth, there will evidently be a great endway pressure on each arbor, which there is not while the teeth are square, or belonging to separate wheels put together. This endway pressure may however be neutralised by again putting two such wheels together with the teeth sloped opposite ways as at *b*. There was a German turret clock on this plan in the 1851 Exhibition, which certainly went with a very small weight; and small clocks with the single helix tooth had been made in England many years before by Macdowall (see p. 100) which also

a Fig. 54. *b*

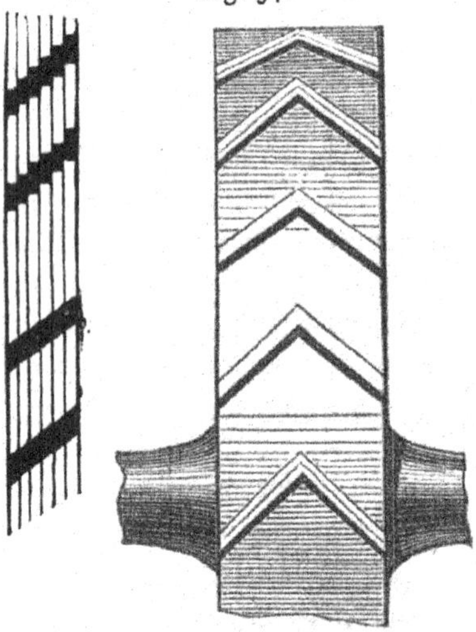

required less force than usual, from the smallness of the friction in the teeth. I do not know that the advantage is worth the expense; but as this construction is very little known, I explain it in case it may be turned to any use hereafter.

Epicycloidal teeth.—It may be said without exaggeration, I believe, that all the teeth now used in machinery are constructed either as epicycloids or hypocycloids, and the meaning of those words is this:—If you roll a circle A G P *on* another circle A R Y, the curve R P traced by any point P in the rolling circle is called an epicycloid to the circle A R Y; and if you roll the small circle A G P *within* a larger than itself, such as A Q X, the curve P Q traced by P is a hypocycloid to that circle. And it is remarkable that if the tracing circle is exactly half the size of the one in which it rolls, the hypocycloid P Q is a straight line, and part of the diameter of the large circle, and therefore teeth so described are called radial teeth.

Fig. 55.

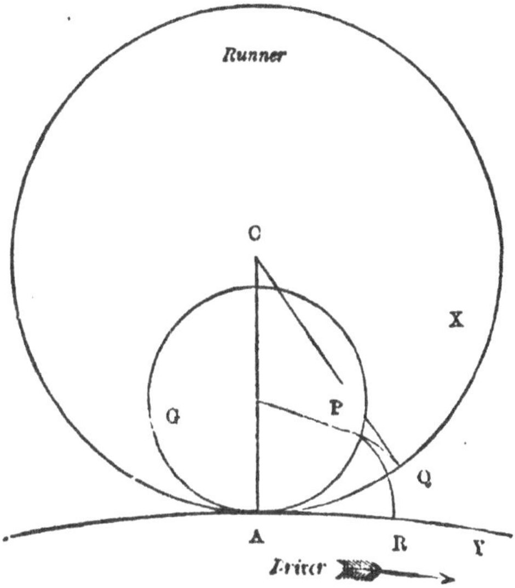

Now suppose A R Y is the circumference of what is called the *geometrical* or *pitch* circle of a wheel which is intended to drive another, and A Q X the pitch circle of the wheel to be driven, which is generally called the 'follower,' but which I think it better to call the *runner*, as followers do not usually run before their driver; then it is easy to see that the arc A P of the tracing circle is equal both to A R and to A Q, and also that the epicycloid is always more convex than the hypocycloid, and therefore that the point P in the tracing circle is always the point of contact between two teeth so traced, and the velocity of the two wheels is always the same as if their pitch circles rolled upon each other without any teeth at all; and therefore it is constant in all positions of the teeth. It is hardly necessary to observe that the teeth of the driver, to act after the line of centres, must be wholly outside its pitch circle, and those of the runner wholly within. The part of a tooth within the pitch circle is generally called its *flank* or *root*, and the part outside is called the *point*, or the *addendum*, and sometimes the *curve*, because the flank is generally made radial, *i.e.* a hypocycloid described by a circle of half the diameter of the pitch circle. For it is further to be observed, that although the points of the driver and the flanks of the runner must be traced with the same circle, it is not the least necessary that the points and the flanks of the same teeth should be traced with the same circle.

In clockwork the wheels always drive and the pinions run, except the 12 hour wheel in the dial-work, and the winding wheels and pinions if there are any. It can be proved, as you may see in Professor Willis's *Principles of Mechanism*, but the proof is too long to give here, that no pinion of less than 11 leaves (except of a kind which I shall describe presently) can be driven entirely after the line of centres. A pinion of 10 can very nearly; and there is so much difference between the force required to drive pinions of 8 and those of higher numbers, that some spring clocks with Macdowall's escapement which answered perfectly with pinions of 10 or 12, failed with the common pinions of 8, for want of force to drive the two extra wheels in the train. Professor Willis gave the following table of the lowest numbers which will work together with all the action after the line of centres:—

Driver	54	30	24	20	17	15	14	13	12	11	10	9	8	7	6
Runner	11	12	13	14	15	16	17	18	19	21	23	27	35	32	176

The practical inference from this is, that if you use these numbers, or any higher ones, together, the driving teeth require no flanks and the running ones no points: indeed if you mean to prevent any action before the line of centres, the runners obviously must have no points, because if they have they will be geometrically identical with the teeth of a pinion intended to drive the wheel after the line of centres when reversed. Suppose for instance, what is nearly the

Fig. 56.

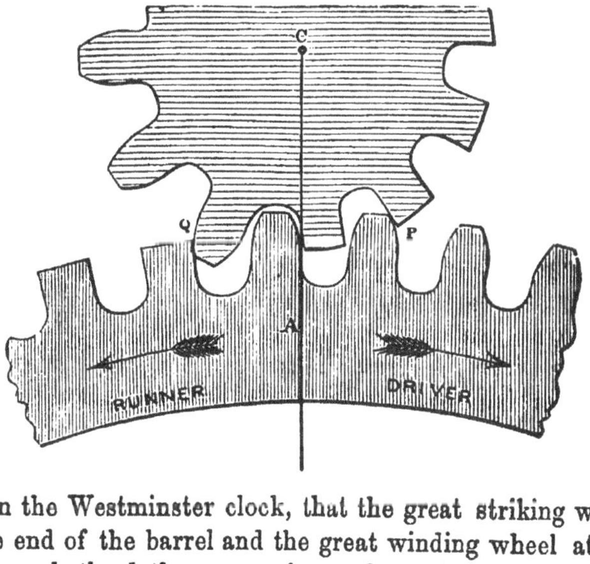

case in the Westminster clock, that the great striking wheel at one end of the barrel and the great winding wheel at the other are both of the same size and number of teeth, and that their pinions are also the same; then as the striking wheel always drives, but the winding wheel is always driven by its pinion, the striking pinion and the winding wheel ought to have no points to their teeth, and the sections of the two wheels and pinions would be as in fig. 56, the right hand representing the striking part and the left the winding, and the action being in both cases, you observe, after the line of centres A C, as the arrows indicate.

It is evident that the same wheel cannot properly drive two unequal pinions with radial teeth. Whenever the same wheel has to drive two such pinions, the flanks of the pinion teeth and the points of the wheel teeth must be traced with the same circle, and that circle must not be larger than half the size of the smaller pinion, or else it will make the teeth of that pinion weaker at the roots than even radial teeth are, which are of course narrower at the bottom than the top, and therefore are a weak form, especially in small pinions. A case of this kind occurs in every clock on the patterns I have described at pp. 183 and 196, and in the Westminster clock, and in short wherever the second wheel in the going train does not turn in an hour, and the dial work is driven independently. In this case, the 40 teeth of the hour wheel (or whatever the number may be) require no points, and should be hypocycloids described by a tracing circle of half the diameter of the other pinion of 10 or 12 which is driven by the great wheel, if the teeth of that pinion are radial.

Lantern pinions.—But there is another, perfectly different kind of pinion, which is much better for small numbers than radial teeth or leaves, viz. what is called a lantern pinion, and in old books, a 'trundle.' These two figures (*t.o.*) will show its construction better than any explanation. I believe it is the oldest form of pinion in the world, but it had almost (if not quite) fallen into disuse in England, when it was restored in Dent's turret clocks about 30 years ago. They work with much less friction than common leaved pinions of low numbers, when driven, the run upon them being less and the action wholly after the line of centres, and the shape of the wheel teeth requiring less

Fig. 57.

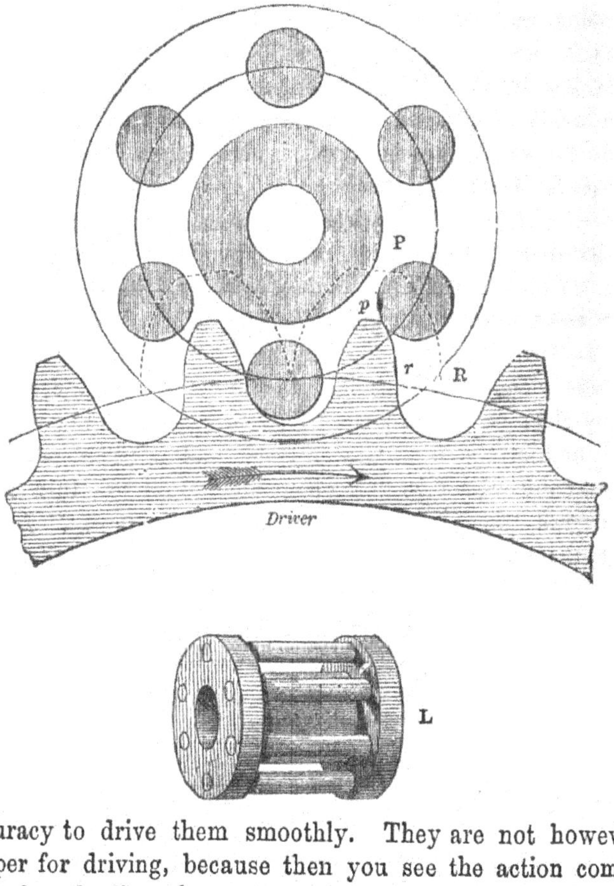

accuracy to drive them smoothly. They are not however proper for driving, because then you see the action comes all before the line of centres. In some French turret clocks the winding pinions are nevertheless wrongly made as lanterns, and the pins themselves pivotted instead of rivetted in their sockets so as to turn while they are working, which makes the working loose and shaky and the pinion itself very much weaker than when the pins are fast, and saves very little in friction besides.

For the purpose of geometrical construction, we may first consider the pins as being of infinitely small thickness, and then the teeth which drive them would be of the dotted form P R in fig. 57, being epicycloids traced on the wheel with a circle the full size of the pitch circle of the pinion. Then in order to get the shape of the teeth for pins of the actual size, you must gage off half the breadth of the pin from each side of the tooth, which reduces it to *p r*, and you may leave on just as much point as will keep hold of the departing pin P until another tooth has got well hold of the next pin just as it crosses the line of centres. This operation of reducing the theoretical to the actual tooth is practically equivalent to tracing the tooth with a smaller circle: how much smaller, will depend on the number, *i.e.* on the thickness, of the pins, in proportion to the size of the pinion. I find that a lantern of 8 or 10 requires a tooth which fits a leaved pinion of the same number so nearly that I can see no difference in the curves on a pattern as large as 9 inches diameter; and even with 12 the difference is very small; although a theoretical lantern pinion with pins of no thickness requires the same shape of teeth as a radial pinion of twice its size. I have no doubt that a lantern of 8 runs as easily as a leaved pinion of 12, and of course it requires only ⅔ the number of teeth in the wheel, and is also itself stronger, and much less liable to break, both in hardening and in working afterwards.

I may however repeat the caution that cast iron wheels do not work so well with steel pinions, which lanterns necessarily are (or rather, they soon wear them out), as with cast iron; and therefore if the great wheel only is of iron and the smaller ones of brass or gun metal, the pinions should be made of cast iron or steel accordingly. Also it should be borne in mind that you cannot draw out an arbor with a lantern pinion endways, by unscrewing the front bush only; and therefore they should not be used where you cannot conveniently get at the back bush to take it off. These pinions are used in all the American clocks and also in the cheap German or 'Dutch' clocks, both of which, it is well known, will go with an amount of dust in their insides which would stop a clock with leaved pinions completely. But the English clockmakers will not use them in small clocks; and as English small clocks are not yet made in factories, as large ones are, and as they are everywhere else in the world, the men who make them up have the power of obstructing every such improvement, and exercise it by immediately charging a higher price for every deviation from the common form, or for everything which they fancy is going to be applied to some new use.

The leaved pinions of English house clocks are made out of 'pinion wire,' which is in fact a very long pinion drawn through a hole like wire, and the leaves are turned off to form the arbor and pivots (see p. 21). The American and Dutch clocks prove clearly enough that lantern pinions can be made at least as cheap as others; and if any man of skill, capital, and determination would follow the example of Mr. Hobbs in locks, and set to work to manufacture clocks in a factory of his own, we should soon see this and other improvements made, and the clock trade recovered out of the hands of foreigners, to whom it has been in a great measure sent away by this combination of workmen, who will ruin this and every trade in the kingdom if they are allowed to have their own short-sighted way.

Inasmuch as English clocks are thus made by hand, and therefore probably no two are exactly alike, it is necessary to have a tool for marking the places where the holes for the pivots of the wheels are to be drilled, to bring them to the proper depth for working. This *depthing tool* is something like two vices framed together parallel to each other, each carrying a thick sliding pin sideways through each of its jaws. These 4 pins have a small hole or 'centre' at the inner end and a point at the outer. The pivots of the pair of wheels to be fitted are put in the 4 centred ends of the pins, and the machine is adjusted by trial till the wheels are felt to work together comfortably. Then the points at the outer ends of each pair of pins indicate the distance of the centres of the wheels, and may be used to mark that distance on the frame-plate. If you cannot get two wheels to work together without shake (so long as they are driven the same way) by any adjustment of their depth, the teeth are wrongly cut in one of them at least, and they have no business to be used.

Internal wheels are wheels with the teeth turned inwards, in which case the rim must evidently be raised in a different plane from the spokes. They are never used in clock-work, except for some special purpose, such as Sir G. Airy's maintaining power in the Exchange clock (p. 151) or in a sun and planet wheel for the same purpose, neither of which are desirable, and in some train remontoires (p. 234). Whenever they have to be used they are best made with pins driven into the flat rim, especially if they have to be driven and not to drive. Otherwise the same rules apply to the construction of internal teeth as to external ones. They almost necessarily involve the fixing of the wheel or pinion which works into them on a stud instead of on an arbor with pivots, which is generally objectionable and ought never to be allowed if it can be helped, either for wheels or pulleys.

Bevelled wheels.—When the direction of motion has to be changed, as it often has in leading off to dial-work in turret clocks, bevelled wheels are used (as shown in p. 230). They may be made to act through any other angle just as well as 90°. The only condition to be satisfied is that the teeth should in every respect converge to the point where the two arbors would meet if prolonged, as if they were on the faces of two cones. Nevertheless in a great many bevelled wheels, the sides of the teeth do not converge properly, the spaces being made of the same width all along, and so the teeth converge too much, and have no contact at all except just at their outer edges. The reason of this is that if the sides of two adjacent teeth are cut in the common way with the same cutter, the breadth of the cut is the same throughout, and not narrower at the inside than the outside, and therefore the teeth evidently taper too much. Fortunately there is seldom much pressure on the bevelled wheels in clocks, and therefore the defect is not very material: but still it is one, and its frequent occurrence is another reason why the bevelled wheels should be large in diameter rather than in thickness; for the larger they are the less pressure there is on the teeth, and the less any inaccuracy of cutting is felt (as it is in wheels of all shapes), and also the less shake of the hands and wheels there will be for any given amount of shake in the teeth.

Skew-bevelled wheels.—It may be worth while to know that bevelled wheels can be made with oblique teeth, to work with their arbors not in the same plane, provided they have only to work one way; but the friction is very great if they work what may be called the wrong way, and even in the right way the friction is more than in the usual conical wheels. I have never myself seen any clock where it was necessary to resort to this construction, which may be found in books on machinery, and therefore I only mention it in case anybody may have occasion to resort to it.

Cams may be defined as teeth which have to raise a lever, or a sliding rod, and not a succession of teeth, and therefore each cam must work up to its end, and drop the lever there, whereas in wheels a second pair of teeth may and always should come into action before the preceding pair have quite separated. The simplest form of cams to raise a lever is that shown in any of the pictures of striking work of house clocks, figs. 33 to 36, viz. a set of pins stuck into the side of a wheel, which catch the lever at some distance from its end and work up to the end and then let it drop off. This does well enough for very light hammers requiring only a small clock weight, but is the worst plan that could be invented for large ones. If you take the trouble to draw a wheel with 8 pins in it, each pin acting on the lever through about 36° (leaving the difference between that and 45° for the clearance) you will see that the angular lift of the lever towards the end of its motion is only one third of what it is at the beginning, and therefore ⅔ of the clock weight is wasted during part of the lift. And that is by no means all; for if you look at p. 211 you will see that, as turret clock hammers are usually fixed, the weight of the hammer acts more vertically or requires more force to lift it at the beginning of its motion, where the pin has least leverage or power, than at the end, where it has most; and besides all this, the loss of power by friction in driving through short levers is much greater than through long ones with less angular motion. Under all these circumstances it is no wonder that the weight required to make a large clock strike is often three or four times the theoretical 'duty' of the clock, or the equivalent of the hammer × its lift × as many blows as it strikes for once winding; *i.e.* that ⅔ or ¾ of the force is wasted in bad leverage and friction: whereas in the Westminster striking part little more than ¼ of the theoretical duty of the clock weight is lost in friction, leverage, and the necessary clearance or drop for the lever.

The first condition therefore which the striking cams ought to satisfy is they should begin to act at the end of the lever. It is not necessary that the action should be quite at the end all the way through, provided it is so at the beginning, where the lift is hardest for the clock, and at the end of the action so as to let the lever drop suddenly; for which reason also rollers are worse than half round pins, besides the reasons just now given against using pins at all. The curve which does keep the cam acting on the end of the lever throughout, and as a tangent to itself, without any scraping, is called in mathematics the *tractrix*. If you wish to describe it, the way is this. Set a smooth round board of the size of the cam wheel with stiffish paper pasted over it (to efface the grain of the wood) on a pin through the centre on a horizontal table; and set also on the table, on another pin, a model of the intended lever, with a vertical pencil at its intended point, so as to press upon the paper, the lever being weighted a little. Set the lever on the line of centres and turn the wheel; it will drag the pencil over its surface in a curve which is the *tractrix*, unless it has been disturbed by some inequality of friction between the paper and the pencil, or at the axis of the lever; which is so likely to happen that you cannot safely rely on this construction, unless you find that a good many of the curves so traced agree with each other.

It fortunately happens however, that there is an epicycloid which agrees with this so nearly that it may be used without sensible error. Suppose r is the radius of the circle which forms the bottom of the cams (*i.e.* their theoretical pitch circle, allowing nothing for clearing the end of the lever) and l the length of the lever. The lever will work as a tangent on its end throughout (without any appreciable error for such length of cam as is ever wanted) on epicycloidal cams traced with a circle whose diameter = $\sqrt{r^2+rl}-r$. Thus if $r=8$ in. and $l=4$ in. the radius of the tracing circle will = ·95 in. Another advantage of these cams is that you may cut them off at any length, provided only you keep the lever of the proper length; if you alter that you will get scraping friction, and soon wear out either the cams or the lever.

If you prefer having circles to deal with instead of epicycloids, there is another form of cam, which was suggested to me by Mr. Effingham Lawrence (another horological lawyer, like Mr. Bloxam, and all the members but one of that jury in the 1851 Exhibition, both English and foreign), and which acts quite as well as the epicycloid or tractrix, provided you take care not to alter the length of the cam: *i.e.* you must not put more or fewer cams on the wheel than they are designed for, and you must take care that the proper distance of centres of the wheel and lever is preserved. In fig. 58, C A L is the line of centres, and A B the space for one cam on their pitch circle; by which I mean the space occupied in lifting, for you see a little space is left below the line of centres before the next cam begins, to prevent the lever dropping onto the cam itself, which shakes the clock most injuriously. A P is the arc of the lever. Draw A T, which is a tangent to the two circles at A, and B T a tangent to the cam circle at B. That point T will also evidently be the place where a tangent to the circle A P at P would meet the others; or in other words, T is the centre of a circle B P, to which the lever itself will be a tangent both at the beginning and the end of the lift, although the contact

Fig. 58.

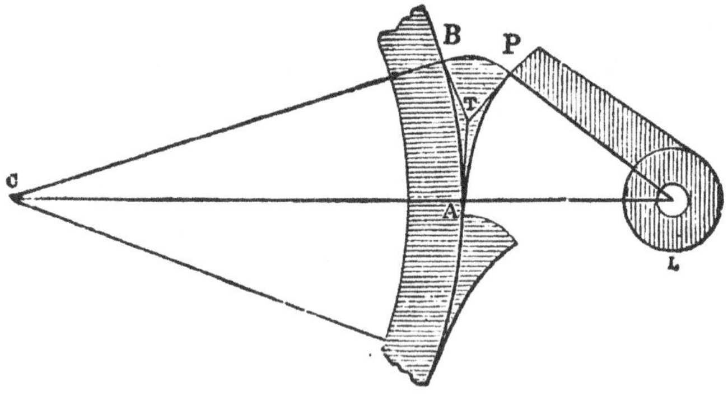

will be a little way from the end during some intermediate part of the action. The backs of the cams must be cut out a little deeper than down to the pitch circle, to let the lever drop freely; and it is important to remember that the end of the lever itself should not be left sharp, or it will cut off the ends of the cams if they are not very hard, and perhaps break them if they are. I know that by experience.

It does not occur to me that cams can ever be required in clock-work for lifting a vertical rod sliding like a stamper. If any such case should occur, involutes of the wheel circle would be the right shape for the cams.* There may also be cases where it would be worth while to pull down a long striking rod by pins in the wheel catching a square hook at the end of the rod, and dragging it on with a little sideway motion until it is struck off the pins by a horizontal stop: this would avoid all the lever work and all friction except at the striking off of the hook.

Oil for clocks.—I believe it is now generally understood that sweet oil is the worst that can be used for machinery, large or small, except when it is purified in certain ways not known to the public; and then it is too expensive for use in large or common clocks. For them purified sperm oil, such as is now made wholesale for other machinery, is quite good enough. Common neat's-foot oil may also be purified into a very good oil, which will hardly freeze here. It can only be done in cold weather; it should be shaken in a bottle with water until it becomes a thick white soup, and then left to stand, and the fine oil that gradually comes to the top skimmed off, taking care to get none of the thick. If it is done in a warm temperature, oil appears at the top as fine, which has no business there, and will not remain fine in cold weather.

* This is no exception to the epicycloidal rule, for an involute is in fact an epicycloid traced by a circle of infinite size, *i.e.* by a straight line.

I have found two advantages in using the animal and sperm over vegetable oil, which some persons may be glad to learn. Working in iron with the common olive oil dirties your hands so that it is very difficult to clean them, whereas animal or sperm oil helps to clean them. It also preserves iron from rust far better and longer, and does not turn green on brass, *i.e.* does not produce verdigris. I understand that petroleum is better still for preventing rust. It is difficult to make amateur clock-cleaners understand or remember that putting oil to pinions driven by brass or gun-metal wheels wears them out; but this does not apply to cast iron; and all pivots should be oiled, and acting surfaces generally. Oil must never be put to the beat pins of a gravity escapement, or it will make them stick to the pendulum quite enough to affect its rate; but it should be to the pallet pivots. I have known the want of it, both there and in the common pallet pivots, affect the arc sensibly.

It must be remembered that oil has always a tendency to run away from small points of teeth, the ends of pins, &c., to the thicker parts of the wheel. In some French clocks the teeth of the dead scapewheel are accordingly made with a kind of lump at the end; but this wastes more space in the clearance or drop, and it is never done in English clocks, so far as I have seen; nor do I know anything that will answer—except putting on fresh oil when it is wanted. Dirty oil should always be cleaned off before putting fresh oil on. When it has got very thick it sometimes requires soda in warm water to remove it.

WATCHES AND CHRONOMETERS.

The early history of watches is no less obscure than that of clocks, though many centuries later. Tradition assigns the invention to the town of Nuremberg, in the fifteenth century. According to a once famous book, Derham's 'Artificial Clockmaker,' of 1714, and Beckmann's 'History of Inventions,' they appear to have existed in the time of our Henry VIII., and of Louis XI., of France, and especially of the Emperor Charles V., who used to keep many watches going, as near together as he could. Erasmus had a watch given him by a Polish nobleman, early in the sixteenth century. And by Queen Elizabeth's time they had become pretty common, and are alluded to by Shakspeare, in *Twelfth Night*—'I frown the while, and perchance wind up my watch.' The London Company of Clockmakers was incorporated in 1631; and not long after that, the invention of the spring balance, the distinctive element of watches, was contended for between Huyghens and Dr. Hooke, who both claimed the invention of the pendulum for clocks (see p. 30). It seems clear that Hooke first ennunciated the discovery of the isochronism of springs in the short sentence, *Ut tensio sic vis;* or the force varies as the degree of tension: which rule however we shall see presently has two rather curious exceptions.

The main-spring of a watch is a thin ribbon of steel coiled up in a barrel round a strong spindle, to which one end of the spring is fixed, the other end being fixed to the barrel. When it is wound up the coils lie close together upon the spindle or arbor, and as the spring runs down, the coils separate from the arbor, and lie close to the barrel. The simplest construction, still used in most of the foreign watches, is for the barrel arbor to be the winding arbor, having a ratchet wheel squared onto it, held by a click on the frame or plate of the watch, and the great wheel is on the barrel itself; and in this case, as I explained at p. 145, with reference to spring clocks, no temporary maintaining power or 'going barrel' is required to keep the watch going while you are winding it up. But it is obvious that as the force of the spring is greater when it is tightly wound up than when it is loose, the force of the train will be very far from constant throughout the day, although that may not affect the going of the watch from one day to another. On the other hand, there is found to be a very singular exception to that rule of Dr. Hooke's, stated just now, inasmuch as there is a position of the spring coiled in a barrel in this way, in which there is no material variation of its force for a few turns. And certainly some of the foreign watches made in this way go very well.

Even if a watch does go rather faster at one time of the day than another (which it is not certain that such watches do), it is of no consequence, provided they go uniformly from day to day; and there has been strong testimony published in the *Horological Journal*, that they do, when well made, of course. I have seen watches with tapered main-springs (though the foreign ones are not tapered), in which you could not perceive any difference in the force, by the usual testing apparatus of a weighted lever, whether the spring was wholly wound up or not. And if this is so, the addition of any more machinery, being superfluous, is mischievous, and only increases the expense and size of the watch, and risk of the chain breaking, perhaps the most common of all accidents. There is evidently a tendency in watch-making now to dispense with such machinery, except for marine chronometers, which are required to go uniformly at all times of day, and not merely from day to day. Accordingly, both in Switzerland and America, which are gradually stealing away our common watch trade, as well as that in many kinds of clocks, the fusee and chain (which I will describe presently) are almost universally omitted.

But in that case there is another risk of breaking, for when the spring breaks, the barrel recoils violently from the sudden removal of the pressure, and often breaks some teeth of the great wheel fixed upon it. Partly to avoid this, and also for general improvement in strength and arrangement of the train, the American watches, as made by the 'Howard Company,' under a patent by a Mr. Reed of 1857, have the spring contained in an immovable barrel, which is a solid part of the upper plate of the frame (calling the dial side the lower). The outer end of the spring hooks to the barrel as usual, and the inner end to the winding arbor, on which the great wheel rides, with a spring maintaining power, just like those of a clock described at p. 149. For safety also, the main-spring is held by two clicks, much stronger than the single one generally put into fusee watches, and the maintaining spring is also very strong. Fig. 59 is a plan of this, and fig. 60 a section of the whole watch. It may be observed also that a watch of this kind turns the same way in winding as one with a fusee, whereas the common foreign watches turn the opposite way: not that there is any real importance in this, except that the absence of uniformity sometimes causes people to wind a new watch forcibly the wrong way.

Fig. 59.

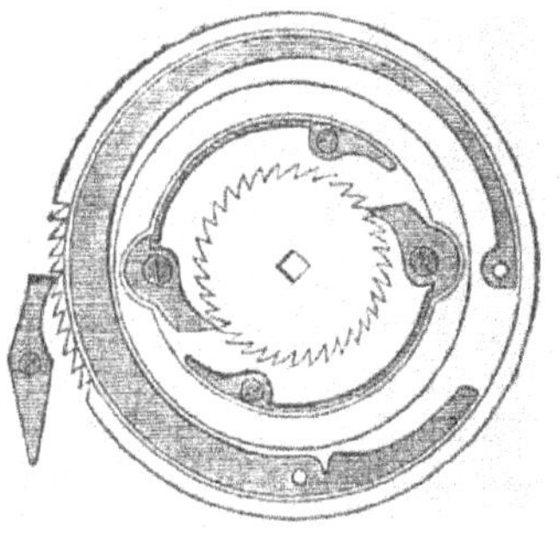

To prevent that, Breguet invented what is called the **tipsy key,** not that that adjective belongs to the 'winder,' in the clockmaker's sense of a key, but in the more human sense.

The handle is connected with the pipe by a pair of ratchets, which would be called *clutches* in large machinery, which can pass each other one way, but not the other; and they are kept together, and then hardly visible, by a spiral spring, which gives way, and lets one pass over the other if the handle is turned the wrong way. The best shape for a winding key is like a pencil with a rough handle, to enable you to twirl it more easily between your fingers. You thereby apply less force, and are less likely to do mischief by wrong, or over winding; and moreover you keep the watch steady, instead of turning it in your left hand, and agitating the balance at every turn.

Fig. 60.

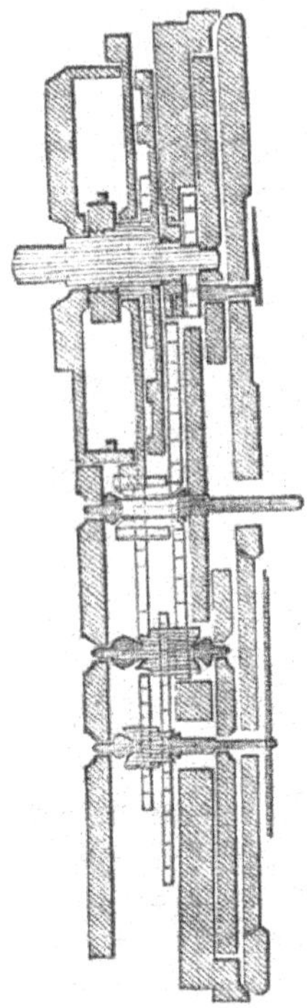

If you compare figure 60 with 61, you will see that in the latter, which is the common English fusee and chain watch, the third wheel of the train has to be sunk in the plate below the centre wheel, whereas in the American watch (fig. 60) it stands free and clear. There are a few other things in this arrangement which had better be postponed till other parts of the watch have been described, and we will proceed with the description of the common English watch construction, and therein first of the

Fusee.—That piece is a hollow-sided cone, which you see in this picture of a chronometer or English watch movement, with a chain round it and the barrel, and the great wheel is no longer on the barrel, but on this conical piece called the fusee. When the spring is wound up and its force is greatest, the chain acts on the small end of the fusee, and therefore with the smallest leverage; and as the spring unwinds, the chain acts on a thicker part of the fusee, and it can be, and in good watches is, so adjusted that the force on the train and escapement is constant. It was suggested by Mudge, the inventor of the lever escapement in the form now used in 99 out of 100 English

Fig. 61.

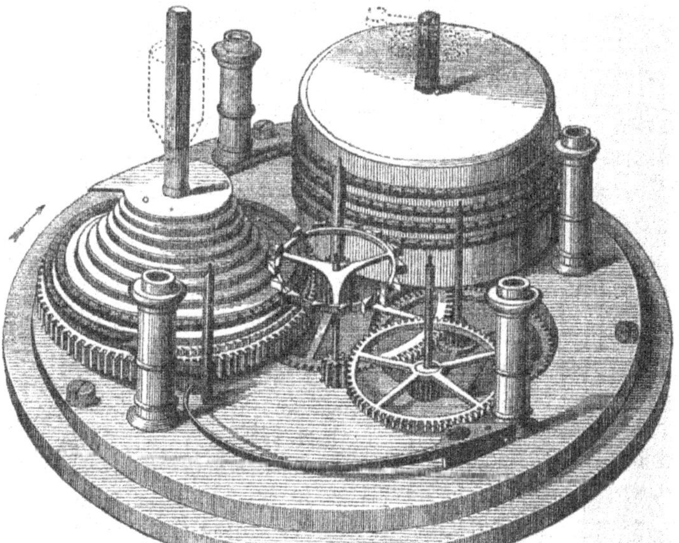

watches, that the usual position of the chain is wrong: and so it is; for you see in fig. 61 that it acts on the opposite side of the fusee to the centre pinion, and consequently the pressure and friction on the fusee pivots (which are necessarily large ones) is the *sum* of the force of the spring on the fusee and of the great wheel on the pinion; whereas if the spring acted on the same side as the pinion, it would only be the *difference*. I confess I know no reason why the common arrangement should be adhered to, except that it is the common one, which is generally considered reason enough for anything bad.

The train of wheels in a watch is much the same as in a clock, except that the scapewheel is not the wheel which turns in a minute and carries the second hand, but is another faster wheel. In a pocket lever watch the balance generally beats in 2-9ths of a second, and in a chronometer either in that time or in ¼ sec. The scapewheel generally has 15 teeth, and therefore turns in 6 seconds, or something near it. In a good lever watch the pinions are generally 7, 8, 8, 10, and the wheels 63, 60, 64, 75; in a pocket chronometer the pinions are 8, 10, 10, 12, and the wheels 80, 75, 80, 72; in box or marine chronometers the numbers are still higher, the pinions being 10, 10, 12, 14, and the wheels 80, 80, 90, 90. Box chronometers are generally made to go rather more than 2 days, though they are wound up every day, and they have a small hand on what is called the 'up and down' circle in the dial, indicating how far the barrel has run down. In these watch trains it must be observed that where a slow wheel has fewer teeth than a quick one, the teeth must be larger, or the wheels could not be put into the frame. I think no pinion so low as 7 should be admitted, and I cannot understand why 9 should be a prohibited number in clocks and watches, as it seems to be, or at any rate was until I introduced it frequently in turret clocks.

Common watches have the balance above the upper plate. What are called **three-quarter plate** watches have a large piece of that plate cut away, and the balance lies about level with the rest of the plate, and most of the wheels have their upper pivots in cocks screwed to the lower plate. This saves thickness in the watch.

Winding stops.—A watch, or a spring clock, with a

fusee, is stopped from being overwound by a long tooth which you see in fig. 61, sticking out from the thin end of the fusee. There is a spring lever with a hook fixed to the frame with a little play on its pivot, so that when the chain comes to that end, or the fusee is full, it pushes the lever just far enough for its hook to catch the tooth, and so stops the winding. In foreign watches without a fusee, a thing called the *Geneva stop* is used: it consists of a small wheel on the barrel-arbor, with only one tooth in it, and the rest of the circumference filled up blank; this tooth works into the teeth of another loose wheel, with the hand on it in fig. 62, which has only as many teeth as the turns that the barrel has to make in winding, and has also a blank space in its circumference. The one-toothed wheel turns the loose wheel through the space of one tooth for every turn of the barrel, and when those teeth are all past, the one-tooth jams against the blank in the loose wheel and lets the barrel turn no more, and so stops the winding. These two wheels are above the upper plate. The same thing might be put on a fusee arbor, but the spring stop is preferred.

The 'up and down dial,' just now described, on the usual construction, involves an additional pair of wheels. But Mr. A. L. Dennison of America, now settled here, who is said to have been the originator of watch-making by machinery (which at last is taking root here, at Messrs. Rotherham's well-known watch factory at Coventry), patented a modification of the Geneva

Fig. 62.

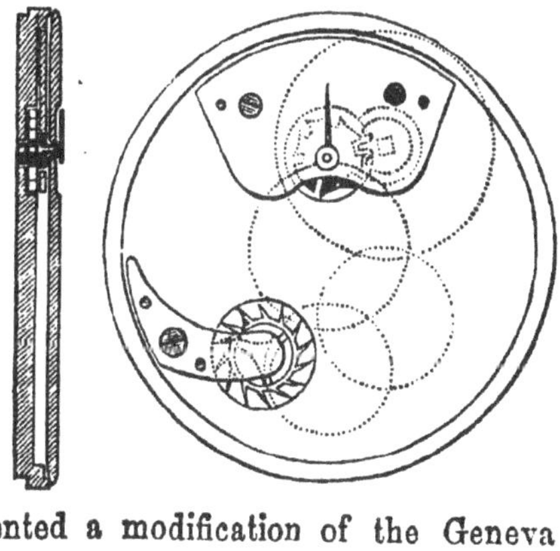

stop which enables this kind of indicator to be added without any more wheels. It consists merely of adding a short arbor and hand to the loose or second or star wheel of the Geneva stop, which is applied to watches on the American plan described at p. 292. These pieces are shown in this figure 62, of which the right hand part is the plan and the other the section. This is not of so much value in a common watch as in a chronometer, but is convenient to be able to see at a glance whether you have wound your watch up fully, without the risk of straining it in trying.

The dial wheels of a watch are more like those of a turret clock than of a house clock in the division of the numbers of the teeth, as there is no occasion for the intermediate wheel called N in pages 134, 164, to turn in an hour, as it does in house clocks to discharge the striking, and even in silent clocks for uniformity. The hand sockets are also only held on by friction without any spring. In other respects the train of a watch is substantially the same as that of a small clock until we reach the escapement, except that there is one more wheel in the train, for the reason given just now. The dial pinions in a good watch are 12 and 14, and the wheels 42 and 48; in a box chronometer they are 14 and 18, and 54 and 56.

Balance spring.—The rest of the machinery of a watch only differs in size from that of a clock, until we come to the escapement and the balance. We shall consider the various escapements presently, but the object of them all is the same as in clocks, to produce an escape of teeth by the vibration of a balance, the time of which depends on its own moment of inertia and a coil of very thin steel wire called the balance spring, and sometimes popularly the hair-spring from its thinness, and sometimes (very absurdly) the pendulum-spring, though it is of the very essence of a pendulum to act solely by gravity and the essence of a balance to be free from the action of gravity, or else the 'balance' is not balanced and the watch will go differently in different positions. The spring by which a pendulum is hung produces no sensible effect on its time, and is only used to avoid the friction of any other mode of suspension.

The balance-spring is fixed at its outer end to a stud in the watch, and at its inner to the balance arbor or staff, and at the neutral position of the spring the balance is at the middle of the escape. In some kinds of escapements therefore the balance cannot stand still if the watch is wound up; and in all of them the spring tends to bring the balance back again after it has gone a certain distance, depending on the force of the escapement and the cleanness of the pivots and teeth on one hand, and the strength of the spring on the other, and then the balance swings an equal distance beyond zero the other way. Whereas in the original clocks with a balance but no spring (p. 28) it was only the recoil of the escapement that brought the balance back again, and such a balance could not have been used with any kind of dead escapement, because it would never have returned. Moreover, considering that a pendulum moves by gravity, through an arc of only 4° or 5° with a variation never amounting to 30′ in a good clock, while the balance swings independently of gravity through an arc of sometimes 400° and sometimes no more than 100°, it is evident that a watch escapement must be a very different thing from that of a clock.

When I say that a balance is independent of gravity, you must remember the distinction between mere mass, which we denoted by the letter M in treating of pendulums, and mass acting as a force by means of its weight, *i.e.* by the earth's attraction, which we called Mg. The mass and the moment of inertia of a balance have quite as much to do with its motion as they have in a pendulum, but the force which makes it vibrate or return from the impulse given in the escapement is not gravity, which must be entirely excluded, for the reason just now given. It was to this balance spring that Hooke's law of the force varying as the tension or space moved through (which always implies isochronism) was meant to apply, and does apply pretty generally; but not invariably, because it is found that not every length of a given spiral spring is quite isochronous, but only certain lengths, which have to be determined by experiment.

The time of vibration therefore depends on the moment of inertia of the balance directly and the force of the balance spring inversely; and in a given spring the force varies inversely as the length. Consequently you can quicken the vibration either by reducing the moment of inertia or the size of the balance, or by shortening the effective length of the spring. The latter method is used in all common watches, and the former in chronometers and watches in which extreme accuracy is aimed at.

Regulation of a common watch, to make it go faster or slower, is done by a lever or index SPT (fig. 63), which turns on a ring set on the watch plate (through which the *staff* or arbor of the balance has to pass from the inside to the outside of the watch frame), and it has two small pins at P which embrace the spring, one end of the spring being fixed to the frame at R and the other end to the balance at S. It is evident that as you turn the regulator to the right you shorten the length of the acting part of the spring and so make the vibrations faster, and if you move it to the left, slower. If the regulator has been moved as far toward *fast* as it can go and the watch still looses, the spring must be taken up altogether at R; and then in order that the balance may still be in the middle position of the escapement when the spring is neutral, the piece S by which it is attached to the balance is itself a ring which fits tightly round the staff and can be moved when the balance is taken out, to alter the position and length of the spring.

It must have occurred to everybody who has had to regulate a good watch for very small errors that there is a want of some better method than the common one both for moving the regulator and for seeing how much you move it. Occasionally, but very rarely, the index is made to move by a tangent screw, turnable by the watch key, and I am surprised that good watches are not always made in this way. But I have lately seen another, in the American watches referred to at p. 296, which is still more delicate, and also easier to make. The index is resisted one way by a slight spring fixed to the plate, and a common screw, in the same direction as a tangent screw, acts upon it the other way. This avoids the shake of a tangent screw, and you know that however little you turn the screw you move the index some distance. In these cases you must learn by trial how much half a turn of the screw accelerates or retards the watch per day, and after that you can regulate it to the utmost nicety—so far as its errors are amenable to regulation.

Fig. 63.

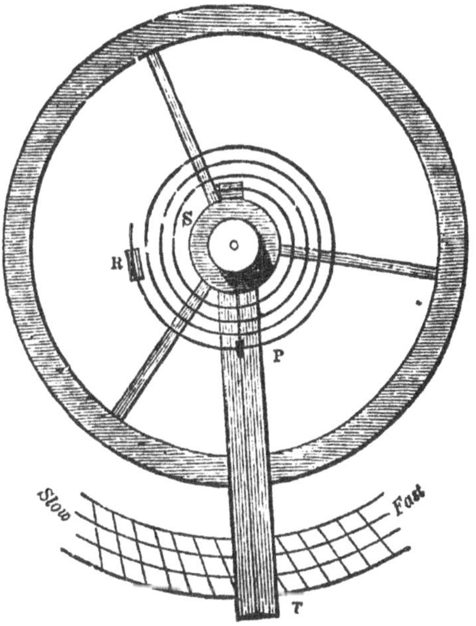

Another method suggested by the late Mr. Dent is shown

in fig. 63. Instead of being made with a point, the index has fine bevelled edges lying quite over the index plate, which is to be made with oblique divisions and two or three cross lines, like the scale generally engraved on a dividing rule: this enables the eye to measure much smaller divisions than could be either seen or cut upon a degree plate of the common form. But this does not remove the difficulty of moving it a very little, and the screw is manifestly better. The old plan of a segment of a wheel turned by a pinion with the index fixed to it (fig. 64) is better than the common one, and I do not know why it is disused.

Fig. 64.

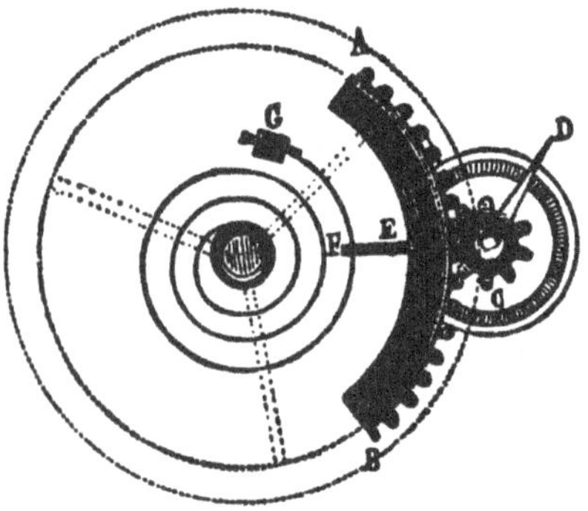

This mode of regulating by shortening the effective length of the spring is not accurate enough for chronometers for another reason, viz. that all lengths of the spring are not quite isochronous for different arcs of vibration, and therefore if you have got the spring adjusted for an isochronous length, it will become unisochronous if you shorten it a little; and moreover a spring moving in that way, partly held fast at the ends and passing loosely through curb pins at P, is not so steady as one held fast only at the ends. Chronometer balances are therefore regulated by *timing-screws*, which are screws with heavy heads set in the rim of the balance: screwing them in of course diminishes the moment of inertia and quickens the balance, and vice versâ. In chronometers also the spring is generally made as a cylindrical spiral and not a flat one. Some chronometer makers have doubted whether there is any advantage in the cylindrical form, but it is now almost universally adopted, and therefore I suppose the balance of experience is in favour of it. It is now said that the best form is one that combines both, or a cylinder returned into a flat spiral both at top and bottom.

But the quality of the spring is probably of more importance than its form. A medal of the highest class was awarded in the 1851 Exhibition to M. Lutz of Geneva for some balance springs, made by a secret method, which bore being pulled out nearly straight, and laid on a hot plate without suffering any change of form, which was not the case with any others which were then submitted to us. I see that aluminium bronze has been successfully used in Saxony for balance springs; but I know nothing more of them.

Glass balance springs.—These were suggested by Berthoud but have been very little used. The rate of a chronometer with a glass spring, which Dent sent to Greenwich many years ago, was very good; and they have the advantage of varying so much less than steel or any other metal in elasticity that they require very little compensation of the balance. I do not know why they have not been more used, unless it be from the fear of breaking them when anything is done to the watch. It appears that with them, as with steel springs, the watch always gains after a new spring has gone a few months, as if it acquired more elasticity by working; and this is analogous to what takes place in bells, which become more sonorous, *i.e.* more elastic, after a few months ringing.

Timing for position.—As a watch sometimes lies flat on a table, and sometimes vertical, and not always in the same position in your pocket, it is necessary that the balance should keep the same time in all positions, both horizontal, and with either iii, vi, ix, or xii, upwards. With a heavy balance it is impossible to get the same arc of vibration when the watch is vertical and horizontal, because the friction of the pivots is much greater when they are acting on their sides than on the point of one of them. If you take a small chimney-piece clock with a balance and hold it sideways so that the balance becomes vertical and its staff horizontal, you will see the vibration diminish very much, and then the least want of isochronism in the spring will set the rate wrong. Marine chronometers, which are only very large watches, are therefore always set horizontally in a box, in *gimbals*, as in this drawing, which keep the watch horizontal, even when the box is tilted by the ship: if the box is tilted at E or W, it turns on the outer pivots N S of the gimbal ring, and if the N or S side is tilted, then the box and ring together turn on the inner pivots at W and E leaving the watch steady. The level of the pivots should be only just enough above the centre of gravity of the watch to make it keep its level, for if the c. g. is much below the points of support the watch will swing when they are moved.

Fig. 65.

Ship time-pieces.—There is another kind of ship-clocks made to hang against a wall, not pretending to the accuracy of chronometers, but a very neat and convenient and cheap form of time-piece for other places as well as ships. They are a large 8-day lever watch with the balance staff and the scapewheel arbor vertical, and therefore the third wheel in the train a *contrate* or crown-wheel as at p. 28. Like other things, they are of very different qualities with the same external aspect. They do not work well without a fusee, though many of them are made so for cheapness, and the contrate wheel arbor ought to have an end-stop to the pivot which tends to push away from the scapewheel pinion and work upon its shoulders, and the two vertical arbors ought still more to have their lower pivots resting on stops, either of hard steel or jewels, to take the weight and friction off their shoulders. End stops are sometimes put to the horizontal arbors of highly finished clocks; but as there is no end pressure on them, I think it is hardly worth while there, though it is in the two cases I have just mentioned. Chimney-piece clocks with balances (the only ones which housemaids will allow to go, unless they are too heavy to move) ought always to have them compensated, for otherwise the great changes of temperature to which they are exposed will make it impossible for them to go right, as I shall now explain.

COMPENSATION OF BALANCES.

Watch balances, or rather the springs on which their time depends, vary much more with heat and cold than pendulums, and therefore compensation is still more essential if you expect great accuracy of rate; and if it were not that watches are kept at a pretty even temperature during the day by near contact with men's bodies, their variations in hot and cold weather would be enormous: women's watches being generally worn loose, go worse than men's for that reason. A small portable clock with a balance uncompensated goes twenty times worse than the commonest pendulum clock. The balance itself also expands a little with heat, as a pendulum does, but the effect of that is small compared with the variation of the force of the spring. It appears from some experiments made by Berthoud in 1773, and others by Dent communicated to the British Association in 1833, which nearly agreed, that the loss of time due to expansion of the balance for a rise of temperature from 32° to 100° is not a 5th of that due to the loss of elasticity and elongation of the spring, and that the two together amount to no less than 6½ minutes a day; and a variation of 33° would produce a variation of an hour in 3 weeks; whereas we saw that a common iron wire pendulum would only lose 10 sec. a day for such an increase of heat, and a wooden one not more than 1½ seconds.

The first watch compensation was made by Harrison, who has been mentioned several times already as the inventor of various horological improvements; and he received the first parliamentary reward for improvements in chronometers with a view to finding longitude at sea. His method is quite disused now, and indeed he was himself dissatisfied with it, and suggested that the compensation ought to be done in the balance and not by any contrivance for altering the effective length of the spring, which was the principle of his own and all the early compensations. For this purpose he put the curb pins (P in fig. 63 p. 300) on a compound bar, of which one side was made of brass and the other of steel. As brass expands more than steel, the bar bends to the steel side when the heat increases, and thus the curb pins were moved along the spring, as they are by hand in the common regulator. It is not necessary to dwell on any of the various modifications of this plan, as they are all now abandoned for the balance of compound bars, which appears to have been first made, as well as a mercurial compensation balance, by Julien le Roy, a celebrated French clockmaker, but afterwards much improved by the first Arnold, Earnshaw, and various other English makers. A great variety of these early contrivances may be found in Rees's Cyclopædia by those who are curious about them.

The plan which is now always used, with some modifications in certain cases, as I shall explain afterwards, is exhibited in this drawing: *tat'* is the main bar of the balance, with the timing screws *t t'* for regulation at the ends; and *t b*, *t' b'* are two compound bars, of which the outside is brass and the inside steel, carrying weights *b b'*, which may be screwed on at different places. As the heat increases, those bars with their weights bend inwards and diminish the moment of inertia of the balance. The only secure way of making these balances is to cut them out of a solid steel disc round which melted brass is run which *brazes* itself fast on. This plan of uniting the metals was introduced by Earnshaw, it appears, besides various other improvements in the construction of chronometers, in which very little alteration has been made in nearly 100 years. The principal expense of a really compensated balance is in the time required for adjusting it, which can only be done by trial. Many watches are sold with balances only constructed for compensation but never adjusted, and therefore under or over compensated, or with no regularity in the action of the compensation; and some still worse, with a mere sham compensation, resembling a compensated balance only in appearance, sometimes not even cut through.

Fig. 66.

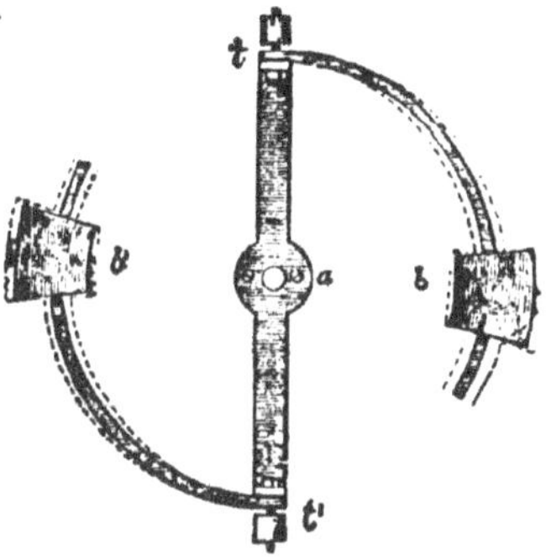

The chronometrical thermometer is simply a watch with a balance compensated the wrong way, *i.e.* with the brass inside and the steel outside, so as to increase the retardation from heat and the acceleration from cold. The use of it is to measure the quantity of heat or cold received during any given period without recording the actual degree of heat or cold at any particular time. It is therefore used at the Greenwich Observatory for trying the rates of compensated chronometers under great variations of temperature.

Secondary compensation.—When chronometers had been brought to great perfection, so as to go with scarcely any sensible variation of rate while they were kept within moderate limits of temperature, it was observed that they always lost if the temperature either rose or fell beyond those limits; and on the other hand, if the compensation was adjusted for two extreme temperatures, then the watch always gained at mean ones. I believe it has never been disputed that old Mr. Dent was the first person to explain the cause of this error, in the *Nautical Magazine* in 1833; and he gave the following illustration of it: The diminution of the force of the spring proceeds uniformly in proportion to the increase of heat, and may therefore be represented by a straight line inclined at some angle to another straight line which is divided into degrees of temperature. But the inertia of a compound balance such as I have described cannot be made to decrease quite as fast as the heat increases; and therefore its rate of variation can only be represented by a curve, and can only coincide with the straight line representing the variation of force of the spring in two points, either the two extremes, or two means, or one mean and one extreme point: in other words, the compensation can only be exact for some two temperatures for which you may choose to adjust it.

The same thing may be shown mathematically as follows: Let r be the distance of the compensation weights from the staff or axis of the balance, and we may call them both together M, and for this purpose we have nothing to do with the rest of the balance. Let dr be the increase of distance of the weights for some given decrease of heat. Then the new moment of inertia of the balance will be $M(r^2 + 2r\,dr + dr^2)$, and the ratio of the new inertia to the old will be

$1 + \frac{2\,dr}{r} + \left(\frac{dr}{r}\right)^2$; and now the term $\left(\frac{dr}{r}\right)^2$ is too large to be disregarded as it may be in the similar formula for pendulums, because *dr* must be much larger in proportion to *r* than it is in a pendulum, as we have seen already. Again, the ratio of the moment of inertia for an equal increase of heat to its amount of inertia in the middle state, will be $1 - \frac{2\,dr}{r} + \left(\frac{dr}{r}\right)^2$, assuming that equal successive increments of heat produce equal variations of *r*, which however is not quite the case, as it is in pendulums. Consequently the increase of moment of inertia, for a given rise of temperature is less than its decrease for an equal fall by $2\left(\frac{dr}{r}\right)^2$, or the compensation fails to that extent in one of the three states of cold, middle, or hot temperature.

The correction of this error is called the secondary or auxiliary compensation, and it is the point to which I think every chronometer invention for the last thirty-five years has been directed, chiefly because the chronometers are now exposed to such extreme and unnatural variations of temperature in the Greenwich trials—nearly as much as 100°, that the attention of the makers seems to be withdrawn from everything else; and every year's tables display an increased number of contrivances for this purpose, but no increased accuracy in the recorded rates of the best of them.

Greenwich trials.—When the rates of some chronometers were returned to Greenwich after one of the Arctic expeditions, the Astronomer Royal reported that they had been *kept so warm* that they afforded no test of the relative value of the different modes of compensation employed in them. I think it might also have been inferred from that fact that the trials at home in such unnatural variations of temperature are only calculated to test a single and comparatively immaterial quality of the chronometers, and to frighten away from public trial every invention for the general improvement of the instrument in other respects. Any nautical man who wants chronometers for real use would rather have one that would not vary a second a week in such temperatures as it is likely to be exposed to in any voyage in a ship's cabin, than one which may possibly go better than it at some extreme and improbable variation of temperature, but worse in all ordinary and probable variations.

Mr. Hartnupp's report of the trials of above 1000 chronometers at Liverpool, in 1872, says that the best of them, with the ordinary compensation, generally gain 6 sec. a day more at 70° than at 55° or 85°, if the compensation is adjusted for these two temperatures, and that those which have the same rate at 55° and 70°, or at 85° and 70°, lose 1·5 sec., for a change of 15° below or above the temperatures for which they are compensated. He adds, that the connection between rate and temperature remains constant for a long time.

Eiffe's compensation balance was the first invention for this purpose which was disclosed; and a reward of 300*l*. was therefore given him by the Admiralty on the recommendation of the late Astronomer Royal. At the very time while this invention was under trial at Greenwich, Mr. Molineux took out a patent for one precisely similar; which is only one of the usual proofs that when the time is ripe for an invention, it is almost sure to be made by several people at once, of whom one gets rewarded—or suffers, by a patent. Mr. Eiffe described several methods for effecting the secondary compensation; the one which he chiefly relied on, and which has been followed by some other makers, is a balance in which the compensation bar, or a screw in it, is made to reach another in the form of a spring with a small weight upon it, when it has bent inwards to a certain extent, and it carries that other with it, so as to diminish the moment of inertia still more than the single compound bar. It seems an obvious objection to this, that it is discontinuous, *i.e.*, that the secondary compensation comes into action suddenly at one change of temperature only. Nevertheless it appears to answer as well as or better than many others which are free from that objection.

Dent's balance.—That objection was obviated in Dent's balance, of which also there were several forms: this is the one which he proposed as the best in his pamphlet on the subject. The flat cross-bar *r r'* is itself a compensation bar with the brass below and the steel above, so that if the compensation weights *b b'* were set upon upright stems from the bar they would be bent in towards the axis when the heat increases. But they are in fact set upon some other bent compensation bars *s t*, *s t*[1], of which the brass is inside the bend, and consequently the weights approach the axis more than if they were set on stems of fixed length; and as they can be set anywhere on the stems the compensation is thus adjustable. The reason why the bent bars stand outwards across the principal bar is to leave room for the balance spring, which is attached to the smaller cross-bar *d* shown in fig. 67, in which the spring is omitted to avoid confusion. This compensation has the advantage of being continuous.

Fig. 67.

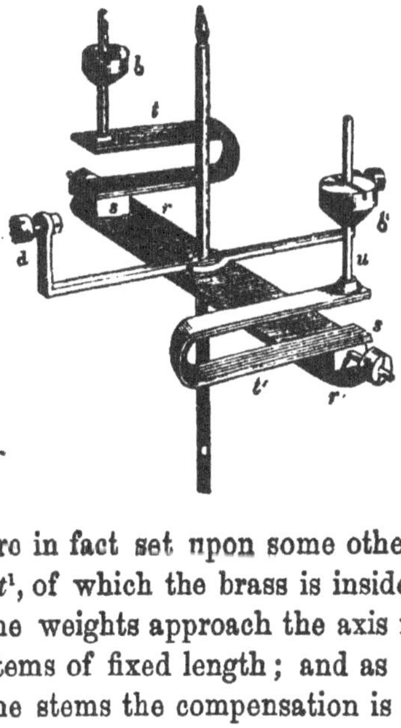

Loseby's balance was an ingenious alteration of a previous mercurial compensation of Le Roy into this form.

D D are the weights on the usual compound bars B B, for the primary or principal compensation. Besides, there are two small bent thermometers with the bulbs at F, and the tubes at E, into which the mercury runs as the heat increases, and so more of the weight of the balance is carried inwards than is due to the mere bending of the primary bars. The tubes are sealed with a little air included. C C are the usual timing screws independent of the compensation. The action here is equally continuous with Dent's, and Loseby's

Fig. 68.

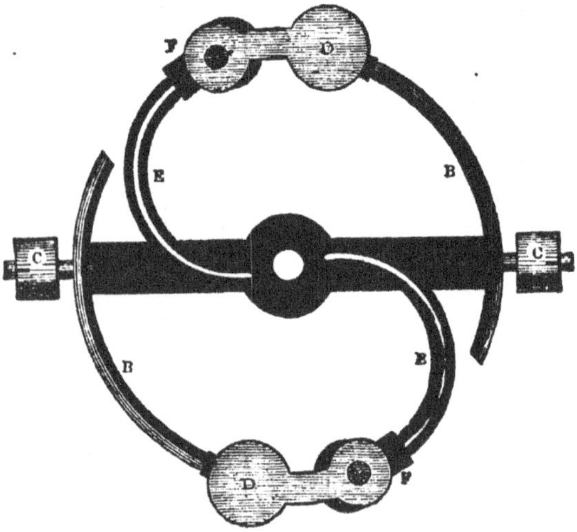

chronometers generally got a very high place in the Greenwich lists during the seven years in which he sent them there; and he would have done wisely to be content with that success. But he applied no less than four times for a reward for his invention, and the Astronomer Royal four times reported against his claim, on the ground that his balance was neither the first to do what was wanted, nor was proved to be the best. Mr. Loseby then sent to the Admiralty a long memorial, containing a comparison of the rates of his chronometers with others in the Greenwich lists for 5 years, to prove that his was the best; and the last horological paper to which old Mr. Dent put his hand before his death was a counter statement, showing that a proper analysis of those very trials exhibited Loseby's compensation to be decidedly inferior to Dent's own, and not at all superior to Eiffe's or several others of which the constructions were not disclosed.

After his death in 1853 the controversy was carried on in the Society of Arts' Journal by Mr. Loseby and me, as he had assumed that I wrote Dent's paper for him, and so I extended the analysis to the whole seven years during which Loseby's chronometers were at Greenwich, and with the same result. The nature of the analysis is simple enough, viz. this: If you divide the 6 months of published rates in each year into 3 periods, one containing all the coldest weeks, another all the hottest, and another the mean temperature weeks—and whether you make the division into equal periods, or into two periods of 6 weeks of extreme temperatures, and 13 or 14 of mean, or make it just where the register shows that the greatest breaks of temperature actually occurred—the result was always the same, that Loseby's compensation was the most successful in only one of the seven years, three other makers' each once, and Dent's three times. Sir G. Airy, the Astronomer Royal, was therefore clearly right in rejecting Mr. Loseby's claim to a public reward, whatever may have been his personal skill in preparing a single chronometer for trial in a year, and so, on the whole, getting a high place in the list.

Kullberg's compensation balance has been so successful in the Greenwich trials that I add a description of it to those mentioned before. The drawing in the next page shows its shape, but not the peculiarity of its construction; for the brass is attached to the top of the cross-bar A C, but to the bottom of the curved rims C B, the effect of which is that the ends of the cross-bar are carried downwards by heat, and the weights P upwards, and more inwards than if the cross-bar did not bend, as it does also in Dent's balance; D D are the timing screws. The principle of it is better illustrated by the second sketch, of half the balance seen edgeways and with its action enormously exaggerated. The thick line is the steel and the thin one the brass. The principle of its action is the same as Dent's, but it is simpler and stiffer

Fig. 69.

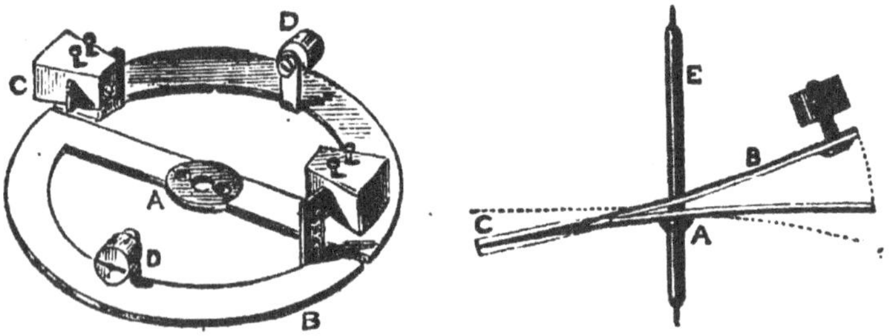

I doubt whether it is necessary to have a curved rim at all, and whether sufficient compensation could not be got by making the pieces C B straight, each lying close by the staff A E. Mr. Kullberg has also made balances of the common form but with much wider rims hollowed outwards, looking like a pulley with a very flat groove. The expansion sideways of the outer brass rim tends to flatten it, and so makes it easier to bend inwards under the expansion of the brass end ways; but he says he has not yet succeeded in getting sufficient secondary compensation in this way; as was the case also with

Dent's prismatic balance.—This was also invented by the first Mr. Dent and improved by the second, for the purpose of effecting the primary and secondary compensation together, or a sufficiently near approximation to it, by a simpler construction than any of the others. The steel part of the balance is the usual flat bar; but the brass is an obtuse-angled bent prism. This invention was founded on the fact that a bar of that form bends more easily from the angle than towards it, and therefore the compound bar bends in more easily and farther than it does outwards, for equal degrees of heat. It has the advantage of making the balance stronger also. It was not supposed that this would make the secondary compensation complete for the extreme temperatures of the annual Greenwich trials; but there seems no doubt that it is better than the common flat bar compensation.

Fig. 70.

There have been sundry other inventions for secondary compensation. Indeed nearly every maker in the Greenwich lists professes to have one of his own. The late Astronomer Royal invented one in 1875 and went so far as to get an Admiralty order that it should be used; but the trade pronouncéd it a failure, and the order was first suspended and then withdrawn. It is difficult to compare the general performance of the best clocks with that of chronometers under severe trials. The best recorded rates of chronometers seem to be a daily variation of not more than half a second during the time of trial; which, however, might come to a good deal if continued. I am inclined to say that the rate of a very good watch in ordinary use, and a lever quite as much as a chronometer according to my experience, is sometimes nearly equal to that of the best clocks for some months together. For ordinary use no secondary compensation is at all necessary.

WATCH ESCAPEMENTS.

There is a greater variety of escapements which may all be said to be in use in watches than there is in clocks, omitting in both cases merely fancy ones of which only a few specimens have been made rather for show than for use, and which I do not think it worth while to describe, either for clocks or watches.

Vertical escapement.—The original watch escapement, which has remained in use for about two centuries, but is almost gone, exactly corresponds to the old crown-wheel escapement in clocks described at p. 28. This figure (71) shows the mode of arranging it in a watch. There are in fact two crown wheels in it, for the wheel with its arbor vertical, the minute wheel of the train, is one, and the scapewheel itself is another. The plain rimmed wheel is the

Fig 71.

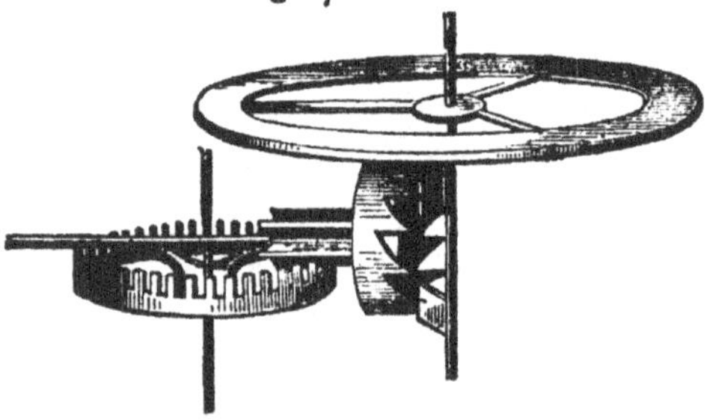

balance. It has all the properties of the recoil escapements in clocks, and is very inferior in accuracy to any of the others, and now has hardly the advantage even of cheapness over the lever, which is the standard English watch escapement.

A very simple vertical escapement, *i.e.* one involving a crown wheel, was invented (or perhaps re-invented) for cheap watches by the late Mr. John Gray, a well-known dentist. The teeth lie dead on a thick verge or cylinder except at the moment when they pass through notches cut across it; and they give an impulse in passing. It involves no more friction than the horizontal escapement, the commonest in foreign watches. The crown wheel prevents the watch from being thin, but one which Mr. Gray lent me for trial was not thicker than most silver watches, and went very well.

Lever escapement.—This in its original form was invented by Berthoud about 100 years ago, and according to the picture of it in Rees's Cyclopædia it is identical in all but the position of some of the parts with the *rack-lever* escapement, of which Litherland of Liverpool used to be the great maker: the first watch I ever had was one of that kind. If you suppose the crutch of the dead escapement of a clock to end in a piece of a wheel, of radius equal to the crutch, and working into a pinion set on the arbor of a balance, just like the regulator in fig. 64 at p. 301, that is the rack-lever escapement. The objections to it are the friction of the rack, *i.e.* of the wheel and pinion, and the dead friction on the pallets.

These were almost entirely got rid of by the modification of it which was invented by Mudge about thirty years afterwards, and which used to be called the *detached lever*, but is now generally called the lever escapement simply, since the rack has gone out of use; and the term 'detached' is now applied to another class of escapements. It is a curious fact in the history of watchmaking, that a parliamentary committee in 1793 thought fit to award to this Mr. Mudge (or his son for him) 3000*l.*, the same sum as Arnold and Earnshaw had had from the Board of Longitude, in opposition to the opinion of that Board, for a chronometer escapement which was not worth a farthing, and indeed turned out worth a good deal less than nothing to his son, who spent a considerable sum in making them. However he well deserved his 3000*l.* for the invention of this lever escapement, of which 100 times more are now made in England than of all the other escapements together, and which is quite equal to the best in accuracy of performance. The following is the construction of it (see next page).

The scapewheel and pallets are precisely those of a clock dead escapement; but the pallets A B are set on a lever which turns on their arbor C and has a notch at the end, into which a pin P in a small disc on the verge of the balance works, being in fact a single tooth of the old rack-lever pinion. The teeth only just lock on the dead part of the pallets, and the pin and the notch are so arranged that as soon as the escape has taken place the pin slips out of the notch, and so the balance is detached from the lever during the remainder of its swing. When it comes back again the

Fig. 72.

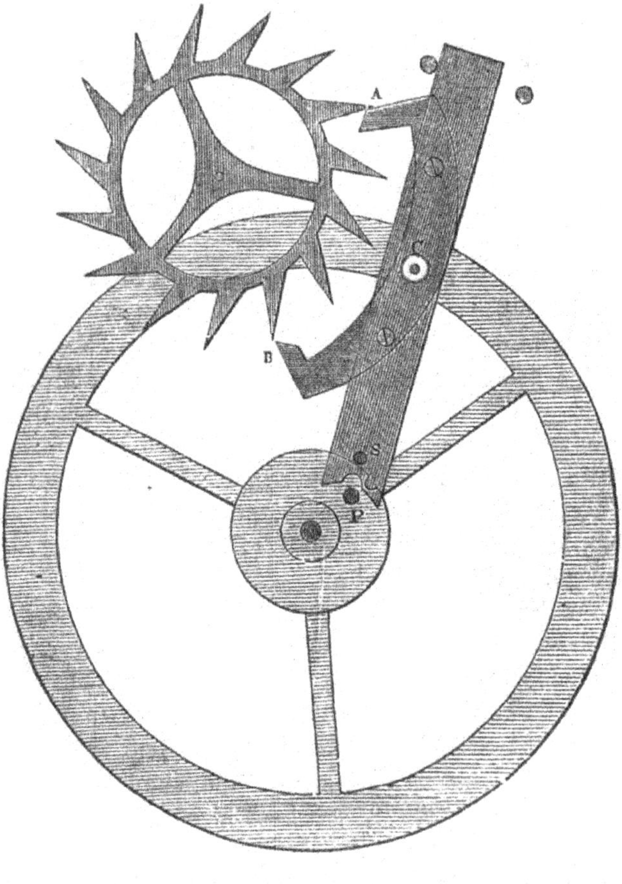

pin re-enters the notch, moves the lever just enough to send the tooth onto the impulse face of the pallets, and then the scapewheel acts on the lever and balance until that tooth has escaped and another tooth has dropped onto the dead face of the other pallet, when the pin passes out of the notch in the other direction and the balance is again free. The dead faces have a little recoil the wrong way, to prevent the risk of the teeth slipping off while the balance is free; and besides that, there is a 'guard-pin' S on the lever, which moves through a notch in the 'roller' at the same time that P moves through the lever notch: S has no actual contact, but it is there to prevent the lever from shaking back and making a false escape while the balance is free. The pallets are always jewelled, except in very cheap watches, and the staff or verge of the balance ought always to have jewelled pivot holes, and in good watches the staff of the lever also has.

It should be noticed that you cannot let a lever watch run down, for repairs or any other purpose, by merely taking out the balance, because the lever will stop it. This is sometimes inconvenient; and another part of Mr. A. L. Dennison's plan is to set the upper pivot of either scapewheel or lever in a bridge or cock, which can be taken off so as to let the train run down. The same remark applies to the chronometer escapement, but not to the vertical, horizontal, or duplex, in all of which the staff of the balance does the escaping directly on the wheel.

The balances of lever watches are generally made to vibrate about three quarters of a circle, or 270° from zero, when the watch is clean and lying flat, which always diminishes when they are in the pocket, and also as the oil thickens. Nevertheless, under a sudden jerk or twist, the balance sometimes swings quite round, and then the pin strikes the lever again, and occasionally (though very seldom) hard enough to break itself. Several inventions have been made to enable the lever to give way a little when that happens, but I cannot find that any of them have been successful, however promising they may have been in appearance, and on merely trying an experiment Several of them are described in the *Horological Journal*, ii. 53, xii. 80, and a very ingenious one in April 1882, by Mr. Hillgreen, dispensing with the lever altogether by putting the pallets on the balance arbor, not rigidly, but connected by the balance spring, so that the balance may swing as far as it likes. But as no escapement can be pronounced successful until it has been proved by long experience, I only thus refer to it.

Horizontal or cylinder escapement.—It seems strange

Fig. 73.

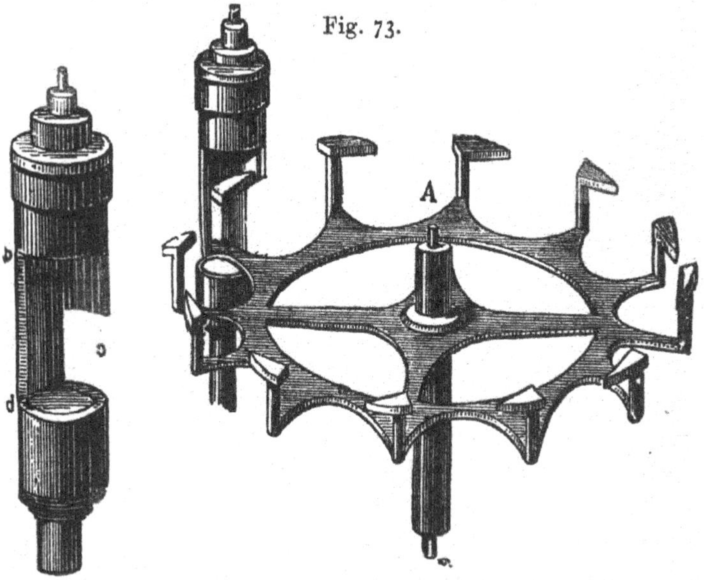

that Graham, the inventor of the dead escapement, should not have been the author of its adaptation to watches, and should have invented instead a very different one, in appearance at least, which is now almost as universally used in foreign watches as the lever, which you see originally came from France, is in English ones. The verge of the balance in fig. 73 is expanded into a comparatively large hollow cylinder in the middle, large enough to hold both a tooth of the scapewheel and a short stem on which each tooth stands; and about 150° also of the side of the cylinder is cut away, leaving only the shaded portion *a b* in fig. 73, or A E in fig. 74; in which the tooth B*b* has just escaped, giving the impulse to the balance by its oblique face acting on the edge of the cylinder as it passes out. The point of the next tooth A*a* then falls on the outside of the cylinder, just as a tooth in a clock escapement falls on the dead face of the pallets, and it rests there till the balance returns, and then the outside of that tooth begins to give the impulse until it escapes into the

Fig. 74.

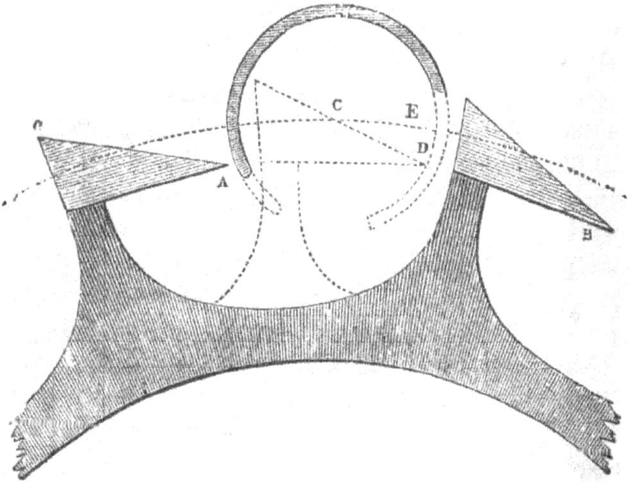

inside of the cylinder and is stopped there, as at D, till the balance returns again. The inferiority of this to the lever escapement is evident, inasmuch as the balance is always subject to the friction and pressure of the teeth; and moreover if the staff of the balance gets broken by a fall it is expensive to replace with a new one, which it is not in the lever. In the best watches the cylinder is made of a ruby, and the wheel is generally made of steel, instead of brass as in the other escapements.

There is a French escapement rather like this, called the *virgule escapement*, in which the teeth are in the plane of the wheel as usual, with small pins rising from them near the points, which act on a hollow cylinder, smaller than in the horizontal escapement. At one beat of the balance a pin only passes from the outside to the inside of the cylinder without giving any impulse. At the other beat the emerging tooth acts upon a long impulse face or pallet added to the cylinder and gives the impulse. They are very little made. Most of the foreign watches have the horizontal escapement.

Duplex escapement.—This is probably so called because the scapewheel has two sets of teeth, one for the locking and the other for the impulse. The inventor of it is not known. Its action is peculiar and requires some attention to understand it. It is even more distinctly than the virgule, a single beat escapement, for the scapewheel only moves sensibly at every other beat, *i.e.* at every vibration of the balance in one direction only: not that this makes any difference in the time of revolution of the scapewheel, because in all double beat escapements only half a tooth-space passes the pallets at each beat, whereas in the single beat escapements a whole space passes, or one tooth runs immediately into the place occupied by the previous one: the effect of which is that in a duplex or a chronometer escapement the scapewheel almost looks as if it did not move at all. V*v* is the verge of the balance, which is made of a ruby and has a nick in it through which the long teeth of the scapewheel can pass: one of them is represented as just escaping, and at the same moment one of the impulse pins S is just ready to act upon the tooth or pallet P on the verge, and so to give the impulse to the balance. By the time S has got to *t* the next locking tooth B has fallen upon the verge and is stopped there as shown by the dotted line *bv*. When the balance returns, the nick slips past the tooth *bv*, and the balance goes on till the pallet has got into the position *p*, with the nick quite beyond the tooth, and then the balance comes back again in the impulse direction; the tooth enters the nick and the escape begins, the impulse pins moving from *s* to S and from *t* to T without giving any impulse, except the small amount which the balance receives from the long tooth acting on the nick in the verge until it escapes in the position here shown, as before.

Fig. 75.

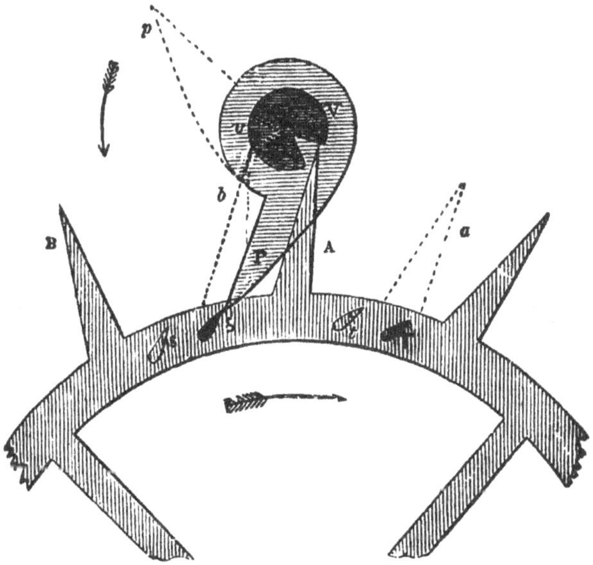

Therefore the balance here also is never free, but the cylinder on which the friction takes place is smaller than in the horizontal escapement, and what is of still more consequence, the impulse is given directly across the line of centres in the most favourable way, which constitutes the great merit of the escapement, and makes it rank next to the chronometer or completely detached escapement, which I shall next describe. But it requires great accuracy of construction, and is liable to stop from any sudden twist of the watch which prevents the balance from once swinging far enough for the nick to clear the tooth; and on the whole it seems to be going out of use, as being neither so cheap nor so safe as the lever, and not so good as the chronometer escapement, which is used in all watches requiring very great accuracy. But that also is liable to stop in common wearing, and some levers go quite as well, and I prefer them on the whole.

Chronometer, or detached escapement.--The principle of this escapement, as of the lever, seems to have been invented in France, by Julien le Roy; but also, like the lever, it acquired its now standard form in England, under the improvements of the first Arnold, who died in 1799, and Earnshaw, the latter of whom appears to have beaten with his ordinary chronometers the picked ones both of Arnold and his other rivals of that time, such as Brockbank, Emery, and others. The second Arnold, who died in 1842, was as different from the real chronometer man, as many sons are from eminent fathers, and owed his reputation first to his name, and afterwards to his partnership with old Mr. Dent, who told me that the business had fallen so low as to be a losing one when he joined it for a few years.

I do not, however, see any material difference in principle between Arnold and Earnshaw's escapements, except that Arnold's detent D T V (fig. 76) turned upon a pivot at D with a spiral spring round it, while Earnshaw's is set upon a straight spring as in this drawing, which diminishes the friction. That is doubtless a great advantage, but I cannot help thinking that Earnshaw's personal skill had a good deal to do with the superiority of his chronometers.

It should also be known that Earnshaw was the first watchmaker who had sense enough to set at defiance the vulgar and ignorant prejudice for 'high finish' of the non-acting surfaces, and to leave them 'in the gray,' as it is called. But so long as smooth work which everybody can see is easier than accurate work which few people can judge of, watches, like other things, will be got up for show. Old Mr. Dent used to say, 'We must work for the fools.' Nevertheless, in large clocks we have got rid of this folly to a great extent, and convinced people that painted iron can go better and last longer than polished brass, when the work is properly constructed.

Fig. 76.

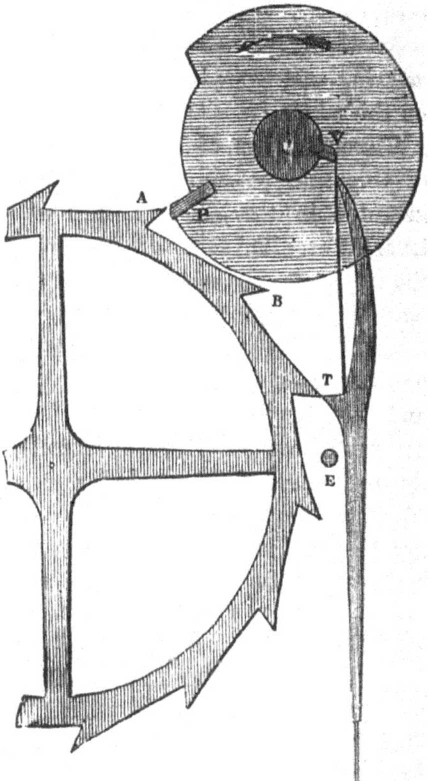

This drawing (fig. 76) shows the form in which the chronometer escapement has now been made for many years. The verge has a small tooth V upon it which pushes aside a lever or *detent* V T D as it passes in one direction, but can pass the other way without moving the detent, by merely pushing aside a very slight spring T V set on the end of the detent, which is therefore called the passing spring. The detent is attached to the watch frame by a spring at D, like a stiff pendulum spring, so as to avoid the friction of any pivots, as its motion is very small, and it has a stop T against which the teeth of the scapewheel are stopped as soon as the escape has taken place. It then rests itself against a pin E. The impulse is given exactly as in the duplex escapement, by the scapewheel teeth A acting directly on the pallet P projecting from the verge, although their form is rather different. The impulse begins as soon as the tooth V has unlocked the detent: it is shown exactly in that position in the figure. The detent stop is a little undercut for safety, as the pallets are in the lever escapement, and it is always made of a jewel in good chronometers, and so is the impulse pallet. The chronometer escapement has been made on the duplex plan, of long teeth for the locking, and short for the impulse, but that has never been generally adopted.

Fig. 77.

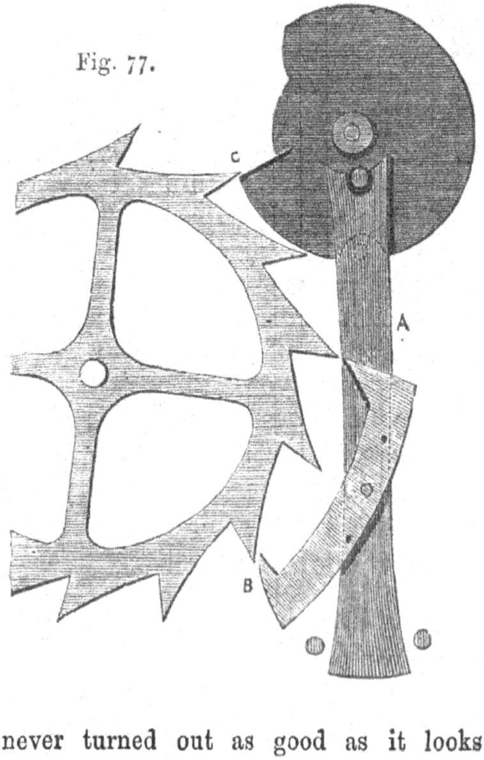

Lever chronometer escapement (fig. 77).—This is a combination of the lever and the detached escapement, which has been several times re-invented, although—or because—it has never turned out as good as it looks.

Therefore it may be useful to preserve its epitaph as a warning. Its action is this. The pallets A B (fig. 77), which look like the lever escapement, only just lock, and have no impulse. The balance is here represented at the moment of unlocking A and the impulse going to begin at C, exactly as in the common chronometer. By the time the impulse is finished the tooth now between A and B will arrive at pallet B and be stopped there. As the balance returns it will unlock B, which lets the wheel run a very little, just enough to carry that tooth past the pallet into the position shown in the figure, and transfers the locking to A again; and this is what corresponds to the passing of the passing spring in the chronometer. Things are then again in the condition ready for the unlocking of A and giving the impulse as before. This, like the French virgule escapement, wastes a little of the motion of the wheel in passing from one locking to the other without any impulse.

Tourbillons.—I saw in the *Horological Journal* that what is called the tourbillon escapement was reported to have done the best in 3 years' trials of chronometers at the Neufchâtel Observatory. It is not however really an escapement at all, but an additional piece of machinery which may be used with one escapement as well as another, like a train remontoire. The scapewheel and balance are set in a frame which revolves every minute, or thereabouts; and the object is to make the watch correct its own 'errors of position' by making the balance pass through all positions frequently. Consequently it would not be of much use for marine chronometers, which are always kept horizontal by being set in gimbals. The tourbillon frame is driven by the third wheel of the train: what would be the fourth wheel is fixed by its rim to the watch plate a little above it, the tourbillon arbor going right through them: the pinion of the scapewheel rides round the fixed wheel as a 'planet,' as the tourbillon revolves; which is the same thing as if the fixed wheel revolved and the tourbillon frame stood still, except that the scapewheel makes one more revolution for every turn of the tourbillon, or as many as if the fourth wheel teeth were increased by the number of the scapewheel pinion, that being the effect of sun and planet wheels, just as there is one more sidereal than solar days in a year (see p. 3).

Remontoires. — The escapement for which Mudge received the parliamentary reward I have already spoken of, was on the remontoire principle, and there have been others on the same principle. But they have never come to any good, and I do not believe they ever will in watches, although they are very useful in large clocks with any except a gravity escapement of good construction. But the conditions are essentially different in so many respects that no inference can be drawn from one which can be of any use for the other. It is sufficient to say that the friction of the train and dial work is the great source of variation in large clocks, and that the pendulum arc varies very little; while the variation of the arc of the balance in a watch is very large indeed in different states of the oil and in different positions; and nothing is more requisite to impress on horological inventors than this, that nothing more complicated than what is now in use has the smallest chance of being adopted. And for that reason I do not think it is worth while to increase the size of this book by describing any other of the numerous escapements that have been invented at various times.

Repeaters—or striking-watches, have so much gone out of use that no English workmen now even profess to make them. All that are sold at all here come from Switzerland. In watches as in clocks, time-keepers which profess to perform many feats generally perform them all ill, and frequently require mending to make them go at all. I shall therefore merely indicate the general nature of the machinery of repeaters: full descriptions of it may be found in Reid's book and in Rees's Cyclopædia. Watches were never made (so far as I know) to strike spontaneously: but pushing in the handle knob is made to wind up the spring of a striking part to such an extent as the position of the dial-work allows it to go, as I have already described under the striking work of clocks at p. 174. The watch then strikes the hour either on a very flat bell or on a ring surrounding the works, like those spiral springs on which small clocks sometimes strike, and the quarters, and perhaps a half-quarter afterwards, on two other rings, or on another less sonorous part of the hour-bell.

KEYLESS WATCHES, &c.

A watch may be made to wind without a key in several ways. One plan is to put a kind of gathering click to the handle knob, which pushes in and takes hold of a ratchet set on the barrel, or the fusee if there is one, and winds it up as you pull the handle out again. But this was very liable to get out of order, and was also objectionable because it pumped air into the watch, which produced condensation of moisture; and the following plan (fig. 78) was invented by a foreigner and adopted by Dent and some other makers: *d* is a wheel set on a ratchet on the barrel arbor, so that it will only turn the barrel the right way (there is not room to introduce this machinery in fusee watches of the common size); *c* in the left hand figure is an intermediate oblique bevelled wheel between *d* and a pinion *b* on the handle. It is evident therefore that if you turn the handle *a* the right way you will wind up the watch, and if you turn it the wrong way you will do no harm.

But besides this you can set the hands by the handle; for there is a small wheel *e* on the hand arbor with another *f* by the side of it on a lever *fgh*, by which that intermediate wheel can be thrown into gear with *d* as well as *e*, the lever coming through the side of the watch-case ; and then it is clear that by turning the handle either way you can turn the hands. If you have to turn the same way as serves to wind the watch you do also wind it a little (and therefore if

Fig. 78.

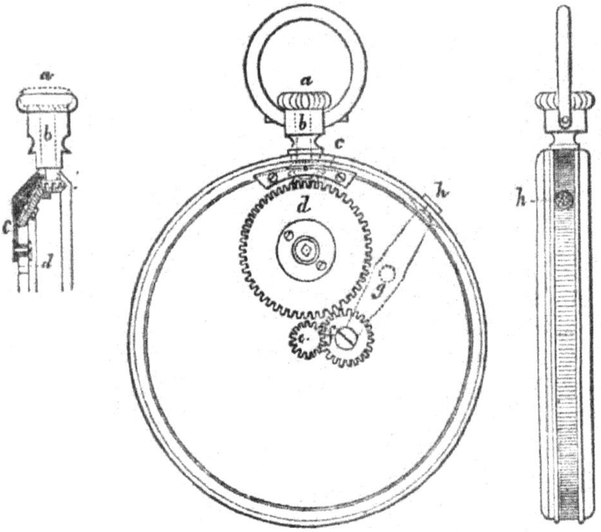

it is fully wound you cannot set the hands that way) ; but if the other way, then you do not move the barrel, as the wheel *d* slips on the ratchet.

Another keyless watch, by Mr. Kulberg, described imperfectly in the *Horological Journal* of April 1869, appears to be now more generally used than that just described. I cannot afford space here for more than a statement of its principle, and a fuller description would be of no particular use to anybody. The wheel *d* in the last figure is driven by the pinion *b* in the pendant, without the oblique bevelled wheel, and that wheel *d* (for setting) drives another, and that other the centre or cannon pinion of the minute hand, in much the same way as in

Dent's when pushed into gear. But the winding is done differently. The wheel *d* is set on what is called a platform, having a sideway motion something like a remontoire frame in a large clock; and the first thing turning the knob and wheel does is to move the platform a little (when it is not pushed into gear for setting) so as to move itself into gear with the fusee, which it then proceeds to wind. The platform is kept out of the way generally by a spring, so that neither the fusee nor the centre pinion is touched by either of the wheels on the platform.

Mr. A. L. Dennison patented another keyless watch, which is fully described in his Specification, No. 356 of 1872. There are also other methods (see *Horological Journal*, April 1874). The advantages of these modes of winding and hand-setting are that the watch has never to be opened, which lets air and dust in, and so the back requires no hinge, which never works quite air-tight, but snaps on as a separate piece; and also that the inner case or 'dome' is saved.

Self-winding watch.—Napoleon I. had a watch which wound itself up as he walked, by means of a weighted lever with a slight spring under it, which danced up and down at every step, and had a click taking into a ratchet on the barrel.

Pedometer.—A similar lever may be made to drive a train like a watch train, but without any escapement, and then it in fact counts the number of your steps and indicates them on a dial. You can adjust it for the number of steps which you usually take in a mile, and then it measures the distance you walk, in a rough and approximate way; but it ought to be understood that it is really nothing but a step-counter, and unless it is properly adjusted, and you are walking at the rate for which it is set, it is worth nothing for measuring distances accurately.

Stop watches.—It is sometimes convenient to have the means of marking a short interval between two observations with a watch, or to mark the exact time of an observation without looking off the thing you are watching. Several contrivances have been invented for this, most or all of them involving some kind of duplication of the seconds hand. In one there are two seconds hands on concentric arbors connected by a very weak spiral spring, and when you push in a pin one of them is stopped, while the other will go on for some seconds without the connecting spring having force enough to stop the watch. But this is clearly objectionable, and a better plan is to have the two hands or their arbors connected by a sort of eccentric or heart-shaped piece acted on by a spring which brings them together again either forward or backward, through whichever is less than half a revolution. Several watches of different constructions, on this *split-seconds* plan were exhibited in 1851, and others have been invented since, called 'chronographs,' and other names.

One of them is of this kind, so far as I can describe it here. Pushing in a pin for a moment drives on a ratchet wheel with a few square teeth half a tooth-space; and that raises (as we may say) a spring lever, which carries a pinion with a disc or 'roller' on it (which is always going with the train) into frictional contact with another disc on the arbor of the extra seconds hand, which is thus set going with the train. Pushing in the pin a second time drives the ratchet wheel another half space, and so lets the lever fall again, into a space between two teeth, and takes the discs out of contact, and *might* leave the hand standing at whatever point it has reached: which might be useful for some purposes, but is not in fact done; because it is of more consequence to have the hand returned to o, and you can look where it has reached before returning it. That is done thus:—the second movement of the pin also brings a 'jumper' spring to bear on the heart-shaped piece which is fixed on the hand arbor (as before described), and so sends it backward or forward to o according as it is left before or after 30 sec., or half way round the dial.

In watches of this kind, at least in some shown to me by Lund & Blockley of Pall Mall, the stop-seconds hand is central, which gives it the benefit of the full size of the dial, and enables the space of each ordinary minute to be subdivided into fractions of a second corresponding to the time of vibration of the balance. For this reason also a single beat escapement, like the chronometer or lever-chronometer or duplex, is not so good for these split-seconds as a double beat one, such as the lever or horizontal or the old vertical, in which the scapewheel moves equally for every beat of the balance, and not only for alternate ones. The falling of a small time-ball in a suitable frame (see p. 162) is easily made to push in the pin the first time, and when you take up the watch afterwards you see by the difference between the two seconds hands how much it is before or after Greenwich time.

There is also another perfectly different plan, which enables you to make a mark on the dial at the exact time when you push in the pin at the time of observation. In fig. 79, D D is the dial of a large watch, with the seconds hand E A B in

Fig. 79.

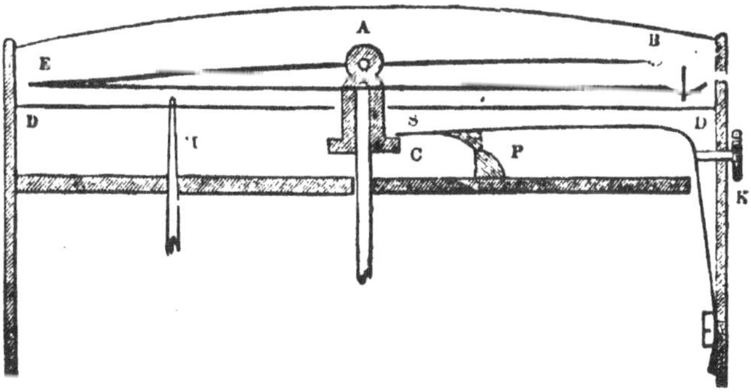

the middle: the hand is double, and the lower piece of it ends in a little spoon with some thick ink in it and a hole in the bottom through which a point from the upper hand

E A B can pass and make a mark on the dial. That hand is pulled down very suddenly by a lever D P C which slips over a stop P of a shape difficult to describe, being pushed in by the knob K, and is immediately thrown out of contact again with the link A C, by means of which it pulls down the hand.

Dials of watches and of small clocks are either made of gold or silver (which soon tarnishes) or of copper covered with enamel, which is a kind of glass. Such dials with black hands are more distinct than any other kind, and if the black figures are burnt in with black enamel, the dials would be everlasting and never want painting. I noticed the absurdity of gold hands on gilt dials before, at p. 137. Some of the public clocks in Paris have enamelled dials, which are far more expensive than white glass ones would be.

Watch cases.—I am not aware that there is anything else of a rudimentary character, or belonging to the principles of watchmaking, which requires notice. Minute details of watchmaking can be learnt by nothing but experience; whereas clockmaking is easily learnt by any person of mechanical ability. Case-making is not horology, and I have nothing to say about it, except that the cases of what are called hunting watches, which fly open with a spring when you press the handle, cannot be so close against air and dirt, as those which snap tight together. Persons who are afraid of breaking their watch-glasses, may be tolerably safe with 'half hunting watches,' which have only a small and strong glass in the middle of the cap, which may then fit tight, and need never be opened except when the hands want altering. The closeness of the case makes a great difference in the time a watch will go without cleaning. They generally want at least cleaning about every two years; though very good ones, with all the escapement work jewelled and well-made, will go 4 or 5 years with no material variation of rate, which I think may be regarded as one of the greatest triumphs of mechanical art. At the same time it should be remembered that letting either watches or clocks go too long without being cleaned and oiled is very bad economy; for as soon as the oil is all gone wearing out of the pivots begins.

American watch-factories.—In the *Horological Journal* for April 1869, and January 1873, there are accounts of some of these factories, where watches are made by machinery, so that every piece will fit every watch of the same pattern; on the same principle as Hobbs's locks. There can be no doubt in the mind of any one who understands machinery that this is the best, as well as the cheapest way of making machines which require precision and uniformity. Adjustments will after all have to be made by hand, and a machine which has always to be in motion is not quite on a level with a lock. The degree to which machine-making of machinery can be carried cannot be defined *à priori*. To a certain extent the same thing is done at the celebrated watch-factory of Messrs. Rotherham at Coventry, and also at Prescott in Lancashire, where watch 'movements,' *i.e.* the train set in the frame, are chiefly made. I can give no description of the American machinery here, but its elements are stamping plates and the holes in them and the wheels, and then cutting the teeth of many wheels together. Although labour is dearer in America than here, this machinery enables them to undersell English watches of the same quality, as the Swiss also do with cheaper labour and more organization, though with less use of machinery; and if our English makers do not bestir themselves they will lose the trade in all but the best watches, as they have already lost that of both cheap and ornamental clocks.

BELLS.

THERE is still less to say of the history of bell-founding than of clock-making. Indeed it is hardly a progressive art; for as soon as the right shape and composition are discovered, and the means of making a sound casting, there is nothing more to do, except to see that that is done. Small bells, probably not cast but hammered, and of gold, are, we know from the Bible, as old as the time of Moses; and some such old bells of sheet copper and even sheet iron are said to have been found at Cologne and in Ireland. The Romans also had them apparently. They can only be very thin, and of a very inferior sound to our thicker cast bells. The nearest modern approach to them is the Chinese gong, which is made exactly of our bell-metal, but hammered, which that metal oddly enough admits of when it is heated and cooled suddenly in water. Gongs, like thin bells, sound very ill except when you are near them, and have less power than the same weight of metal cast into a bell of a much higher note. The following is the short history of cast bells given in some of the Encyclopædias:—

'The large bells now used in churches are said to have been invented by Paulinus, bishop of Nola in Campania, about the year 400. They were probably introduced into England very soon after. They are mentioned by Bede about the close of the seventh century. Turketul, abbot of Croyland, who died about 870, gave a very large bell to that abbey; and his successor Egelric cast a ring of six others. Pope John XIII. consecrated a very large new cast bell in the Lateran Church in 968.'

It is said in *Otte's Glockenkunde*, the latest German book on the subject, but one which does not contain much information of a practical kind, that bell-founding flourished in all the monasteries in the twelfth century, and that there were travelling bell-founders who went about casting bells where they were wanted. I suspect this practice went on very much later; for it is impossible to believe that there were regular bell foundries in anything like the number of places from which the 'legends' on still existing bells testify that they have come. The great clock-bell on the south-west tower at Canterbury was recast in the cathedral yard as lately as 1762. Indeed the carrying of very large bells along the roads of old times would have been a more serious affair than casting them. A great deal of curious matter of all kinds, historical, practical, and artistic, has been collected by the Rev. H. T. Ellacombe in a quarto volume on 'The Bells of Devon and Bells of the Church' generally, who rings peals himself at above the age of eighty. I shall confine myself to the practical.

The founders of church bells in England have for many years been very few. The most famous bell-founding family on record were the Rudhalls of Gloucester, who are known to have flourished there from the time of Henry VIII., and some people think still earlier, till about 60 years ago. They cast the Westminster Abbey and Chester Cathedral bells, and the Magdalen bells at Oxford (of which the large ones are too thin), and a great many others. Farther back the Purdues of Salisbury seem to have been equally famous for a long time. One of them cast the great single clock bell, called Peter, of Exeter, and some of the bells of that the largest ringing peal in the kingdom, though now exceeded by those of Manchester and Bradford, which are only chimed by machinery—a very inferior mode of sounding bells. Miles

Gray was the great founder of the eastern counties in the seventeenth century: the famous tenor bell of the noble church of Lavenham is his, which is considered the best in England for its weight, 24 cwt. The firm of Watts, Eayres and Arnold existed in Leicester and St. Neots for two centuries, and is now represented by Taylor, of Loughborough, whose foundry now has more than the old celebrity of the Rudhalls and Purdues, and the more modern Whitechapel foundry of Phelps, Lester, Pack, and Chapman, and several generations of Mearses, until the last destroyed its reputation by a multitude of very bad bells, and finally their Big Ben exploit, before spoken of. The only other founders of bells or peals worth notice now are the Warners, of Cripplegate, whose Big Ben I. was very nearly as large and heavy as Taylor's 'Great Paul,' but very inferior to it in sound and casting. They made the quarter bells at Westminster, of which the largest is about 4 tons, and several others for Leeds, and other town-hall clocks, of about the same size, and the whole, or the greater part, of several large peals at Doncaster and Kensington parish churches, and St. Nicholas, Aberdeen, Fredericton Cathedral, and many smaller ones. Gillett and Bland, the clockmakers at Croydon, who have been mentioned before, now make their own clock bells, and, I believe, ringing peals also, but I have not heard any of them. There is a nice new peal of eight at St. Andrew's, Wells Street, by Lewis, an organ-builder also, at Shepherd's Lane, Brixton, but too thin, the tenor being E flat, and only 21 cwt., which should be F. The firm of Moore, Holmes, and Mackenzie, at Redenhall, Norfolk, have a patent for making the clapper of a 'raised' bell fall out of contact as soon as it is struck; but I know nothing more of them; nor of several others who advertise as bell-founders, which may mean anything. Tom of Oxford, by Hodson in 1680, is decidedly the worst of all our great bells, though he also made good bells of moderate size close by at Merton College. The tradition that Tom came from

Oseney Abbey is a fiction; and even if some predecessor did, it is a melancholy fact there is no personal identity and very often no resemblance between an old bell and its recast successor. The Oseney bells were once called the best in England, and were moved to Christchurch Cathedral, but have been all recast long ago.

Great Tom might be recast for nothing into a very much more powerful and an infinitely better bell (for it hardly could be worse) by reducing the weight about a third, though, if the College does recast it, they are not likely so to reduce it for the sake of a hundred pounds or so. I wonder none of the Christchurch men have managed to get it cracked before now, and so compelled the College to recast it. Mears's great bad bell of York ought to be treated in the same way; and if that were reduced a third in weight, which is a little more than a note in sound, it would still keep its rank as the third largest bell in England. It is all but useless now. Every now and then there is a movement for getting something done with it, and getting the clock made to strike on it (which the present clock is not strong enough to do), but it always dies away again. Whatever is the price of copper, and therefore of bell-metal, the founders invariably charge two guineas a cwt. for re-casting, *i.e.* they give you back at that price as much metal as they receive; and as the price of new bells generally fluctuates round seven guineas, you may expect to get a new bell about five-sevenths of the weight of the old for nothing.

Bryant of Hertford was a celebrated maker of both clocks and bells early in this century. There are some nice peals of his at Waltham Abbey, Saffron Walden, and St. Alkmund's, Shrewsbury, a town full of peals, both bad and good. The best peal of 5 bells I ever heard, at Castle Camps in Cambridgeshire, was by Dobson, who lived at Downham but failed and died in the Charter House. He is said not to have been equally successful with larger bells.

All these have disappeared, besides many smaller ones; among which I ought to mention Daniel Hedderley of Bawtry, as the founder of the fine old peal at Doncaster Church, which was inferior to none, but is closely imitated both in weight and tone by the new one of 1858 by Messrs. Warner, which will be given presently.

The art of bell-founding is evidently no better understood abroad than here. There were some beautifully cast bells from various places on the continent in the 1851 and 1862 Exhibitions, but they were nothing remarkable in sound, and the best in appearance are by no means always the best in sound. Notwithstanding the laudation of a Belgian foundry by a musical amateur of no experience in bell-making, the Boston bells cast there (p. 210), and the heavier peal which the Duke of Westminster was advised by another musician to get there for his chimes at Eaton, are by no means equal to many modern English bells of corresponding size, and the small ones are a decided failure. That however is from another cause which will be noticed afterwards, but it proves that the Belgian founders have no special secret. The art had reached its lowest point about 30 years ago, when three bad peals in succession were cast for the Royal Exchange. I doubt if there was a fairly good peal of bells cast from about 1830, or perhaps earlier, till the Doncaster one in 1858. Certainly the previous Doncaster peal of 1835 by Mears, which was burnt with the church in 1853, was very inferior to the older one of 1722, and to many others of the last century and the earlier part of this.

When we began the Westminster bell business in 1855, I found there was as good as nothing of a practical kind to be learnt from books, and what little there was was contradictory, and some of it evidently wrong, and not always right even on the simple arithmetical relation of the musical notes to the different sizes of *similar* bells (using that word in its mathematical sense, of all the proportions varying alike). Accordingly, when I had to undertake the designing of those bells, for the reason which I had to state at p. 269, I stipulated for the power to make experiments as to thickness, shape and composition; and in every one of them I found reason to alter the traditional practice of the founders more or less. The result has been that nearly as good bells can now be made as ever, and they would be quite as good if it were possible to get copper of the old quality; but we had a convincing proof that it is not, in the fact that several pieces of old bells that were analysed at the School of Mines contained fully ·25 of tin, or tin in the proportion of 1 to 3 copper, while our experiments showed that modern copper will bear no such quantity of tin without being too brittle to be safe for ringing. This is only one of a thousand proofs that the effect of science upon 'common things' is simply to make them worse than they used to be. No doubt in this case the reason is that more copper is got out of the ore. So also they have found out a way of spoiling lead by robbing it of the little silver which nature has put into it as a protection from being so easily convertible into sugar of lead, which is poison, by the action of soft water with anything that turns it acid, and especially if it is laid on oak, which the lead of old church roofs always was. Lincoln's Inn Hall had to be re-roofed with slate because the lead on oak had decayed into holes in less than 20 years, while the old lead on many church roofs of oak has lasted for 500 years.

It is surprising how little even musicians bear in mind the distinction between making bells in tune with each other, which a set of cast iron pots might be, and making them individually good in tone: and the best musician in the world is no judge of that unless he knows by experience what sort of tone is attainable by good bells of something like the same size as those he has to judge of. Mears's two first peals at the Exchange were duly certified by musicians of repute, and I dare say they were in perfect tune, but most of them were thoroughly bad bells—so bad that after being twice paid for they were condemned to be recast again in spite of the musical certificates. The fault of the present peal by the late (not the present) Mr. Taylor is that they are too thin, as I warned him that they would be if he aimed at getting those notes with that weight of metal. I suspect also that the composition is too soft, or has too little tin, as that was a very common fault of modern bells until the Westminster experiments.

Fortunately a peal of bells not quite in tune can be tuned, although the tone or quality of a bad bell cannot be mended. They are made flatter by turning a little off the inside of the sound-bow or thickest part; and they can be sharpened a very little by cutting off the edge so as to reduce the diameter of the mouth; but this blunting of the edge is very apt to spoil the bell, and it is seldom done now.

Notes of bells.—The whole theory of the designing of bells to produce the required musical notes is deduced from this mathematical law—that the number of vibrations in a second, in *similar* bells varies as $\frac{(\text{thickness})^2}{\text{diameter}}$; or in other words, the depth of the notes or the time of vibration varies as $\frac{\text{diameter}}{(\text{thickness})^2}$. Consequently, if you want to make (a very bad thing) a peal of bells all of the same absolute thickness (not the same proportionate thickness), their other dimensions must be as the square roots of a set of numbers in the inverse ratio of the vibrations belonging to the proposed notes. But if the thickness itself varies with the diameter, then the sizes will be simply as those numbers; and therefore all the dimensions of 'a peal of 8 tuneable bells,' according to the old phrase, which means a peal sounding the 8 notes of the diatonic scale, will be in this proportion—

$$1,\quad \frac{8}{9},\quad \frac{4}{5},\quad \frac{3}{4},\quad \frac{2}{3},\quad \frac{3}{5},\quad \frac{8}{15},\quad \frac{1}{2};\ \text{or}$$

$$60,\quad 53\tfrac{1}{3},\quad 48,\quad 45,\quad 40,\quad 36,\quad 32,\quad 30;$$

and so on for a larger number of bells, each being half the size of the octave bell below it. These are the lowest numbers which will represent the inverse ratio of the vibrations of the 8 notes without more fractions; and they are easy to remember as a standard, from which any others may be deduced, these being the diameters in inches of a peal of bells in the key of D flat, of the best weight or thickness for such a peal, supposing that the small bells were not made thicker for their size, and therefore larger for their notes, than the large ones, as they always are, to prevent them from being overpowered when they are all rung together, as will be seen from the peals which I shall give presently.

It may save some trouble to observe that in designing peals of bells we have nothing to do with what is called musical *temperament*, which was long a vexed question in organ tuning, inasmuch as a peal of bells is always played or rung in the same key. That question would arise if you took the 7th and 6th of a peal of 8 bells (counting, remember, from the smallest) to make them the 8th and 7th of a smaller peal; for in that case you see the 53⅓ and 48 inches would not be in the right proportion, of 9 to 8, but the 53⅓ ought to be 54; though both would be called E flat, and the error is too slight to be perceived except by very good ears. The same occurs in playing the Westminster and Cambridge, or the Doncaster quarters on the 2nd, 3rd, 4th, and 7th of a peal of 8, as described at p. 203; for the sizes of those bells will be, as we saw just now, 32, 36, 40, 53⅓, whereas the 36 in. ought to be only 35·55 to make it exactly in tune as the second of a peal of 6 or of 10 bells. Without regard to temperament, bells (on any given scale of thickness) two whole notes apart are always as 4 to 5 in diameter; 1½ note apart, as 5 to 6: 3, as 3 to 4; but an interval of 3 notes (*i.e.* 4 bells) of the diatonic scale always includes one half tone, such as B C or E F. Three bells must have no half-tone interval.

Weight of bells.—The weights of *similar* bells, *i.e.* of those in which the thickness and all the dimensions keep the same proportion to each other, vary as the cubes of the diameters, or any of the other dimensions; and therefore the weights of a peal of 8 such bells would be in this proportion (making the tenor 100 for facility of calculation)—

100, 70·23, 51·2, 42·2, 29·63, 21·6, 15·18, 12·5.

The Westminster bells, which would be the treble, second, third, sixth, and tenor of a peal of ten, are very nearly in this proportion; for in them the thickness does vary as the other dimensions, except that the smallest was made a little thicker than the others for its size, in order that its sound might be strong enough. Bells 1½ note apart are as 58 to 100 in weight; and half a note nearly as 5 to 6.

But we have still to ascertain what is the proper weight for any given size or note. From the practice of some of the modern bell-founders you might suppose that it may be almost any weight you please, at least within very wide limits, as one sees bells of about half the weight of old ones and yet of the same note; and then people are surprised that they sound worse.

Postponing for the present the consideration of bells to ring in peals, any one who takes the trouble to compare the weights and the cubes of the diameters in the list of the principal large bells of Europe at the end of this book, will see that the bell-founders of all ages and countries have agreed in fixing rather narrow limits for the variations of weight in proportion to the diameter of their bells. And this is by no means from any blind following of each other; for there is a good deal of variety in the shape and the distribution of thickness over the different parts of the bell, although it all ends in this near agreement of the proportions of weight and size; which would be nearer still, but for the foreign bells being generally taller than ours.

Taking 6 ft. diameter as a convenient standard to reduce them to, you will find that the *least* weight for a bell of 72 inches would be 72 cwt., which is easy to remember; or in smaller figures, 9 cwt. (= 1008 lbs.) for 3 ft. diameter. And such a bell will be nearer B flat than any other note, according to the proposed universal pitch, in which A has 880 vibrations in a second, or that number multiplied or divided by some power of 2. The diameter of bells on that scale is about 13 times the thickness of the sound-bow or thickest part; for if some of them are rather thinner there, they are thicker above the sound-bow, and so the total weight is the same. But nearly all the great European bells, as you may see from comparing their sizes and weights, are very considerably heavier than that scale, and the average *modulus* for them may be taken at 4 tons for 6 ft. diameter, or 10 cwt. for 3 ft., and in that case they will be nearly a note higher. The Westminster bells are very nearly on this thicker scale, or their diameter is nearly 12 times their thickness; and so is the new Great Paul bell: indeed apparently on a still heavier scale, but it is somewhat taller for its width than the other great English bells. Some of the bells in the list (if their recorded weights are right) are still heavier than this, which I shall generally call the 12, and the other the 13 scale; which last gives the before-mentioned size of 5 ft. for a D flat bell, weighing about 42 cwt., instead of 30 or less, as D bells are often made now, but never with my consent.

This 13 scale agrees more nearly in weight than any other simple proportion with that used by the old bell-founders in the large bells of peals, which were made rather thinner than large single bells, and the small ones thicker, to prevent them from being overpowered by the large ones. The tenor of the great peal at Exeter weighs only a 16th less than its weight on the 13 scale of the Doncaster peal; and 'the great bell of Bow,' and the tenor of York Minster, and the similar one at Sherborne (lately cracked and recast) are still nearer to it; although any one measuring their thickness at the sound-bow only might fancy they were on a thinner scale. The Exeter one, for instance, is only $\frac{d}{15}$ at the sound-bow, but the waist is 2 inches thick, or $= \frac{d}{36}$, and not $\frac{d}{45}$ as some modern founders would make it, and the weight is that of a rather thick bell. I am satisfied however that no sound-bow ought to be so thin as that proportion, and that even thickening the waist does not compensate for it. There is a fulness and softness in the sound of a thick bell which a thin one never has. The old bell-founders evidently knew that it is a law of nature that a given weight of bell-metal is only capable of sounding a very narrow range of notes with good effect; and if you infringe that law and make your bells thinner for the sake of getting deeper notes out of them, you are as certain as usual, in fighting with laws of nature, to pay for it by a more than equivalent loss in the quality of the tone.

And it happens that this loss is even greater now than it would have been a century ago, on account of the difference in the quality of the copper which I have already spoken of. It is less tough in working, capable of holding less tin without becoming too brittle, and apparently incapable of a certain softness of sound which even thin old bells sometimes have, but thin new ones never. Indeed it must be admitted that some of the finest old tenors are what we should now call thin. They would probably have been better still if they had been thicker. As a peal of bells is a luxury, meant to give pleasure to those who listen to them, and not a necessary of life, it is astonishing that people will go on raising large subscriptions for them without taking the least tronble to ascertain that they get what is really the best thing for their money, and forgetting that you pay for the same weight of metal whether the bells are thick or thin, only in one case you get them of the right notes for their weight, and in the other case wrong, and so make the bells bad instead of good.

The following is a table of diameters and weights of an octave of bells, including all the half-notes, on the 13 scale of thickness, and also on the 12 scale; from which you may easily interpolate a 12½ scale; and larger or smaller bells than these may be at once deduced from those an octave above or below them in the same scale. Single bells, and especially small ones, should always be on the 12 scale, or very near it; but the tenors of peals on the 13 scale. The smallest bells of large peals are generally still thicker, sometimes as thick as a 10th of their diameter, and often an 11th; and therefore they exceed any weights deduced from this table, as you may see in the large peals which will be given presently:—

13 Scale.		Diameter.	12 Scale.	
cwt.		inches.		cwt.
72	Bfl.	72	C	80
60	B	67½	Dfl.	66
51	C	64	D	56
42	Dfl.	60	Efl.	46½
38	D	57½	E	41
31	Efl.	54	F	34
26	E	51	Gfl.	28½
21	F	48	G	24
18	Gfl.	45	Afl.	21
15	G	43	A	17
12½	Afl.	40·5	Bfl.	14
11	A	38·4	B	12
9	Bfl.	36	C	10

Until about 14 years ago the largest ringing peals in England, and therefore in the world, was that of Exeter Cathedral, which is the largest still, and those of York Minster, Bow Church, and St. Saviour's, Southwark, otherwise called St. Mary Overy, which were all practically of the same size. The Bow peal is the best of the three, the new York one, which was cast by Mears after the fire of 1840, caused by a clock-maker leaving a candle burning, being very inferior to the old ones from the same foundry and patterns. The Southwark peal is half a note lower, being a little thinner, and for that reason worse.

EXETER CATHEDRAL BELLS, MOSTLY CAST IN 1676.

cwt.	qr.	lb.	ft.	in.		
67	1	20	5	11½	Bfl.	10
46	3	14	5	4½	C	9
38	1	16	4	11	D	8
30	1	12	4	7	Efl.	7
21	0	0	4	1	F	6
15	0	0	3	10	G	5
12	2	0	3	4½	A	4
10	1	2	3	1½	Bfl.	3
9	3	20	3	0	C	2
8	3	20	2	9½	D	1
260	2	20				

BOW CHURCH, 1762, AND OLD YORK MINSTER, 1765.

	ft.	in.	cwt.	qr.	lb.
C.	5	4½	53	0	25
D.	4	9½	34	2	6
E.	4	3½	26	0	13
F.	4	0½	21	0	23
G.	3	8	16	0	4
A.	3	5	13	2	22
B.	3	2½	12	0	7
C.	3	0	10	0	0
D.	2	10	9	1	5
E.	2	8½	8	3	7
			205	0	0

The two largest modern ringing peals are those of St. Paul's and Worcester Cathedrals, both by Taylor, of the patterns and composition which I arrived at as the best after the experiments made for the Westminster Bells, modified a little by some later ones, as I shall explain farther on. The St. Paul's peal is on the whole better than Exeter, of which some of the bells are bad; and the Worcester peal is quite equal, if not superior to that of Bow.

St. Paul's, 1878.	in.	cwt.	qr.	lb.		Worcester, 1869.	in.	cwt.	qr.	lb.	ins. thick.
Bfl. .	. 69	62	0	0	12	Dfl. .	. 63	50	0	0	4·8
C .	. 61¼	44	2	0	11	Efl. .	. 56	34	2	12	4·07
D .	. 55¼	30	2	22	10	F .	. 50·4	26	1	8	3·7
Efl. .	. 52½	28	0	7	9	Gfl. .	. 47¼	21	2	11	3·44
F .	. 47⅝	22	1	18	8	Afl. .	. 42½	15	2	11	3·1
G .	. 43¼	16	2	21	7	Bfl. .	. 38½	12	0	0	3
A .	. 39⅝	14	0	4	6	C .	. 36	11	0	24	3
Bfl. .	. 38⅝	13	2	14	5	Dfl. .	. 35	10	1	21	2·9
C .	. 36⅜	11	3	21	4	Efl. .	. 32½	8	3	0	2·8
Dfl. .	. 34	10	0	3	3	F .	. 30½	7	3	10	2·7
Efl. .	. 32½	9	1	15	2	Gfl. .	. 29½	7	0	22	2·44
F .	. 31	8	1	16	1	Afl. .	. 28	6	3	19	0
	Total	271	3	1			Total	212	2	2	

All the 5 largest bells at St. Paul's are only on the 14 scale of thickness, and they would have been better on the 13 scale, but the tenor inconveniently heavy to ring. On the other hand the 9 smallest are all practically 3 in. thick, which also I do not approve of nearly so much as the graduated thickness of Worcester, except the treble there, which gave a great deal of trouble from the highness of the note, and is not satisfactory. But the two trebles there are simply a mistake, and the peal of 10 sounds a vast deal better than the peal of 12, as is always the case. It is not so bad at St. Paul's, because the notes are lower and the bells heavier and slower. But even there the 12 sound confused, and inferior to 10. At Worcester three extra bells have been added to make some more half-notes, D A C, for the chimes, which I omit in the list of the peal. In both cases I shall notice the great single bells separately. The Rochdale Town Hall has a peal for chimes only, almost identical with Worcester. It should be noticed that the addition of an odd 13th bell only half a note below the treble, instead of the whole one, makes a useful small peal of 8, omitting the 4 largest of the great peal, subject to the difficulty of making a satisfactory treble of a higher note than F or F sh. at the highest. That bell at Worcester is a fair one, though the G sh. treble is not. The upper 6 of a peal of 10 make a proper peal of 6.

The largest modern peals, all by Taylor, are those of Manchester and Bradford Town Halls, and of St. Paul's Cathedral; but the Manchester peal cannot be rung, only chimed by machinery. The St. Paul's peal is really a more powerful and better one than Exeter, though that is rather larger; but some of the bells are too thin and otherwise inferior. Sir Christopher Wren, very unlike most modern architects, who will not condescend to learn anything of such matters, but consider themselves qualified to give orders for anything that is wanted, whether architectural or not, had prepared a tower capable of bearing such a peal in full swing with perfect safety. It does not even shake sensibly under the ringing, which is the case nowhere else that I know of with a moderately heavy peal. The next largest modern ringing peal is that of Worcester Cathedral, especially without the two trebles. The St. Paul's peal, being heavier and slower, bears them better; but even there—and everywhere—it is impossible to hear the 12 as distinctly or as pleasantly as 10. For playing tunes there is no objection to any additional bells, as half-notes outside the diatonic scale, which alone is fit for ringing. There are two other chiming peals still larger, which I ought to give, at the Bradford and Manchester Town Halls (see next page). Both of those also are by Taylor. This Manchester peal is the best proportioned of them all, except that even here perhaps the small bells vary too little. You see the large bells are on a thicker scale than St. Paul's, while the small ones are thinner and lighter for the same notes, but quite heavy enough, as I proved by the Doncaster peal (p. 351).

It is a pity that Town Councils will not make up their

BRADFORD, 1873.

	in.	cwt.	qr.	lb.	
A . .	77¼	87	0	0	12
B . .	67	59	2	0	11
Csh .	59⅜	41	2	9	10
D . .	56	38	0	10	9
E . .	50	24	1	14	8
Fsh .	45⅜	18	3	14	7
G . .	43½	15	3	25	odd
Gsh .	41½	13	3	23	6
A . .	39½	12	2	22	5
B . .	35¼	9	0	22	4
Csh .	33¾	8	2	17	3
D . .	33	8	0	11	2
E . .	30¾	7	3	2	1
Total	tons	17	3	10	

In the Manchester peal the 10 numbered bells are hung for ringing. The tenor has been recast rather larger than at first, after being cracked by the clock hammer, which weighed nearly a 30th of the bell, which is decidedly too much.

MANCHESTER, 1876.

	in.	ringing number.	cwt.	qr.	lb.
G . .	91½	..	162	3	0
A . .	80	..	100	2	0
B . .	70	..	71	0	0
C . .	65¾	10th	52	0	0
Csh .	61¼	..	43	2	0
D . .	58¾	9	39	0	0
Dsh .	55	..	31	2	0
E . .	52⅜	8	27	0	4
F . .	50¾	7	23	0	11
Fsh .	47¾	..	21	1	7
G . .	45¼	6	17	1	7
Gsh .	43¼	..	16	0	6
A . .	41½	5	14	1	3
B . .	37⅝	4	10	0	14
C . .	36⅜	3	9	3	14
Csh .	34⅝	..	8	3	0
D . .	33¾	2	8	2	9
Dsh .	32	..	7	3	14
E . .	31½	1	7	2	7
F . .	33½	..	7	1	5
Fsh .	29½	..	6	3	4
Total		tons 34	5	2	3

minds to have peals of bells before instéad of after their towers are built, and that architects cannot learn that bells sound worse for being crowded together, and better the higher up they are: in other words, that bell chambers ought to be as large and as high as possible, as they were in old cathedrals and large churches. Now they seem to resort to every possible device to make them small, and the towers so weak that even if they are large they will not bear a peal of large bells, as was the case with that of St. John's Chapel at Cambridge, which is as large inside as Worcester Cathedral tower. The Bradford tower is no wider than that of a moderate old village church tower, and so is that of Eaton. Rochdale is still smaller.

I have given them all in flats or sharps as they were sent to me, though there is no real difference between bells called D flat and C sharp, except that D flat sometimes means that the bell is nearer D than C. Of course the notes very seldom happen to be exactly those of the received standard pitch, and there is no advantage in their being so.

I next give two modern peals of eight, of successive sizes. They would differ by only half a note and not (nominally) by a whole one, but that the 'sweeps' were rather different, and I don't believe they really differ by a note. But as I have said throughout, notes are of no consequence provided only the bells are in tune among themselves. The Doncaster peal was cast by Messrs. Warner at the same time as the quarter bells of Westminster. The Croydon one, by Taylor in 1870, is substantially the same, but a little heavier, especially in the smaller bells, which the bell-founders are always struggling to increase, with no benefit that I can see. The Burton peal is at St. Paul's church there, which was built by Mr. Bass, substantially from my design, as well as the bells. Ossett is near Wakefield:—

Doncaster, 1858, Warner.	in.	cwt.	qr.	lb.		Burton and Ossett, Taylor.	in.	cwt.	qr.	lb.
Efl.	54	30	1	0	8	F	51	25	3	21
F	48	21	0	24	7	G	45	19	0	0
G	43¼	15	1	10	6	A	40½	13	1	9
Afl.	41	13	0	0	5	Bfl.	38½	11	3	14
Bfl.	37	9	0	0	4	C	35	9	3	9
C	34	8	0	10	3	D	32½	7	1	25
D	32¼	7	0	11	2	E	30½	7	1	9
Efl.	31	6	2	5	1	F	27½	6	2	4
Total		110	2	12		Total		101	1	10

It is not worth while to insert any smaller peals of 8: indeed no peal of 8 with a tenor less than about 4 feet diameter, and 21 cwt., is worth having. A nice peal of that size, by Warner, was hung at Hunslet Church, Leeds, a few years ago, of the total weight of 85 cwt. There are plenty of peals of 8 with E tenors weighing 17 cwt., and even less, made for foolish people who insist on having bells of the deepest possible note and for the smallest possible price, and lazy ringers like them better than the good old-fashioned heavy bells, but they are miserable things.

I next give two new peals of 8 and 6, one unusually heavy for its size for this reason. The three largest bells were made at first without the intention of adding any others, and I determined to have them heavy for their size to sound well alone. They were accordingly made even thicker than a 13th. Afterwards the other three were added, and then a separate bell to make the Cambridge and Westminster quarters (see p. 209), and then a treble to make a ringing peal of 8 was given. That church at Headingley was also designed and partly built by me. The Gainford peal is a particularly good one, made by Taylor for a fine old church which has been admirably restored. I consider that the model size for an ordinary peal of 6, and very nearly as small as they will bear without running into objectionably high notes. There is a very nice peal by Taylor at Coddington, near Malvern, half a note higher than this, which is quite the smallest that will do :—

St. Chad's, Headingley.						Gainford, 1865.				
	in.	cwt.	qr.	lb.			in.	cwt.	qr.	lb.
G . . .	46	19	0	15	6	Afl. . .	40½	12	0	11
A . . .	41	14	0	0	5	Bfl. . .	36	8	3	18
B . . .	37½	10	0	11	4	C . . .	32½	8	0	18
C . . .	36	9	3	24	3	Dfl. . .	31	8	0	0
D . . .	34	9	0	9	2	Efl. . .	30	6	2	17
E . . .	32	7	2	19	1	F . . .	29	6	1	14
Fsh . .	30	7	0	21						
G . . .	29									
Total		77	0	15		Total		50	0	22

The tenor of the beautiful little peal of 5 at Castle Camps, cast by Dobson in 1828, is 40 in. and 11 cwt., and the treble

5 cwt. 1 qr.; but nobody gets peals of 5 now, for they cost very little less than 6, and sound much worse.

Although some very good old bells are thinner than that 13th of the diameter which I have prescribed as the least thickness to be allowed, and no such bells can be made of modern tin and copper, as I have already said, yet the *best* modern bells on the 13 scale are better than any old ones on the 15 scale. A fashion arose in the last century of making the trebles enormously thick, even with the tenors thin, and some of the bell-founders of the present time require constant repression to prevent them doing so. I was assured that the trebles of the Doncaster peal would never be heard with the large bells on my scale of thickness; but they are remarkably distinct. The treble is on the $d = 11\ t$ scale. In peals of 10 they may go up as high as a 10th, but anything beyond that makes the bell too thick to ring—I mean to sound, not to swing. I know a peal of which the middle bells alone are good for anything, because the tenor is too thin and the treble too thick to sound over the middle ones. The rule that I always adopt is to prescribe the 13 scale for half the bells in the peal, and let the others increase in proportionate thickness gradually up to some maximum, which I usually define by limiting the diameter of the treble. And I have several times advised people not to pay the founders for any weight due to excess above the prescribed diameters, which is easy to calculate. There is not the least difficulty in casting to any prescribed diameters.

The 4 largest bells at Doncaster were 'maidens,' *i.e.* they came out not only of the prescribed diameter, but thickness too, so that they required no tuning: the smaller ones were intentionally made just too thick, so that all the tuning might fall on them, and they required very little. A good many of the Worcester bells are also maiden; and every now and then, but very rarely, a whole peal, for instance that of

St. Andrew's, Wells Street, is so; but I always regard them with some suspicion that one or two bells have been left a little out of tune for the sake of calling it a maiden peal.

Sizes of peals.—I have given the minimum for peals of 8 and 6. The smallest tenor suitable for 10 bells is D flat, of 5 feet diameter and 42 cwt., or D at the very highest; for the same reason that F, 48 in. of 21 cwt., is the lightest tenor for a good peal of 8; viz. that if you go much higher you run into a G sharp treble, which, for some reason that neither I nor the bell-founders have discovered, though the fact is certain, never sounds well together with large bells. At that point some change takes place in the character of the sound, and bells above and below it do not sound homogeneous. Peals of 12 I have already said I disapprove of altogether; and it is nothing but the vanity of having them which induces ringers to cry out for them, and subscribers to find money for them. It is almost impossible, with the very best ringing, to distinguish the bells in them, and the best ringing is very difficult to get. If you will have 12 bells the tenor should not be higher than C, for the same reason as I gave just now, and even that makes the treble G, though that is not quite so hopeless as G sharp. A B-bell would probably be the best for such a peal, and anything below that, if of proper strength, roars over the others in a way which does not produce a good effect. Perhaps even the Bradford chimes would have been better without the A tenor, though no bell is too big for a clock to strike on. The tenors of Exeter and St. Paul's, you see, are B flat; but both would be on a better scale of thickness if they had been B with the same weight, or more like the B bell at Manchester, which is not rung however. The Southwark B tenor is decidedly too thin.

I am speaking here only of ringing peals, in which the time of each round is governed by the weight of the largest bell. The only limit to the number of chiming bells, with as many notes as you please beyond the diatonic scale, is that difficulty of getting small bells above C to mix well with the large ones, which is painfully apparent both at Boston Church and Eaton Hall, where alone it has been tried in England; and as both these peals are entirely or chiefly Belgian, the defect cannot be attributed to any inferiority of the English founders. The largest bells in both peals are good enough.

Shape of bells.—Hitherto I have assumed that the bells are to be of the well known and universal shape of church bells, as shown accurately enough for this purpose at page 212. And I am convinced it is the right shape, notwithstanding the multitude of public and private assurances which we had that it is wrong, and that the hemispherical form, or something like it, is right. The persons who kept making these suggestions evidently did not know that large and small bells require different shapes. Why it is so, I do not pretend to explain. But the hemispherical form had been used for ages in small clock and house bells up to the largest size at which it can properly be used. I had tried myself for several years, and with several founders, to get one made as large as 9 inches diameter that would sound well in a house clock, and I was obliged to give it up and be content with one of 7 inches, which seems to be the limit of their musical capacity. There was a very large one in the 1851 Exhibition, but it was obliged to be struck with a muffled hammer: otherwise the sound would have condemned it at once.

I know that bells of 3 or 4 cwt. of that shape are made for cemeteries, for which their horribly doleful sound is appropriate enough; and as they have the advantage of not being heard nearly so far as bells of the common shape, it is perhaps still more appropriate. But those are not the qualities usually sought for in either ringing bells or clock bells.

On the other hand, small bells of the usual church form, such as musical hand bells, require to be thinner and straighter in the side, or more conical than larger ones, which are bad if the sides are less hollow than at pp. 210 and 358. I exhibited a 6 in. bell-metal model of the Westminster bells at the Royal Institution in 1857, and it sounded worse than a common door bell, or a railway hand bell. But above 2 feet in diameter, all these peculiarities vanish; and according to my observation and experience, and the universal practice of the bell-founders of all ages, the long-established shape and proportion of church bells appears to be equally right for a bell of 4 cwt. and of 220 tons, the probable weight of the largest Russian bell. I may just mention here that the form of a very prolate hemispheroid, which the Chinese and Indian bells have, is so manifestly bad, that no one need hear them twice to know that that at any rate is wrong.

But when you have to design bells for construction it is necessary to go somewhat farther than this general conclusion; inasmuch as what we may fairly enough call the established form of bells, when speaking popularly, or comparing it with a very different form, will be found to have considerable variations of its own—considerable at least to the eye of a person who knows what a great difference in tone may be produced by an apparently small difference in shape. This, and the thickness, and the composition of the metal, were the three great points to be settled in the experiments and observations, which, as I have already said, were made before and during the two years which were occupied in casting the Westminster bells. It would be tedious and useless to describe the different variations that were tried. They ended in our coming to the conclusion that the best shape was something between the most common English pattern and the usual foreign one. It is indeed a good deal nearer to the English than to the continental pattern, except the Russian, which I was surprised to find from a section given in *Lyall's Russia*, after the first Westminster bell was made, agrees very nearly with that pattern.

After trying and observing the effect of a great many patterns, and without any *à priori* theory in favour of any particular curve, I saw that the one which we all thought the best in effect was very like an ellipse in section, though not the same ellipse as had been previously, and is still used by some of the English founders. And after further experiments with slightly varying shapes, I came to the conclusion that the following is the best for large bells on the 13 scale of thickness. Thicker ones only require a slight modification of the outside curve, and it is not worth while to complicate the description by going into more details for them. (See next page, fig. 80.)

Consider the diameter of the bell mouth divided into 24 equal parts. Then the inside sweep is simply the quadrant of an ellipse, whose semiaxis major A C is 14, and minor B C, 6 'parts;' in which I shall now give all the measures. You see at once the arrangement and lengths of all the vertical and horizontal lines in the figure. To draw this ellipse, mark B at 14 above D, and with radius 14 from centre B mark S, and also H (beyond the size of the page), in the line A C prolonged. S and H are the foci of the ellipse; which is drawn in the well known way, by sticking pins in a table at S and H, through the ends of a thread made just long enough to reach over S B H, and running a pencil along it, keeping it stretched. The pencil will then trace out the elliptic quadrant A P 8 B, the 8 indicating 8 parts from A, for a purpose I shall mention presently.

The outside can obviously be made of no single curve, but must be compounded of several curves in some empirical way to produce what we find to be the best proportions of thickness throughout. As the thickness of the waist of the bell is to be a third of the sound-bow P Q, which is a 13th of the diameter, *b* must be a 39th, or practically 2 thirds of a 'part,' outside of B. It is necessary to put the minor

Fig. 80.

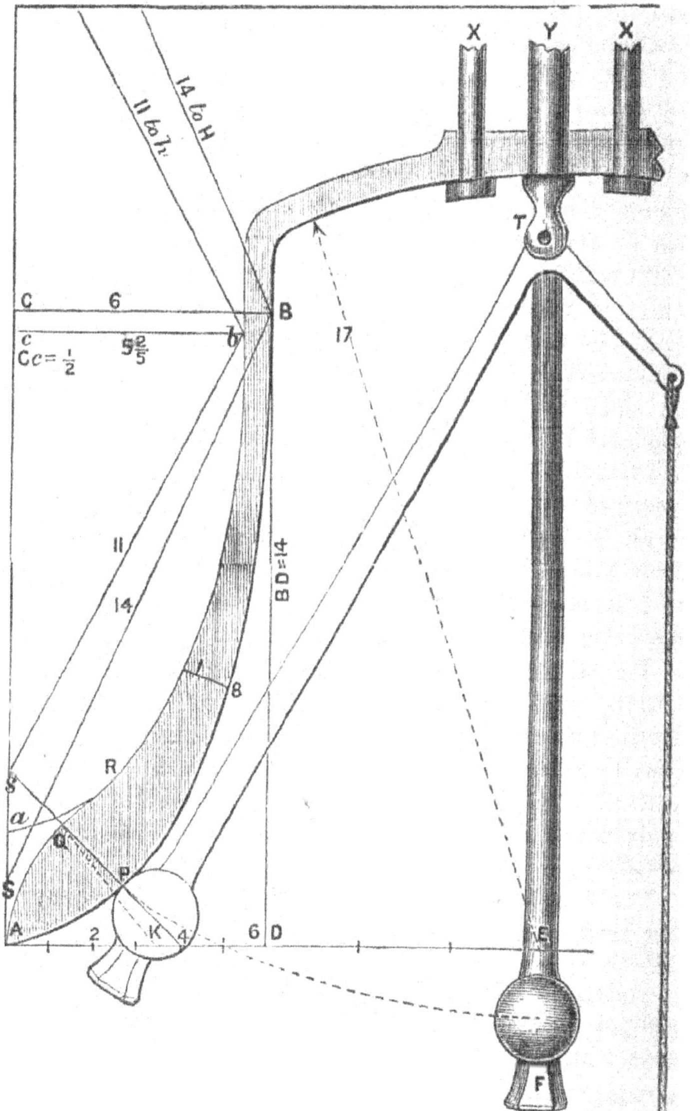
X
Y
X
14 to H
11 to h
T
C
6
B
c
b
Cc = 1/2
5 2/5
17
11
14
BD=14
8
R
s
a
Q
P
S
A
2
K
4
6
D
E
F

axis *c b* of this ellipse ½ p. below C B, to make it come right below. From centre *b* with radius 11 mark *s* and *h* for the foci of another ellipse, and then describe the quadrant *a* R *b*, in the same way as before. The lower part *a* R is useless, and that part of the bell curve is made up thus: —Draw *s* Q P 4 to the point 4 in the base, and mark off P Q = a 13th of the diameter. Then, with a radius of 3½ or 4 (for there is hardly any difference), draw the circular arc A Q, letting the centre K come where it happens. The remaining little bit Q R is easily filled up by hand. This makes the thickness = 1 at 8 from A.

The lowest bells are somewhat taller than the elliptic quadrant which forms the inside sweep, and the remainder is simply added as a cylinder. The top is drawn as a circular arc with a radius = 16½ to 18, from E the centre of the base, small and thick bells being usually made taller than large ones. This one is drawn with radius 17 for simplicity. I have drawn the crown without *canons*, or ears for hanging in the old fashion, for a reason which I shall give afterwards; and have accordingly added two of the 4 or 6 bolts X X, for hanging the bell to the stock; and Y the larger bolt to which the clapper is hung at T, which I have inserted to show its length, and to remind people that the *pit*, or frame to hold a swinging bell, must be a good deal longer than twice the height of the bell, as I have known it forgotten, and the frame spoilt in consequence; and if the clapper catches the frame the bell is almost sure to be cracked. The pit should be at least 5-3rds of the diameter of the bell; and for tall bells, or bells hung long, or with the gudgeons above the crown, the pit must be longer still. The clapper should strike at the thickest part of the sound-bow. The tail F, called the *flight*, is almost always requisite to make the clapper fly properly, and its axis, or pivot at T, must be sensibly below the bell gudgeons, or the clapper will not fly at all. This only has to be considered when the bell is *tucked up* in the stock; for otherwise the clapper axis cannot help being below the gudgeons.

Very thick bells require the sound-bow to be rather higher than in the above figure, and therefore the lip to project a little more outwards, or it becomes too lumpy. In that case, the bit of curve A Q requires drawing with a longer radius than 4. But minor details of this kind must be left to the judgment of the bell-founder. I have given these fundamental rules, because the time may come again, as it did when I undertook to design the Westminster bells, that there may be no means of learning these things, except by going through a course of experiments again, or else accepting what the bell-founders call experience, which means nothing but the way they happen to have been doing things for some time. The then existing 'experience' was manifestly wrong.

This construction makes the sound-bow, and also the part above it, rather fuller outside than had been usual. According to all the sections in books, and in most of the bells that I have seen, you can lay a straight edge against the lip and the top shoulder of the bell; but in the Westminster pattern the straight edge would be thrown out a little beyond the lip, by the protuberance of the sound-bow. I do not profess to give any reason why this pattern should be any better than the more hollow ellipse, of major axis 12, which was the more usual English pattern, instead of 14, or than the foreign pattern, which is not an ellipse at all, but a very much *less* hollow curve, and not ending horizontally at the mouth: all I can say is, that having tried them all, both I and other people who examined them came to the conclusion that this is the best in effect. The greater tallness of the foreign bells, beyond the height of 18 'parts,' at any rate, which has sometimes been copied in English ones, appears to me to be a pure waste of metal in large bells, besides being a serious incumbrance in the increased momentum and centrifugal force of the bell in ringing; and I believe all the bell-founders are of that opinion now. Among other experiments we tried making the 'waist' thinner than a third of the sound-bow, but it produced an unpleasant whistling sound and was plainly wrong.

Composition of bell-metal.—I exhibited in a lecture at the Royal Institution, in 1857, a variety of small bells of different shapes, metals, and alloys, and different proportions of copper and tin, which had been suggested as possible improvements on the usual composition of old bells. It was clear that none of them were improvements, and that something near the proportion of 3 lbs. of copper to 1 of tin had been generally used by the old founders. The tin was sometimes a little more than that, or tin and antimony together, which both produce the same effect of making the alloy brittle, and I understand are difficult to distinguish in the analysis. But it was also clear from experiments that 3 to 1 with modern copper is the turning point between safe and unsafe brittleness, although the sound improves as you increase the tin up to that proportion, and even higher; and that antimony is an inferior substitute for tin. I found it was the modern practice to use much less than this. Accordingly we fixed on 22 to 7 for the Westminster bells, and others made about the same time; none of which have cracked, and we had great difficulty in breaking an experimental bell of that composition. The specific gravity of the sound part of Mr. Mears's bell, as of the previous one, is 8·8, but of the unsound part only 8·32: some thin pieces at the edge of the first bell were as high as 8·94, and very difficult to break. There is no such test of good casting as a high specific gravity; but unfortunately it cannot well be applied to large bells, as it involves weighing them in water: for the specific gravity is the weight in air divided by the difference between the weights in air and in water.

But I have since come to the conclusion, for chemical

reasons, that the proper composition for bells is 13 of copper to 4 of tin, though it is not mentioned in any book, nor came out exactly on the analysis of any old bell-metal. But in old times the doctrine of 'chemical equivalents' or atomic weights was unknown; and the reason why 13 to 4 is the proper proportion is, that it is the only one near 3 to 1 which is in atomic proportions. For the 'atomic weight' of copper is 32, and of tin 59, and $\frac{13}{4} \times 59 = 6 \times 32$; or the mixture of 13 to 4 by weight is a true chemical combination of 6 atoms of copper to 1 of tin, and is written in chemical language Cu_6 Sn (Sn being short for stannum, tin).

I am aware that some persons who know more of chemistry than I do dispute the application of that theory to metallic alloys, on the ground that the metals do not refuse to combine in other proportions, as gases do and some other compositions. But the practical question is whether those proportions do not combine more firmly and with less risk of the alloy becoming unhomogeneous in cooling, or otherwise defective. And I am convinced that they do. Without going into other metals, telescope speculums only differ from bell-metal in containing rather more tin, and no alloy has to undergo such severe tests as they have. Lord Rosse made all his speculums as Cu_4 Sn, or 128 to 59 by weight, and a very little deviation from that proportion runs great risk of spoiling the speculum. I have seen some specimens intentionally varied a little, and certainly the difference between the appearance of a fracture of the metal of atomic proportions, and of one very slightly deviating from it, is remarkable. I was very near prescribing the above-mentioned atomic proportion for the Westminster metal, and indeed I published it as a suggestion in 1856; but it only differs about ·01 from the 22 to 7, which Mears told me agreed with his own practice, and it was therefore inserted in his contract. Although not one bell in 1000 may fail in casting to the extent that Mears's bell did, and small ones probably not at all, because they cool before the metals have time to separate, it is clear that there is a tendency to do it; and indeed I find it is well known at Woolwich as a thing to be provided against in casting large guns of gun-metal, which is about 1 tin to 9·5 copper.

It seems to me at any rate imprudent not to avail ourselves of this undoubted law of nature, especially as there is an atomic combination which happens to be a particularly convenient proportion, being exactly a mean between the 3 to 1 which is just too brittle to be safe, and the 3½ to 1 which the bell-founders like because it is easier to tune, but which is certainly softer and less sonorous than it need be. I always now therefore require large bells to be made of this 13 lbs. of copper to 4 of tin, or 76·5 and 23·5 of the whole alloy; and they should be rejected as unhomogeneous if any part of the bell is proved to be beyond the limits of 77 per cent. of copper or 23 of tin. It must not be supposed however that this would have prevented the porosity of Mr. Mears's bell, which is a quite independent defect, and would have made the bell a bad one, even if it had not also miscarried in composition.

The only other atomic combination within the range of bell-metal is Cu_7 Sn, or 19 copper to 5 tin by weight; but that is too soft, except for small house bells, which are thinner and have clappers much larger in proportion than church bells. Old Tom of Lincoln and the old York Minster bells of 1765, and probably the Bow bells, of nearly the same date and made from the same patterns, contained ·03 of antimony, which has a hardening effect like tin. That is much too large a quantity to have got in by accident; but there was certainly no improvement in the sound from introducing about ·03 of antimony into a small bell. The antimony diminishes the specific gravity of the alloy, which tin does not, though so much lighter than copper by itself.

So far as I have an opinion on the point, it is at present against the antimony, and the bell-founders have the same opinion. Very small quantities of iron, lead, zinc, arsenic, and sulphur sometimes appear in the analysis of bells; but they are mere impurities. I have indeed seen lead and zinc in considerable quantities innocently put down in books as ingredients of bell-metal; but they are mere adulterations; for both those metals are injurious and have no business there at all, especially the lead: a little zinc is sometimes put into small bells, I do not know for what reason.

Steel bells.—I have frequently been asked my opinion of these bells, which are made both in Germany and at Sheffield. And I answer, as I did to the makers who asked me for a testimonial, that if the object of bells is to make the greatest noise for the least money, steel bells are very good ones, but that the less they asked me to say about the quality of the noise the better. It was remarked in the 1862 Exhibition that the sound of the very large ones was rather less horrible than of the smaller ones; and that was the best I heard anybody say of them. I have heard nothing of them since 1862. Moreover they require hammers two or three times as heavy as bell-metal bells: if so, there would be little saved in using them for large clock bells, as the clocks would cost much more.

Silver.—The most inveterate of all popular delusions about bells is the notion that old bells had silver in them, and that all bells would be improved by it. There is not the slightest foundation for that belief. Nevertheless we had some experiments made for the purpose of being quite sure that silver was of no use, either with reference to sound or strength of the metal; several different proportions were tried, beginning with sixpence in a bell of nearly a pound weight, and it was clear that the silver rather did harm than good in both respects. I suppose the delusion has arisen from the ring of shillings and half-crowns, which justifies no such inference, any more than in the case of aluminium, which some people fancied would make very fine bells, until I exhibited one at the Royal Institution, which M. St. Claire Deville of Paris was good enough to cast for the purpose, and the sound was worse than of cast iron. A bell of copper and aluminium was also bad, though the bronze of 9 copper to 1 aluminium is in other respects a very superior metal to either brass or any alloy of copper and tin. No composition for bells has yet been discovered equal to copper and tin in the proportions I have given. There was a large bell of iron and tin in the 1851 Exhibition; but that also was very inferior indeed to bell-metal; and it required an enormous blow to bring out the sound, though it was thin, and was at last cracked thereby.

Moulding.—There are two different ways of making the moulds for bells. As to the internal mould or *core* they are nearly identical, that being made by covering a cone either of brickwork or of cast iron with moulding clay, which is *swept* over into the shape of the inside of the bell by a piece of wood called a *sweep* or *crook* fixed to an axis or spindle set up in the middle of the core. The advantage of the iron core is that it can be lifted up and put into a furnace to dry, instead of lighting a fire inside it. At this point the difference between the two methods begins. The old method is to make a clay bell on the core by means of another crook, and when that is dry to make the outside mould or *cope* on the top of that. The cope has hair and hay-bands, and in large ones, iron bands worked into it, to make it hold together and lift off when it is dry; then the clay bell or *thickness* is knocked to pieces, the cope dropped down again and weighted with earth in the pit where the bells are cast, and the metal poured in at the top through one hole, another being left for the air to come out at.

In the other way there is no *thickness* made, but the cope is an iron case lined with clay, and swept out by an internal sweep to the shape of the outside of the bell. The *wires*, or ornamental rings round the bell, are made in both cases by the second sweep, and the letters, and any other ornaments, are pressed by stamps into the clay of the cope while it is soft. These iron copes can be bolted down to a plate under the core, and therefore do not require to be sunk so deep in the ground, provided only proper care is taken to get a high enough head of melted metal above the bell, or else it is certain to be of bad specific gravity, and probably porous in a casting of any considerable size, as in the top of the first Westminster bell, where the metal ran short. I was told by an old bell-founder that the core should not be strong enough to resist the contraction of the bell in cooling, or the strain injures the tone. Small bells are generally cast in sand, like iron, from models, and not in loam moulds made by sweeps. Bells are always cast mouth downwards, so that the sound-bow, which is by far the most important part, may have the best chance of being sound by having the greatest pressure of metal on it. The importance of this was illustrated by Lord Rosse's experiments with large cast iron crucibles for melting speculum metal in, which were always porous in the bottom, which is their most important part, until he had them cast with their mouths upwards.

Bell-metal melts at a temperature far below that of the copper, as is usual with the alloys of one easy-melting metal. Small bits will melt in a common house fire. It is melted in a reverberatory furnace, in which the fire is at one end of a long shallow trough which holds the metal, and the flame is drawn over it to reach the chimney at the other end, and is reverberated down upon it from a 'bridge' or a low roof over the trough. The different founders use different fuel as well as different ways of moulding. At Whitechapel they confessed that they still use wood. I suppose the old founders used charcoal, as I understand they do in Russia, which is probably far better, because wood contains so much moisture that it takes much longer to get the requisite heat up with it, and it is notoriously a bad thing to keep the metal long melted getting up the heat. At Woolwich they have given up using wood for that reason, and use coal or coke in melting gun-metal, which is bell-metal with much less tin in it. Warner and Taylor use coal. In order to test this as far as possible, I gave to Warner and Mears the same pattern for a bell of about 12 cwt. for the two new churches at Doncaster; and the result was very decidedly in favour of the coal-cast bell; but whether from that cause or some other difference in the general management of the casting, of course I cannot say. Most of the metal of Mears's Westminster bell was nine times as long in the furnace as Warner's (which however I suspect was run too soon), and was also much longer running into the mould than either that bell or Great Paul. I have no doubt that that slow running contributed to its unsoundness. Lord Rosse specially mentions the importance of quick running of large speculums, which again are bell-metal with a higher quantity of tin.

Mending cracked bells.—Whenever a bell of any importance cracks, there invariably follows a flood of suggestions, public and private, for mending it: generally by 'cutting out the crack' in some way or other. And as there exists in some persons what is called colour-blindness by the world at large, so there are persons to whom a bell so divided, and others apparently to whom a bell with a deep crack in it, sounds (at least, they say it does) no worse than when it was whole, though to other people's ears the tone is generally altered several notes, or is made otherwise intolerably bad. Another way of mending, which is continually being reinvented, is by what is called burning the parts together; *i.e.* cutting the crack wide enough to let hot metal be poured in and through, which is done until the constant application of it partially melts the faces of the division, and then the running is stopped and left to cool. This may *perhaps* answer in very thin bells, though the cracking of one of Sir C. Barry's gun-metal hands which were put together in that way is not very encouraging even for much thinner castings; and the contraction, and consequent tension of the metal, would most likely crack the bell again as soon as it is rung. There is no evidence that I can learn, of any bell as large as a common church bell having been successfully treated in this way yet. And if the bell has cracked from any radical defect, as both the Westminster bells did, it is absurd to think of mending it, even if it were otherwise possible; for the same defect would make it crack again.

Fig. 81.

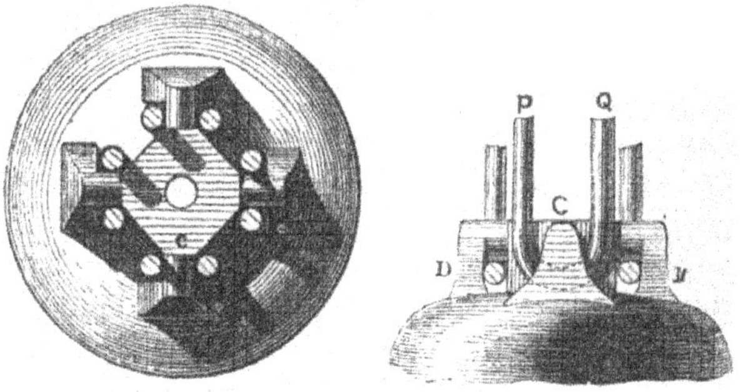

New bell crown.—Church bells used always to be hung by 6 long ears, called 'canons,' which cut a large piece out of the stock, and weakened it very much. They were not set radially, and the iron bolt which carries the clapper was cast into the bell. Consequently when a bell got worn in one place it could not be turned without a new stock and a great deal of trouble. Several plans to cure this had been proposed, but none of them were satisfactory; and I had the Westminster bells and some others made with a top like a mushroom or button embraced by a collar in two pieces, or by bolts with heads of proper shape surrounded by a ring, which may advantageously be connected with the gudgeons or pivots. But the founders complained (unduly, I think) of this ironwork being expensive, and so I invented another crown, with 4 short and thick canons, which will be understood from figure 81. It has been generally adopted by Messrs. Warner. Mr. Taylor uses 6 very short canons, and in many cases, with my approval, none at all, but only a thick crown with 6 bolt-holes through it, as at p. 358. I believe canons are of no use. In all these plans the clapper bolt goes through a round hole in the bell, but a square or octagonal hole in the stock. By placing the bolts (of my plan) each with one leg in the stock, or two of them inside and the other two outside of the stock, you may get 4 different pairs of places for the clapper to strike, with the same stock and without any cutting of it, and the stock itself is far less cut into and weakened than by the usual canons, which are necessarily taller. The Doncaster bells were the first peal made in this way.

On the old plan too, the clapper bolt is always cast into the bell, and must be cut off and a new one stuck on in some way, to turn the bell at all, which is an awkward job, and postpones the turning of bells for years after it ought to be done. And in connexion with that, it is another advantage of this plan, that it enables the clapper bolt to be adjusted both for length and position, which cannot be done when they are cast in. The length of that bolt ought to be more or less according as the bell is hung more or less high in the stock, in order to bring the point of suspension of the clapper below the gudgeons or pivots of the bell, without which it will not swing or strike properly; and again, if the bolt is cast in a little on one side, the clapper will not strike true, and the time between the blows will be unequal, which makes it impossible to ring the bells truly. If some such plan as this had existed in old times, whereby a bell could easily be turned in the stock, many a fine old bell would now be alive which got cracked from being constantly struck in one place, and so worn too thin. These four short and thick canons are moreover much stronger and less liable to crack than the usual six thinner and longer ones.

Opinions differ whether large bells should be what is called 'tucked up in the stock,' or the top of the bell made higher than the 'pivots or gudgeons.' The advantage of it evidently is that it diminishes the centrifugal force, or side-way strain of the bell on the frame; and if friction were out of the way it would of course make the bell easier to raise and ring. But friction is *not* out of the question; and as a bell in swing is in effect a pendulum, and not (as I have even heard bell-founders represent it) a body lifted by a steady pull like a lever, it may very easily happen that a certain amount of friction on the pivots may make it impossible to make the bell pendulum swing through 360° by any practicable force that can be applied to it at the beginning of its motion, which is the only time when the rope acts upon it. The Rev. Mr. Taylor told me, that a bell of about 52 cwt. at Hereford, which he and some other boys used to raise and set (*i.e.* ring till it stands mouth upwards) was made unraisable by them by being re-hung, and at the same time 'tucked up;' and so confident was he of the mistake of this mode of hanging, that he offered to fill Mears's great bad bell at York with beer if *any number* of men could set it, and they never could, as *Browne's History of York Minster* also testifies.

There is moreover a remarkable difference in the facility of tolling, according as the bell is hung low or high. I can toll the two largest bells of Doncaster Church together, weighing 2½ tons, one with each hand, whereas it is difficult to make a high-hung bell toll at all. At the same time, it is true that the largest bells of a peal, where the tenor is above 30 cwt., can hardly be rung with the others, unless they are somewhat tucked up, though it makes them 'rise false,' or with the clapper striking the low side of the bell and lying on it, and the men have to go up and set them right before ringing (and they will keep so), otherwise there is a risk of cracking the bells, and they do not sound so well. The reason why heavy bells in peals must be rather tucked up is that otherwise their momentum is so great that no man can check them, so as to keep time; but this would not apply to a single bell.

I am surprised that very large bells, say above 2 tons, are not hung on friction rollers—*i.e.* so much of the circumference of a friction wheel as = the circumference of the gudgeon or pivot of the bell. Of course brasses must be put to keep the gudgeons in their place, against the side swing. And, by the way, let me say a word here to warn those whom it may concern, that if the smaller bells of a peal jump out of their brasses when they are new, it is because the gudgeons are not sunk deep enough. I have known that happen several times. Of course the founders denied that that was the reason, but as I cured it by getting deeper brasses, the denial was worth nothing.

I recommended rollers of about 2½ in. diameter in annular bushes round the 6 inch pivots of Great Paul, like bicycle bearings on balls. Swinging that bell through only 60° from the vertical (the most that is proposed) is equivalent to sliding 18 tons 6 inches rapidly at every swing. Even tolling it is equivalent to sliding that weight an inch forwards and an inch backwards at every pull. Tolling consequently takes three men now.

Tolling-levers.—The great Worcester bell is hung, by my advice, on wedge-shaped rolling gudgeons, only round instead of sharp, to enable it to be tolled almost without friction, by a long lever; for the tower would not bear it in full swing; and it went so easily that the Rev. H. T. Ellacombe, then 80 years old, tolled it with one hand. But for all practical purposes it answers equally well to toll it by a short lever, or shank, projecting from the top of the clapper, and pulled by a slight rope; which is a very good plan for all bells which the tower is either too small or too weak to bear swinging. If a double lever is used with two ropes, one for each hand, you may ring as rapidly as the bell itself would sound when swinging about 'frame-high,' or half-way up. See such a lever at p. 358 for one side.

A few years ago I introduced a mode of hanging chapel and school bells, *i.e.* single bells from 1 to 4 cwt., or 18 to 28 in. wide, so as to swing lengthways in the plane of the wall, instead of across it. A bell of very moderate weight will soon pull an ordinary wall to pieces if rung in full swing across it. The bells in 'bell gables' can hardly ever be safely rung for that reason, but only tolled. Mr. Taylor has introduced an ingenious plan for ringing small school bells without the risk of the rope being carried over, by putting it to a crank in the iron axis which serves for the stock of such small bells: you may then ring the bell 'over' as often as you like, and the rope always keeps right.

Bell-ropes.—I said just now that some people fancy a bell is hauled up like a dead weight on a lever, and not swung up as a pendulum; and they forget still more, that the measure of the strength required for a rope is that of the man who has to pull it, and has nothing to do with the weight of the bell. A small bell can be 'raised' in a few pulls, while a large one takes many, but in either case the ringer may be exerting his full strength, though in fact no man does in raising small bells; and therefore *some* difference in thickness of the ropes should be allowed. But experienced ringers know too well that inexperienced rope-makers always make the ropes much too thick, though probably not many ringers have reflected on the proposition that it is their force used in pulling, and not the weight of the bells, that determines the thickness of the ropes—allowing reasonably for wear, and for some greater force being used upon the large bells.

The great superiority of tone of bells ringing in full swing over tolling, and even of tolling over striking by a clock hammer, has been often noticed, but never yet accounted for. I think the explanation is this. It is now well known that the note of an approaching source of sound like a railway whistle is sharper than when it is receding, because the velocity of transit is added to, or subtracted from, the velocity of vibration which fixes the notes. (The analogy of light to this has been made use of by Mr. Huggins, in one of the greatest astronomical discoveries of late, to determine the velocity of approach or recession of the stars, as is explained in my Astronomy.) The bell in full swing, while it also vibrates from the blow, is always sending out vibrations of slightly different velocity in each direction, and so varying its own note a little. I remember a gentleman trying to persuade the Institute of Architects that bells sound as well stationary as swinging, and he illustrated it with a common dinner bell; which by the way a servant of experience always swings as far as he can. Everybody exclaimed immediately that he had refuted himself. The motion in tolling is probably too small to vary the note sensibly, but the inevitable variation of a man's pull makes it less monotonous than the unvarying lift of a clock hammer, or of any chiming machinery, a melancholy and miserable substitute for a set of ringers, only justifiable as a precaution against strikes—of the men, I mean. Even that hardly justifies it.

Ellacombe's chiming hammers.—There is however a less objectionable mode of chiming by one man, with a set of separate clappers fixed below the bells, temporarily tied up very near them, so that a small pull of the rope makes the clapper strike. This was invented by Mr. Ellacombe, and is fully described in his large book on bells, and in some smaller ones. Not that even this is to be compared to genuine chiming by tolling the bells themselves, for they are stationary, and those clappers are necessarily too small to bring out the full tone. In this way however, a skilful man can play tunes, whereas in ordinary chiming or ringing a bell can only change one place at a time: e.g. the greatest change that can follow 1 2 3 4 5 6 7 8 is 2 1 4 3 6 5 8 7. I do not profess to teach the art of ringing, or to deal with the interesting and complicated subject of change-ringing, which is quite enough to have a book to itself; and there are several, by Captain J. E. Acland Troyte, and by Mr. Jasper Snowdon, of Old Bank Chambers, Leeds. I will only say that ringing 'the tenor behind,' *i.e.* ringing the changes only on the other bells, always sounds much better than ringing the tenor in the changes, though ringers think more of it as a feat. One man can ring a very heavy bell 'behind,' which would require two to ring in, as its time has then to be altered continually. Also, when two men are required, it tries their wind much less, and it requires only one rope, for one to take the 'fore stroke' only, and the other the 'back stroke' only, though it is hardly ever done.

Stays and sliders.—The stay is a strong piece of wood, about as long as the radius of the wheel, bolted to the stock and pointing upwards, so that when the bell is swung right up, the stay points downwards and rests against a stop there. But if that stop were fixed, the bell could not reach quite the vertical position, at least in one direction of the swing. Therefore the stop is a sliding bar, which lies low enough to clear the wheel and clapper, and it has play enough in its own bed near the stay to allow the bell to go just beyond the vertical both ways; and so a small portion of its weight is borne by the stay and slider when it is 'set,' or so resting, until it is pulled off again by the rope and swung the other way. Sometimes the stay has to be put by the wheel, on account of clock hammers at the other side, but generally it is at the other end of the stock.

After a long struggle with bell-founders and their hangers, I got the practice introduced of striking the bed or board in which the slider runs as an arc of a circle from the gudgeon as a centre, instead of leaving it flat, which the smallest knowledge of mechanics will show anybody must produce a grinding action between the stay and the slider. Moreover, sliders were generally much too heavy, which increased the friction and the cost of breaking. A slider need be little more than a common stick, and can be replaced almost for nothing in a few minutes, whereas the breaking of a stay is a more serious matter, as many a learner of bell-ringing knows. At the lower end (away from the stay) it need only lie loosely in a hole; its weight and slope will keep it there without a pin. Stays can hardly be too strong, and should always go through a strong iron loop, and not be merely bolted to the stock, which weakens them, and gives more leverage for breaking. Boards should be put over the sliders to keep the grease off, which increases rather than diminishes their friction; black lead is the thing to diminish it on wood.

The Gudgeons, or pivots on which the bells swing, are often made too short, under an erroneous idea that that diminishes the friction, which it does not, and may in fact increase it, because if the pressure on each particle of surface is too great, it squeezes out the oil, and the metals come in contact, which it is the object of oil or grease to prevent. Besides that, short gudgeons tend to twist the beams. They should not be less than 2 in. long and 1 in. thick for the smallest bells in a peal, and rather longer and much thicker for the large ones. I think they should be of steel. And it should be specified that the 'brasses' in which they run are to be 19 copper to 5 tin, for the reason given at p. 363. The brasses should be large, to allow for wearing down, and for firm fixing in the beams. All the ironwork should be well painted, except the gudgeons; and the woodwork too, though it hardly ever is, which is very bad economy in the long run. But oak should not be painted very soon.

Clappers should be hung by or on wooden blocks, and not with a mere piece of leather inside an iron strap, which soon wears away, and leaves the two irons in contact, and they are sure to be rusty by that time. Thick bells of course require heavier clappers than thin ones. I remember a very good bell of Warner's remained unsold in the 1851 Exhibition, and long after, merely because it was very thick (copied from the 3rd Bow bell) and had too small a clapper. I happened to try it with a large one which was lying loose, and bought it immediately for a chapel for which I had been asked to get one. All nuts connected with clappers should be either double or keyed on with wire, or they will shake loose and come off.

Iron stocks and frames.—I have seen several peals of bells with cast-iron stocks, with gudgeons let into them; others have the advantage of not shrinking and swelling as wooden stocks do. But they have never come into general use: nor have iron frames, of which some also were made many years ago. Messrs. Moore, Holmes, and Mackenzie, of Redenhall, however, I see by their advertisements, use both iron stocks and a new kind of A-shaped iron frame for small towers, with the smaller half of the peal above the others, the gudgeons being at the top of the A-frames, and those of the larger bells on cross bars lower down. It is certainly a very strong pattern, if you are to have an iron frame. But they are not pleasant to walk on when you want to get to the gudgeons, nor to fix clock and chime hammers to; and if a bell slips in any way and strikes the frame it is pretty sure to get broken. So far as I see at present there does not seem to be the same advantage in iron frames as in iron stocks.

Bell-frames, like most other things nowadays, are generally starved of proper bulk. I do not see my way to laying down any rule for the thicknesses of the beams, as they must vary so much according to the weights of the bells. The beams of the smallest bells ought never to be less than 5 in. wide, and 10 in. deep, and of the large ones for a peal with a tenor about 30 cwt., not less than 9 × 12, and the lower beams should be always a little wider than the upper ones. It is also specially important to have the diagonal trusses very deep, particularly under the heavy bells. Dove-tailing and any cutting away of the beams should be avoided as much as possible, by using 'bed-bolts' and iron T pieces bolted on at all weak corners. The upper beams should also be bolted to the lower by long diagonal bolts, set the opposite way to the trusses.

And now I have to correct one of the most common errors about frame-fixing. It is quite true that the upper beams should not touch the walls, unless the walls are very thick and strong. But if they must touch anywhere, for want of room, they should touch close, and in fact tight, all along the beam, so as to avoid anything like battering. The lower beams should always do so, or be somehow connected with the walls as firmly as possible. The notion that they should not is an extraordinary piece of architectural ignorance. Either the whole frame must float about on the floor under the swing of the bells, or it must be fixed to the floor; and if it is fixed, the horizontal thrust must ultimately go to the walls, pulling the floor or its beams against them anyhow; and it is much better to make this connection firm at once, and avoid all chance of battering.

All the elasticity takes place between the upper frame and the lower, and it acts like carriage springs, and the pressure is diffused over the walls as well as it can be, if the lower frame is tight. The oscillation of towers under ringing is merely due to the elasticity of the whole mass of stonework; and though they sometimes move enough to make the little call-bell from the church into the belfry almost ring by accumulated vibrations, the actual motion of the tower is too small to see, except perhaps with a fixed telescope. The idea of preventing it by building up a frame from the ground within the tower is ludicrous, though some architects do not know it, or that such frames have always at last to be wedged against the walls of the bell-chamber.

Another 'vulgar error' about bell-frames is that bells swinging at right angles to each other tend to correct each other's swinging thrust upon the frame and tower; whereas every mathematician knows that that is exactly the position in which the thrusts cannot the least neutralise each other. Besides that, it has the effect of making the ends of some beams thrust against the gudgeons of other bells, and so occasionally binds them tight. I was assured by a bell-founder's foreman, before making the first frame on the parallel plan (for the cathedral at Fredericton, which also had the first gravity escapement clock), that 'all experience was against such a plan.' I said, I should like to be referred to some frames of that kind which were proved by experience not to answer. His reply was, 'O, nobody ever thought of making such a frame.' That total absence of experience of the very thing in question is what so-called practical men mean by experience—and many other people too.

The following is the plan which I prescribed for Doncaster and the others mentioned at p. 351. All the bells swing north and south: 2, 1, 8, 7, are on the south side, and 3, 4, 5, 6, on the north; all the beams going through, except the one on the west side of 8 and 5, which is divided, and overlapped and bolted together, so as to make the tenor pit wider than the 5th by the thickness of that beam, and the treble pit as much narrower than the 4th. The east and west beams are only notched on to the others and bolted, not halved or cut away as usual. There should always be long bolts nearly vertical, but oblique, holding the upper and lower beams together the opposite way to the diagonal struts which carry the weight, because the mortices always get loose, but the bolts can be tightened from time to time. In this frame the ropes can fall in a very good and wide circle, in one or two ways which will be obvious; and in large towers such as those, there is also room for the entrance; indeed in all but Croydon you can go all round the frame.

But when the tower is only just large enough to hold the frame, it is expedient to adopt some other and inferior plan, in which the bells cannot all swing parallel, in order to avoid having the ropes too close to the walls for the men to stand behind them; for leading them down obliquely in troughs increases the friction and labour of ringing. It may be taken as a rule that a peal of 8 bells hung properly on beams of proper thickness requires a frame nearly 4 times the width of the tenor, both ways; and one of 6 bells requires rather more than 3 times the width of the tenor. In hanging 10 or 12, some space may be saved by making a few of the smaller bells swing at right angles to the others; but it should only be the small ones. When the large ones all swing parallel, they are sure never to be thrusting all one way twice together; and as it is only an accumulation of similar vibrations that could affect the tower, they are certain on the whole to escape that effect, and in a great measure to neutralise each other, though of course not completely.

Clappering.—It is still necessary to warn clergymen and churchwardens against allowing the lazy and pernicious practice of 'clappering,' *i.e.* tying the bell-rope to the clapper, and pulling it instead of the bell. More bells have been cracked in that way than by all other causes together, and there is not the least excuse for it, as any man may learn to toll a bell in one lesson, and the work of doing it is nothing, for the bell moves so little that it seems wonderful that the clapper should strike it; but they both act as pendulums which a very small impulse will keep up, and the blow being elastic loses very little of the force. Bells are consequently much easier to toll when they are not tucked up in the stock, because they act more like a pendulum and less like a grindstone; in fact, it is difficult to set a much tucked-up bell tolling, though easy to keep it up afterwards. The only safe way of allowing bells to be clappered is to fix a separate pulley in the floor under the bell with a thin rope, in such a position that no pulling of the rope over the pulley can hold the clapper against or very near the bell, so that it must depend entirely on the swing given to it by the rope.

Specifications.—As I gave a specification for large clocks, I will do the same for bells, so far as any one model can serve for a variety of cases. I add however some variations and notes to explain them, which will enable anybody with the least understanding of such things to make out a specification for the peal he wants. And first, I remark that specifications should always be silent about notes, and stick to diameters and weights. Most of the bad peals in the kingdom, except those by utterly bad bell-founders, are greatly due to people demanding low notes for low prices; and besides that, there are so many uncertainties about notes, even with bells of very nearly the same dimensions, that it leads to nothing but confusion to specify anything about them. Suppose then a specification is wanted for a peal of ten bells, intended to be as good as possible, and of no unusual weight: it should be as follows:—

1. The founder to make, and hang complete for ringing, a peal of 10 bells in perfect tune; the tenor to be 60 inches wide (or more, if you like, up to 67 or 68). The 4 or 5 largest bells to be a 13th of the diameter in thickness; and the smaller ones to increase gradually up to the treble, which is not to exceed a 10th of its diameter in thickness.

For 8 bells, the 4 largest to be a 13th, increasing up to an 11th in the treble; and for 6 bells, the 3 largest to be a 13th, and the treble may be an 11th, or a 12th.

2. The bells are to be of the shape described in this book as to their 'sweep,' and all to have independent clapper bolts going through a round hole in the crown, and through the stock. The larger bells to have no canons, and the smaller bells low canons, if any. The bells are all to consist of pure copper and tin only, in the proportions of 13 to 4, and are to be guaranteed as perfect castings, homogeneous, and free from porosity or other defects. (I should mention that those superficial marks called seams are of no consequence, though of course a casting looks better without them; but the slightest visible porosity in the smoothest bell is a fatal mark of unsound casting, though a bell-founder should find any number of brass-founders and engineers to swear that it is not, as I know that Mears was prepared to do about Big Ben, before Dr. Percy's examination of the 'taster' cut out of it (see p. 271). Also an inequality in thickness round the bell, such as Great Peter of York has, is equally fatal, and is not a 'perfect casting.')

3. The clappers to be hung with wooden blocks, not iron and leather (for the leather soon wears out). The tenor clapper to be a 90th of the weight of the bell, and the others to increase upwards, to about a 30th for a treble a 10th of its diameter thick, and a 36th for an 11th, and a 40th for a 12th. All nuts connected with the clappers to be keyed on, or else double nuts (or they will shake loose).

4. The brasses to be of 19 copper to 5 tin, and none to have less than 1½ in. of metal on each side of the hole, and the holes to be deep enough to prevent the small bells from jumping, and to have at least 2 inches of metal below the hole. They are to be boxed over with wood to keep out dust. [Metal tops always get torn off and stolen.]

5. The gudgeons to be of steel, and to run from

2 in. to 2½ in. long in the brass, and 1½ in. thick for a bell of 30 to 35 cwt. (It is difficult to define the thickness farther, and the fault is more generally in making them too short than too thin.) They are to be fixed so that they will keep themselves quite true in the stock. (The founders have different ways of doing this, and I do not like to prescribe any one absolutely, but I must say that Taylor's plan is the best I have yet seen. The old one is very rough indeed, and cannot be relied on to keep them either true or firm.)

6. The stays to go through iron loops bolted to the stocks, the bottom only of the stay having a bolt through it, and to be so strong that the slider may break rather than the stay. The sliders to run in circular beds, traced from the gudgeon, which are to be covered over to protect them from grease dropping (which makes wood stick, though it makes metals run).

7. The wheels are all to have at least 4 side stays to keep them true. (The founders will never put more than two unless they are made, and others have always to be added in a year or two.) The wheels to be wide enough and deep enough to prevent the ropes from slipping (a source of frequent accidents, and always arising from bad or untrue wheels; in fact wheels, like everything else now, are generally made too flimsy; but I do not see how to define their dimensions in a specification, nor indeed many other things which can only be judged of by a competent eye when they are done).

8. The frame to be of English oak, and the founders to state the dimensions of the beams. (This will at any rate be useful in comparing tenders, which otherwise are deceptive.) The wood must be perfectly dry before using, and if joints get loose afterwards, the founder will be held responsible for all expense incurred in consequence. The beams are to be cut into by mortices as little as possible, and iron

T pieces, and bed-bolts to be used instead, or in addition. The diagonal struts especially are to be deep, and diagonal bolts inclined the other way are to be used also.

9. The *bottom* of the frame to fit tight against the walls, if it is about the width of the tower (see p. 378), and if not, to be strongly bolted to the great beams below; the position of which the architect should be made to settle with the bell-founder beforehand, if the tower or the floor is new; and also to arrange for having two floors between the ringers and the bells, with gravel or small stones on the lower one, if they are very close together (see p. 385). All the large bells, as far as possible, to swing either north and south or east and west, and not across each other.

10. The tender to state the probable weight of each bell, and maximum of the whole peal (the word 'about' not to be allowed there), and the founder will be paid for no excess beyond that maximum. The tender also to state the cost of frame and hangings separately, besides the price per cwt. for the bells, and all other expenses, as far as they can be estimated. (Recasting of old bells is always charged at two guineas a cwt. whatever is the price of new metal, see p. 338.)

11. The whole of the work to be subject to the approval of some competent person or persons, not being in the trade, to be appointed by the [committee], who may require any experiments to be made which they may think fit for testing the bells.

12. All the iron and woodwork to be painted twice, and the frame also after a year, with Carson's anticorrosive paint. (It is very odd that nothing will induce people to paint their bell-frames. They leave them exposed to damp, and frost, and heat, and see them cracking all over worse and worse yearly, and then they are surprised at having to spend three times as much as the painting would have cost, in repairs after a few years. It is said not to be expedient

to paint oak immediately, and for that reason only I have allowed the year for it.)

Bell-towers.—I proceed to notice a few things relating to bells which are more the business of architects than bell-founders, except that architects nowadays rather think it their business to ignore everything that has to be done in a building after they have left it. I have known plenty of cases where they knew perfectly well that it was intended to have peals of bells and clocks, and yet have built costly towers without condescending to learn, either by reading or inquiry, what provisions should be made, or even what dimensions were required. That they may have the less excuse, and that those who want church bells may have some idea of what their architects ought to do, I will give the following information as the result of my own experience; and I am probably the only person living who has the experience of designing both bells and clocks, and the towers in which they are to be placed.

It follows from what I said before about bell-frames, that the smallest tower fit for even a small peal of 6 bells, should be 11 ft. square inside; and the smallest for a very moderate peal of 8 should be 16. Even these sizes make the ropes hang closer to the walls than they ought to do, and therefore they are much better somewhat larger. Very few *old* towers with 8 bells are so small as 16 ft.; they are much more frequently 20, and those with 6 bells at least 14. In what is called *A Book on Building*, I have shown how much narrower modern towers generally are than old ones; but I am only now dealing with them as campaniles, which in fact they always are, without reference to architectural reasons in the same direction. Of course when 10 or 12 bells are intended, the towers should be larger still, certainly not less than 22 and 24 ft. square inside.

There is also another reason for it. Bells sound much better in a large chamber than a small one. Anyone who has heard the Doncaster bells will hardly believe that the three bells in All Saints' Church, Margaret Street, are repetitions by the same founders of the 1st, 4th, and tenor of Doncaster. But the Doncaster bell-chamber is 23 ft. square, and Margaret Street not more than 14, or just two-fifths of the area. And the same at All Saints, Halifax, with a spire 240 feet high. The bells, a smaller peal than Burton, could not be got into it on one level.

Another important consideration is the windows. I have known two cases, both in Leeds as it happened, where clock bells had to be re-hung some feet higher up than the architects had provided, because they did not know that the bells ought to be above the sills of the windows. Louvres again are a frequent source of trouble, by being put too close or too much overlapping, even if they are on that ugly and now fashionable foreign plan of having a few enormously wide or deep boards sticking out beyond the face of the mullions, which never were in genuine English architecture. It is no use trying to keep out snow, or even small driving rain, and it does no harm if bell hangings are kept painted, and the floor is made waterproof and drained. Louvres just overlapping will keep out ordinary rain, and I am afraid cannot be dispensed with in church towers, though they are at Westminster, where the floor under the bells is flagged. It is the custom in the eastern counties to put small clock bells quite open on the tops of the church towers, instead of making the clock strike on the tenor of the peal, and they are generally heard farther; in fact, a clock bell (if only a clock bell) cannot be too open, as it has no hangings that will spoil with rain if the hammer gudgeons are occasionally oiled—or even if they are not, for a long time.

There must be two floors between the ringers and the bells, or they cannot hear for the noise. In most towers of good size this can easily be done, but in low central towers it is sometimes difficult. The plan to be adopted then is to lay a strong floor about a foot below the bell-floor by straps from the great beams, and cover it a few inches thick with large gravel, or broken stones (not sand), or else to fill up the whole space between the floors with shavings, including a box full of them for the trap door which must be left under the great bell. I need hardly say that the great floor beams ought to come under the beams of the bell-frame as much as possible, and should rest on large corbels, and for heavy peals should be strutted also from corbels lower down. This too is often neglected, and a floor made at random, and then the bell-founder is expected to fix a firm bell-frame on a floor not fit to carry it.

But the worst abomination of all, committed by architects and church restorers, is the destroying of the belfry (*i.e.* ringing chamber) floor for the sake of making a 'lantern' of the tower. This has been done at Hereford Cathedral, Ludlow and Boston churches, Merton College Chapel, St. Alban's Abbey, and divers other places. At St. Alban's, where the space is abundant, they have lately re-hung the bells higher up, and so regained a ringing chamber, but not so at the others. At Howden the ringing is done inconveniently and dangerously from a narrow gallery round the tower, and so it is at Merton Chapel—unless they ring from the ground now, as the tower is practically in the ante-chapel. At Pershore Sir G. Scott did a better thing than a gallery, by the converse of it, a sort of insulated floor, leaving an open space round it except where there is a bridge to reach it. Mr. Cattley and I did much the same at Worcester Cathedral, which has probably the best belfry in England on account of the great width of the tower, 32 ft. inside. The much larger central towers of York and Lincoln do not contain the peals of bells; they are in the smaller western towers. In the central tower which I designed for Mr. Bass, at Burton, there are small spandril windows to light the space under it, as in the grand old church of Hedon, so that the belfry floor can still be at the natural place, about the level of the roofs of the four limbs of the church; and it is 22 feet high. At Doncaster there was height enough above the lower windows of the tower.

But this lanterning mania has destroyed some belfries irreparably, and ringers will not ring long in dark holes about 8 feet high, which were only intended for the intermediate chamber or clock room. That demon of church destruction, under the name of restoration, Wyatt, in the last century and early in this, pulled down the old campanile of Salisbury, and the bells had to be sold because nobody dared ring them in the great overloaded steeple of the cathedral. Well might Pugin say of him, in one of his 'restored' churches, 'Yes, the monster has been here!'

Where it has become impossible to ring bells properly, the best thing to do is at once to fit them up for chiming only, *i.e.* tolling with levers. Though inferior to ringing in some respects, and much less interesting to ringers, it has a sweeter sound, but feebler. The clappers ought to be heavier for it, as they strike much softer: especially when there are only 3 or 4 bells, chiming sounds much better than ringing. The tolling of 3 large bells has a very grand sound, and the ringing of two bells only frame high, *i.e.* swinging up to horizontal, so that they strike quickly, has a pleasant and lively sound. I always admire it at New College Chapel when I visit Oxford.

The windows of bell-chambers, and of every opening in the tower or spire above them, should be completely covered with strong wire netting, which must also be kept in repair, to keep out birds, which otherwise fill the place with sticks and dirt, which caused the second fire at York Minster, and the destruction of the bells. The netting should be of about 18 gauge woven in squares of half an inch, not in long openings tied with thin wire, which soon perishes.

The well-known story of St. Paul's clock being heard at

Windsor striking 13 by a sentinel who was charged with being asleep, is often regarded as fabulous; and so it would be in the present feeble condition of the striking. But Reid says, in his book on clocks, that he heard it there himself, and I have heard an officer quartered there say the same. The Doncaster clock striking on a bell of only 30 cwt. has been heard 11 miles in a very flat country. Nobody has yet solved the problem why a wind which you can hardly feel will make the difference of your hearing bells half a mile or ten miles, according as the wind is with you or against.

The great bells of Europe are all that remains to notice, by which I mean those above any ordinary weight of the tenor of a peal, which may be called 3 tons; for there is none except Exeter which exceeds or even reaches that weight.* There are clock bells of 3 tons at the town halls of Hull (W) and Halifax (T), and one lately made from my specification, with 4 quarter bells, for the post office at Adelaide (T). I cannot vouch for the accuracy of the whole of the following list; indeed I am sure that some of the figures cannot be correct, and I must warn people against newspaper statistics of such things. For instance, even in London, the St. Paul's clock bell still figures in sundry books as 9 ft. wide instead of 6 ft. 9 in., and I have seen the York bell credited with a greater weight than Westminster; and if such exaggerations as those take place about bells whose sizes and weights have been correctly published for years in this and other books, it is evident that little reliance can be placed on others where the weights and sizes are plainly inconsistent. Most of the foreign ones are taken from the small German book on bells by Otte, before noticed. But I satisfied myself that he sometimes uses the measures of one country and sometimes of another. I have received

* I will use the letters M, T, W, to denote the only three English founders who have made large bells in this century, viz.: Mears, Taylor, and Warner.

sections and measures of some bells from architects and others who had measured them, and I shall be thankful for any more, of really large bells. I give the weights of the two great Russian ones by calculation from their size and thickness, which is easily done, from their similarity to the Westminster pattern. I have seen even a larger weight than 8 cwt. given for the clapper of the Paris bell; but the 8 cwt. agrees pretty well with the drawing; but if the drawing I have of the Erfurt clapper is right it cannot be so heavy as Otte gives it. Our clappers are of a better shape than the foreign ones, though generally too light.

I have no doubt that some of these bells are bad ones, or they would have a greater reputation than they have. The Erfurt bell is the most famous of all the very large ones, and probably that of Paris next, though there are two opinions about that; the old Cologne and Lucerne bells I have heard well spoken of. The new and larger Cologne one was cast three times. I have never heard it praised. On the other hand the new Great Paul was cast successfully at once, and it seems to me as good as possible. We have ample proof at home that large bells may be thought not worth ringing after costing a great sum of money, or may sound no distinct note, or may be supposed to be cracked because they are so bad. I heard the Montreal bell before it went out, and thought it quite as bad as York. I am told that the bell of St. Peter's at Rome sounds as ill as it ought to do from its extremely bad flower-pot shape; I have a model of it, and it is loaded also with ornaments in high relief, which are sure to injure the sound; but the shape is such that it is of very little consequence what the decoration is. The Canterbury bell is struck so feebly by a good-for-nothing modern clock that no one would suppose it had any business in this list: nor Tom of Oxford either, if judged only by his weak and abominable sound. See List, over the page.

Great Bells of	Date.	Diameter. ft.	in.	Weight. tns.	cwts.	Note.	Thickness. in.	Clapper or Hammer. lbs.
Moscow (broken)	1734	22	8	220	0	..	23	
Another	1817	18	0	110	0	..	18	
Another	1878	..		28	13			
Novogorod	..	..		31	0			
Cologne	1874	11	2	25	10	C		1530
Rouen (was)	..	11	0	17	17			
Olmutz	..	..		17	18			
Vienna	1711	9	10	17	14	..		1600?
St. Paul's, T.	1882	9	$6\frac{1}{2}$	16	14	E flat		730
Westminster, M.	1858	9	0	13	11	E		766
Sens	..	8	7	15	0?			
Erfurt	1497	8	7	13	15	..	$7\frac{1}{2}$	1120?
Notre Damo	1680	8	7	12	16	..	$7\frac{1}{2}$	900
Montreal, M.	1847	8	7	12	15	F	$8\frac{1}{2}$	413
Madgeburgh	1702	7	10	13	0?			
Schaffhausen	1486	..		11	10			
Cologne, 2nd	1448	7	11	11	3	G		
Breslau	1507	..		11	0			
Amiens	1748	..		11	0			
York, M.	1845	8	4	10	15	F sharp	8	405
Rheims	1570	..		10	9			
Vienna (2nd)	1558	..		10	8			
Bruges	1680	..		10	5			
Lyons	..	..		10	0			
Marseilles	..	..		8	19			
Gorlitz	1516	..		8	5			
St. Peter's, Romo	1786	7	6	8	0			
Sneeberg	..	7	6	7	19			
Hamburgh (St. Nic.)	1876	7	6	6	7			
Manchester, T.	1882	7	$7\frac{1}{2}$	8	3	G	7	458
Nuremberg	1392	..		7	16			
Oxford	1680	7	0	7	12	?		
Lucerne	1636	..		7	12			
Halberstadt	1457	..		7	10			
Antwerp	..	..		7	3			
Brussels	..	..		7	1			

Great Bells of	Date.	Diameter.		Weight.		Note.	Thickness.	Clapper or Hammer.
		ft.	in.	tns.	cwts.			lbs.
Halle	1480	..		6	10			
Munich	1493	7	3	6	5			
Dantzic	1453	..		6	1			
Cologne (3rd)	1449	..		6	0			
Ratisbon	1325	..		5	16			
Magdeburgh (2nd)	1690	6	2?	5	15			
Leipsic	1634	..		5	14			
Breslau	1721	..		5	13			
Brunn	1515	..		5	10			
Ghent	..	..		5	10			
Rodiz	1841	..		5	10			
Chalons	..	..		5	9			
Lincoln, M.	1835	6	10½	5	8	A	6	240
St. Paul's (old)	1716	6	9½	5	4	A & C	..	180
Mariazell	1830	..		5	5			
Dresden	1787	..		5	2			
Bradford, T.	1873	6	5¼	4	7	A	6	
Worcester, T.	1868	6	4½	4	10	B flat	6	246
Rouen (2nd)	..	6	4½	5	9?			
Exeter (Peter)	1675	6	4	5	8	A	5	70
Frankfort	1371	6	4	5	0			
Preston, T.	1868	6	3	4	16	B flat	6	227
Bolton, W.	1872	6	2	4	2	B	6	200
Leeds, W.	1859	6	2	4	1	B	6	200
Valetta	..	6	1	..		B flat		
Amiens (2nd)	1736	6	0	5	0?			
Boulogne	1800	..		4	0			
Westminster (4th)	1857	6	0	3	18	B	5⅞	175
,, (all W.) 3rd qur.	1858	4	6	1	13½	E	4½	80
,, ,, 2nd ,,	1857	4	0	1	6	F sharp	3⅞	60
,, ,, 1st ,,	1857	3	9	1	1	G sharp	3¾	56
Chichester, T.	1877	5	10½	3	14	B flat		142
Exeter (tenor)	1676	5	11½	3	7	B flat	5	
Hotel de Ville, Paris	..	..		3	10			
Canterbury	1762	5	9	3	10	C	5½	
Gloucester	1400	5	8½	3	5	C		

INDEX.

AIR, resistance of to pendulum, 44, 70; barometric effect of, 71.
Air-tight clock-case, 72.
Airy, Sir G. B., 10, 11; his calculations for dead escapements, 81, 82, 85; for gravity escapements, 111; his detached escapement, 105; maintaining power, 151; spring remontoire, 236; conditions for Westminster clock, 266; failure of his alteration of the striking work, 272; his report on the Westminster bells, 271; on wheel teeth, 273.
Alarums, 177.
Aluminium bronze, 94, 283; bad for bell metal, 365; balance springs, 302.
American clocks, 135, 153; watches, 292; mainspring barrel, 292; watch factories, 334.
Anchor pallets, 75.
Antimony in some bells, 363.
Arbors, or axes, of wheels, 20.
Architects ignore bells and clocks, 184, 384; their ignorance of requirements, 216; often destroy belfries, 385, 386.
Arcs of vibration in clocks, 35, 72, 92; watches, 318.
Arsenic in bell metal, 364.
Astronomical clocks, 82, 127; dial work of, 137; Greenwich clock, 105.
'Astronomy without Mathematics,' referred to, 2, 4, 373.
Austrian clocks, 154.

BAILY, F., on barometric compensation, 71, 72; his mistake on pendulum compensation, 66.
Bain's electrical clocks, 156.
Balance spring, 297; glass, 302.
Balance springs of watches, 298, 302.
Balance-wheel clock escapements, 28.
Balances, compensation of, 304, 314.
Barometric compensation, 70.
Barrels, going, 147.
Barry, Sir C., his Westminster clock hands, 221, 256, 368.
Beat, adjustment of, 78; self-acting in French clocks, 154; single-beat escapements, 107, 321.
Beat pins, ivory and wooden, 131; eccentric, 102; spring, 103.
Beckett, Sir E., on 'Clocks,' &c., in the *Encyclopædia Britannica*, 20; meridian dial, 13; bolt and shutter, 150; detached escapement, 106; 'drop bushes' for turret clocks, 191; general plans for them, 183, 193, 196; concave dials, 210; spring remontoire, 236; three-legged dead escapements, 101, 104; four and three legged gravity escapements, 121, 129; half-hour striking, 170; plan for the Westminster clock, 265, *et seq.*; improved stays and sliders, 375; new bell-crown, 368; experiments for the Westminster bells, 339, 340; plan for bell frames, 378; for hanging small bells, 372; for Great Paul, 371; rules for bell construction, 357; for composition of bell metal, 361.
Belfries, their destruction by architects, 386.
•Bell-crown, common, 368; Sir E. Beckett's, 368; Taylor's, 368; mushroom-shape, 368.

Bell-founders, some celebrated ones, 335, 337; foreign, 339.
Bell frames, 377; their arrangement, 378; bottom should fit tight to walls, 377.
Bell-hanging, 370; stays and sliders for, 374; for clocks, 211, 212.
Bell-metal, its composition, 338, 340, 361, 362; impurities of, 364; melting point of, 366; silver useless in, 364; its specific gravity, 361.
Bell-moulding, different methods of, 365.
Bell-ringing, books on, 374.
Bell ropes, 372; generally too thick, 373.
Bell-tolling, 371.
Bell-towers, 384; construction and size for bells, 384; windows should be caged, 387; and have louvres, 385.
Bells, history of famous ones, and their makers, 335; great bells of Europe, 388; list of their weights and sizes, 390; Oxford and York very bad, 337, 338; Chinese and Indian, 356; Cologne, cast three times, 389; St. Peter's, and Montreal, and Great Paul, 371, 389; clappers, 376; frames, 377; gudgeons, 375; cracked, incurable, 367; notes, 341; price for casting and re-casting, 338; proper shape for, 355; hemispherical, bad, 355; rule for construction of large bells, 357; scale of weights, 343, 346; specifications for, 380; stocks, 376; thickness of sound bow, 345, 360; 'tucking up,' 360, 370; tuning, 341; sound best swinging, 373.
Bells in peals, scale of sizes, &c., 346; Boston and Eaton Hall, 210, 355, 339; Bow Church, 347; Bradford, 350; Burton, 351; Doncaster, 351; Exeter, 347; Gainford, 352; Headingley, 352; Manchester, 350; Ossett, 351; Rochdale, 348; St. Paul's, 348; Great Paul, 371; Southwark, 347; Worcester, 348; York Minster, 347; large and small together, bad, 210.
Berthoud on dead escapements, 91; his lever watch escapement, 315.
Bevelled wheels, 283.
Big Ben of Westminster, the first and second, 270, 271.
Bloxam, his escapements, 66, 82, 113, 118; corrected Sir G. Airy, 82, 120; invented the dipleidoscope, 14; on barometric error, 71.
Bobs for pendulums, 31, 43, 54, 57.
Bolt and shutter, 150; Sir E. Beckett's improved, 150.
Bond's telescope-driving clock, 232.
Bottle-jack escapement, 29.
Bow Church bells and York, 347.
Box chronometer, 295.
Bradford Town Hall tower and bells, 216, 350.
Brass-work destroyed by town atmosphere, 241.
Brasses for bell gudgeons, 375, 363.
Breguet's tipsy key, 292.
Bryant, bell founder and clockmaker, 338.
Buffers for large hammers, 213.
Burton and Ossett bells, 351.
Bushes, various kinds of, 191, 192.

CAMBRIDGE and Westminster chimes, 202, 204.
Cams, 284; proper form of, 286; for large clocks, 187; in the Westminster clock, 257.
Candles graduated, for marking time, 17.
Canons of bells, 369.
Cast iron for clock-wheels, 239; and steel pinions, 240, 281.
Cast iron jars best for mercurial pendulums, 68.
Casting of bells, test for good, 361.
Cattley, Rev. R., Worcester chimes, 203; lifting off hammers, 206.
Cemetery bells, 355.
Centre of oscillation of pendulums, 36.
Chance's glass for dials, 222.
Chester cathedral chimes, 203.
Chimes (quarter) at Cambridge and Westminster, 202, 204; Chester and Worcester cathedrals, 203; Doncaster and Scarborough, 203, 204; Headingley, 204; Royal Exchange, 202, 204; St. Alban's cathedral, 207; peculiar, on two bells at York Minster,

206; interval between each quarter, 205.
Chime-tunes, 174, 207; improvements by Gillett and Bland, 208; by Lund and Blockley, 209; Boston and Eaton Hall, &c., 210.
Chiming hammers, Ellacombe's, 373; by machinery, 336.
Chimney-piece clocks, 304.
Chinese bells, 356; gongs, 335.
'Chords' in chime-tunes bad, 210.
Chronograph at Greenwich, 26.
Chronometer escapement, 323; secondary compensation, 307; trials at Greenwich, 308; Liverpool, 309.
Chronometers, 295, 303; rate of, 314.
Chronometrical thermometer, 306.
Chronoscopic dials, 143, 144.
Church clocks (see Turret clocks).
Church towers and architects, 384, 386.
Circular error of pendulum, 34, 49, 113.
Clappers, 359, 376.
'Clappering' cracks bells, 379.
Cleaning watches, 333.
Clepsydræ of Greeks and Romans, 16.
Clock out of beat, to regulate, 78.
Clock-cases, 139.
Clock-room, ventilation of, 227.
Clock time, all regulated by sidereal time, 1.
Clocks, invention of, 18; ancient clocks in England, 19; construction of, 20; arrangement of common, 134; going part of, 132; influence one another, 45; to regulate by sidereal observations, 8; striking, 162; striking one at the hour, 163; half-hour striking, 170; strike and silent, 167; striking quarters, 172; silent beat, 100; cuckoo, 154; English or repeating striking work, 165; foreign or locking-plate method, 167; various mechanism for striking work, 177; American, 135, 153; Austrian, French, German, 154.
Clocks, astronomical, 82, 127, 137; electrical, 156; equation of time, 145; musical, 178; self-winding, 155; spring, 152; tell-tale, 178; year, 138; day of the month, 141.
Clocks with dials for many places, 11.
Cologne bell cast three times, 389.
Compensation of balances, common, 304; secondary, 307; Dent's, 310; Dent's prismatic, 313; Eiffe's, 309; Kullberg's, 312; Loseby's, 310.
Compensation of pendulums, 52, 180; barometric, 70; compound bar, 62; Ellicott's, 64; homogeneous, 64; Smeaton's, 62; wood and lead, 59; wood, zinc, and lead, 61; zinc and steel, 54; mercurial, 65; Baily's mistake on, 66.
Composition of bell-metal, 338, 340, 361, 362, 365.
Concave dials for large clocks, 210.
Conical pendulum, 24; its uses, 26.
Construction of bells, rule for, 357; dials, 216, 219.
Continuous motion clocks, 24, 231.
Cooke, clock and telescope maker, 243; his telescope-driving clock, 232.
Copper for bell-making, 340, 365.
Counterpoises for clock hands, 221.
Crab, double-barrelled, 248.
Cracked bells cannot be mended, 367; clappering causes, 379.
Cranks, different forms of, 213.
Cubit of Pyramid, and Jewish, 41.
Cuckoo clocks, 154.
Cumming's escapement, 115.
Cycloidal theory, 32; cheeks useless, 33.

DAY of the month dials, 140.
Day, sidereal, 1; solar, 2.
De la Rue's telescope-driving clock, 232.
De Vick's clock escapement, 28, 30.
Dead escapements, 80; effects of friction on, 82; mathematical theory of, 85; large and small arcs, 92; length and arrangement of pallets, 96; various materials for pallets, 94; pin wheel, 98; Graham's, 81; Macdowall's, 100; Beckett's three-legged, 102.
Degree plate for a pendulum, 47.
Dennison, A. L., of America, his improved Geneva stop, 297; keyless watch, 330; lever escapement bridge 318.
Dent, E. J., his compensation balance, 310; prismatic balance, 313; regulation of watches, 300; 1851 Exhibition

clock, 182, 238; Royal Exchange clock, 233; Westminster clock, 266, *et seq.*
'Depthing tool' described, 282.
Detached escapements for clocks, 104; Sir G. B. Airy's, 105; Sir E. Beckett's, 106; for watches, 323.
Dial wheels, 223; of watches, 297.
Dial work of house clocks, 135, 137; of large clocks, 216.
Dials, chronoscopic, 144; illuminated, 222; moon, 139; of public clocks, 216; colour and materials for, 218, 219; convex and concave, 220; hands of, 220; figures of, 218; of the Westminster clock, 218.
Dials of watches, 333.
Dipleidoscope, the, 14, 15, 16.
Doncaster bells, 339, 353; quarter chimes, 203, 204.
Double-barrelled crab, 248.
Double three-legged gravity escapement, 128.
Driving and running teeth, 279
'Drop bushes' for large clocks, 191.
Duplex escapement, 321.
Dutch clocks, 153.

EATON Hall bells, 210, 339, 354.
Eiffe's compensation balance, 309.
Electrical clocks, 156; Bain's & Shepherd's, 156; Jones's, 158; Ritchie's, 159.
Ellacombe, Rev. H. T., on the bells of Devon, 336; his chiming hammers, 373.
Ellicott's compensation pendulum, 64.
Elliptical section of a bell, 358.
English fusee watch, 294.
Epicycloidal teeth, 276.
Equation of time, 4; table of the, 6.
Equation of time clocks, 145.
Equatorial telescope at Greenwich, its driving apparatus, 26.
Escapements of clocks, 28, 75; dead, 80; half-dead, 91; detached, 104; pin-wheel, 98; silent, 100; recoil, 77; remontoire, 107; single beat, 107, 321; single pin, 100; three-legged dead, 102.
Escapements, gravity, theory of, 107; Cumming's, 115; Hardy's, 116; Kater's, 116; Gowland's, 117; Gannery's, 117; Bloxam's, 118; Beckett's, 121 to 132.
Escapements of watches, 314; chronometer or detached, 323; duplex, 321; horizontal or cylinder, 319; lever, 315; lever chronometer, 325; vertical, 315; virgule, 321; Gray's 315.
Europe, the great bells of, 388, 390.
Exchange, Royal, chimes, 202, 204, 207; clock, 181, 225, 233; alterations of, 239.
Exeter Cathedral bells, 347.

FIGURES on dials useless, and proper size of, 218.
'Finishing' too much undesirable, 96, 241, 324.
Fly, striking, for turret clocks, 194; for gravity escapements, 123, 130.
Fly-wheel clocks, earliest, 24.
Fly, remontoire, 230, 237.
Foreign bell founding, 339; churches, chimes in some, 208.
Forks of pendulums, 46; with springs, 103.
Four-legged gravity escapement, 121.
Four-wheeled trains, 191, 193.
Frames for bells, and fixing, 377; for turret-clocks, 181, 184.
French clocks, 152; improved kind of, 154, 171, 181.
Friction in dead escapements, 82; in turret clocks, 179, 200.
Friction rollers for bells, 371; for clock barrel, 78, 84.
Fusee of watches, 293; often dispensed with, 291.

GAINFORD peal of bells, 352.
Gannery's escapement, 117.
Geneva stop, 296.
German clocks, 153.
Gillett and Bland's chime machinery, 209, 210; clocks and bells, 182, 337.
Gimbals, chronometer, 303.
Glass balance springs, 302.
Gnomon, its position on sundials, 12.
Going part of clocks, 132.
Going barrels (see Maintaining power).
Gongs, 335.
Gowland's escapement, 117.

Graham's dead escapement, 81; horizontal watch escapement, 319.
Gravity escapements, 107; theory of, 109; Mudge's, 109; Beckett's four-legged, 121; double-three-legged, 128; best for large clocks, 130, 239; require heavy weights, 125; in the Westminster clock, 253; rate of some, 131, 182, 263.
Gravity, specific, of metals, 53.
Gray's vertical watch escapement, 315.
Great Paul of London, 344, 371, 389.
Great Tom of Oxford, 337, 338.
Great wheel, striking from, 187.
Greenwich mean time and local time, 8; table for forty-two towns, 9; barometric pendulum compensation at, 75; trials of chronometers at, 308; sidereal clock escapement, Airy's, 105; equatorial driving clock, 26; revolving pendulum, 27.
Gridiron pendulum, 53.
Gudgeons of bells, 375.
Gun metal, 367; its composition, 241; for clock wheels, 241.

HALF-DEAD escapements, 91; Berthoud's, 91.
Half-hour striking, 170.
Half-hours with rack movement, 171.
'Half-hunting' watches, 333.
Hammers, clock, 210; fixing, 211; lifting off, 205; shape, 211; weight, 215, 216.
Hand-bells, 355.
Hands of clocks and watches, the best colours for, 137, 144; for large clocks, 220; proper construction and fixing, 220; counterpoises, 221; of the Westminster clock, 256.
Hanging bells, 371; on friction rollers, 371; chapel and school bells, 372.
Hardy's escapement, 116.
Harrison, the chronometer maker, 53; his clock escapement, 77; going barrel, 148; gridiron pendulum, 53; inventor of watch compensation, 305.
Headingley, St. Chad's chimes and peal, 204, 352.
Helix teeth, 275.
Hemispherical bells, 355.
Henderson's sidereal and mean time clocks combined, 10.
Holmes, an old clockmaker, 62, 143.
Homogeneous compensation, 64.
Hooke, inventor of the balance-spring of watches, 31, 76, 290.
Horizontal or cylinder escapement, 319.
House clock, common, 134.
'Howard Clock Company,' of America, 292.
Huyghens, his theory of the pendulum, 30; his endless chain, 147; theory of balance-spring for watches, 290.
Hypocycloid and epicycloid, 276.

ILLUMINATED dials, 222.
Indian bells, bad shape, 356.
Iron in bell-metal wrong, 364, 365.
Iron stocks and frames for bells, 376.
Isochronal spring, Loseby's, 91.

JEWELLED pallets, 94, 318.
Jobbing in contracts for public clocks, 242, 244.
Jones's electrical clocks, 158.
Joyce's turret clocks, 182.

KATER, his escapement, 116; mercurial pendulum, 67.
Key, best watch, 293; for clocks, 137.
Keyless watches, 328; Dennison's, 330; Kullberg's, 329.
Keys, winding, 137.
King's Cross clock, 100, 238.
Kullberg's compensation balance, 312; his keyless watch, 329.

LANTERN pinions for clocks, 135, 279, 281.
Lantern towers difficult for bell-ringing, 386.
Large and small arcs, 72, 92.
Large and small clocks, difference between, 179.
Large turret clocks, 197.
Lead, adulteration in bells, 364; decay of, on church roofs, &c., 340.
Leap-year, how it affects the equation of time, 4.
Leaved pinions, 282.
Lepaute's pin-wheel escapement, 98.

Lever escapement, 315; chronometer escapement, 325.
Light-house clocks, 26.
Lincoln Minster clock, 182, 199.
Lines and pulleys for winding up large clocks, 190.
Local time, clocks for, 11.
Locking-plate movement, 167, 172.
Loseby's isochronal spring, 91; balance for watches, 310.
Lunar clocks, 139.
Lund & Blockley's chime machinery, 209; stop watches, 332.
Lutz of Geneva, his balance springs, 302.

MACDOWALL'S single-pin escapement, 100; his spiral drill, &c., 101, 102.
Magnet used for regulating clocks, 52, 75.
Maiden bells, 354.
Mainspring of watches, 290.
Maintaining powers of clocks or going barrels, 147; endless chain, 148; Harrison's, 148; bolt and shutter, 150; Beckett's improved ditto, 150; 'sun and planet' power, 151; Westminster clock plan, 199, 254; going-barrel in watches, 291.
Manchester Town Hall clock and bells, 182, 350.
Marine chronometer, 303.
Mean time, 3; how got from sidereal observations, 5, 8.
Mears, bell-founders, 270, 337, 339, 340, 361; their Westminster bell, 363; Montreal and York, 389; Royal Exchange and York, 269, 339.
Mending cracked bells impossible, 367.
Mercurial compensation, 65; Kater's and Beckett's, 67; cast-iron jars for, 68; barometic, 71.
Meridian dial, 12; method of constructing one, 13.
Metre, the French, a bad measure, 40.
Metronome, the, 41.
Moon dials, 139.
Moore, Holmes, and Mackenzie, bell-founders, 337, 376.
Morrison's universal joint, 226.
Moulding bells, 365.
Mudge's gravity escapement, 109; invented the lever escapement for watches, 294.
Musical clocks, 178; hand-bells, 355.

NEW bell-crown, Sir E. Beckett's, 368.
Notes of bells, 341, 351; should not be prescribed, 380.

OIL for clocks, 288.
Old bell-founders, 337, 338.
Old Tom of Lincoln, 363.
Oscillation, centre of, 36.

PAINTING all ironwork of large clocks, 241; bell frames and wheels, 383.
Pallets, 28; anchor, 75; length of, 96; pin, 100; steel, 94; various materials for, 94; silent, 100.
Peal of large bells, specification for, 380.
Peals of bells, some of the largest, 347.
Pedometer, the, 330.
Pendulum bobs, 31, 43, 54; cocks, 45; springs, 45.
Pendulum, conical or revolving, 24, 26; of the Westminster clock, 44, 253.
Pendulums, invented, 29; centre of oscillation of, 36; circular error, 34, 49; compensation, 52; homogeneous, 64; mercurial, 65; wood and lead, 59; wood, zinc, and lead, 61; for church clocks, 179, 180; cycloidal theory, 33; lengths of, 36; the heaviest in England, 57; regulation of, 49; relation to gravity, 37; shape of, 42; short and slow, 41; springs of, 47; standard lengths, 39; suspension, 44; temporary regulation, 51; importance of firm fixing, 45, 93.
Percy, Dr., his report on the Westminster bell, 270.
Phelps & Co., old bell founders, 337.
Pinions of clocks, 21; lantern, 279; leaved, 282 (see Wheels).
Pin pallets, 100.
Pin-wheel escapement, 98.
Potts, maker of Lincoln Minster clock and others, 182, 199.
Pulleys, 22, 148, 190.

QUARTER chimes (see Chimes); on two bells, 201, 206; should be independent of hour in large clocks, 206; of Westminster clock, 257; rack and repeating movements in small clocks, 173.

RADIUS of oscillation of pendulums, 37.
Ratchet wheels, 22, 149.
Rate of the Westminster clock, 263; of some other gravity escapement clocks, 131, 182; of 1851 Exhibition Clock, 238; of watches, 314.
Recoil anchor pallets, 75; escapements, 77; Harrison's, 77.
Regulation of pendulums, 45, 49; of watches, 299; the old plan, 301.
Reid on train remontoires, 228, 229.
Remontoire escapements, 107, 327.
Remontoire train, 228; bevelled wheel, 229; continuous motion, 231; internal wheel, 233; spring, 236, 237.
Repeaters, 327.
Repeating striking work, 165.
Ritchie's electrical clocks, 159; their pendulums, 160.
Robinson, Dr., on barometic compensation of pendulums, 72.
Rochdale chimes and tower, 348, 350.
Rollers, friction, 371.
Ropes, wire, best for turret clocks, 187; in the Westminster one, 258.
Rosse, Lord, his water clock, 17.
Rudhalls, famous bell-founders formerly, 336.

ST. ALBANS cathedral chimes, 207.
St. Paul's clock heard at Windsor striking thirteen, 387; peal of bells, 348; the great bell, 344, 389.
Sam Slick was Eli Terry, 153.
Sand clocks, 16.
Scape wheels, bad weight of, 95; a very small one, 103.
Secondary compensation, theory of, 307; inventions for, 309–314.
Self-winding clocks, 155; watch, 330.
Shape of bells, 355; pendulums, 42.
Shepherd's electrical clocks, 157.
Ship time-pieces, 303.
Siddons and Meese's clocks, 144.
Sidereal day, 1; time, 1, 2.
Silver, none in bell-metal, 365.
Single-pin escapement, 100.
Single-beat escapements, 107.
Skew-bevelled wheels, 284.
Smeaton's glass pendulums, 62.
Snowdon's books on bell-ringing, 374.
Solar day defined, 2; why it varies, 3.
Sound-bow of bells, 360.
Specific gravity of metals, &c., 53; bells, 361.
Specifications for public clocks, 243; bells, 380.
Spring clocks, 152; remontoires, 236, 237.
Stays and sliders of bells, 374.
Steel bells, bad, 364; pallets, 94.
Stop, Geneva, 296.
Stop, winding, of watches, 295.
Stop-watches, 330.
Stops for weights, 189, 263.
Strike and silent, 167, 177.
Striking clocks, 162; half-hours, 170; accuracy of discharge, 186, 199, 206, 260.
Striking part of house clocks, 162; turret clocks, 187, 246; Westminster clock, 258.
'Sun and planet' maintaining power, 151.
Sundials, 12.
Suspension of pendulums, 44, 93.
Sweep or section of bells, 357, 365.
Swinging, effect of, on the sound of bells, 373.
Swiss watches, 152.

TABLE of the equation of time, 6; expansion for compensated pendulums, 53; difference of forty-two towns from Greenwich time, 9; sizes of public dials, 217; weights and sizes and notes of bells, 343, *et seq.*; of the great bells of Europe, 390, 391; of various quarter chimes, 204.
Taylor, bell-founder, 337, 351, 352.
Taylor, Rev. W., and the Westminster bell, 369; and the York bell, 370.
Teeth of wheels, 273; bevelled wheels, 283; cams, 284; construction of, 286; epicycloidal teeth, 276; helix teeth, 275; internal wheels, 283;

lantern pinions, 279; leaved pinions, 282.
Telescope-driving clocks, 17, 26; Cooke's, Bond's, and De la Rue's, 232.
Tell-tale clocks, 178.
Tenor bell in a peal, proper proportions of, 344, 345.
Thermometer, chronometrical, 306.
Thickness of bells, 345.
Three-legged dead escapement, 102; gravity escapement, 129.
Three-quarter plate watches, 295
Time balls and guns, 162.
Time of striking (quarters), 206.
Time, measures of (see mean, sidereal, local); graduated candles, 17; sand clocks, 16; sun-dials, 12; water-clocks, 16, 155; meridian dials, 12; dipleidoscope, 19.
Timepieces, ship, 303.
Timing screws, 300; timing for position, 302.
Tin in bell-metal, 340, 363, 365.
Tipsy winding-key, 292.
Tolling of bells, 371.
Tolling levers, 371.
Tom of Oxford, a very bad bell, 337.
Tourbillon escapement, 326.
Tractrix curve, to describe for cams, 286.
Train remontoires, 228; bevelled wheel, 229; continuous motion, 231; internal wheel, 233; Reid's, 228; spring, 236, 237.
Train of wheels in clocks, 21, 22; in watches, 295.
Transit instrument, 2.
Troughton's pendulum, 53.
Troyte on bell-ringing, 374.
Trumpet and cuckoo clocks, 154.
'Tucking-up' of bells, 370.
Tunes, chime, 174, 207; on church bells, 208.
Turret clocks, 179; best position for, 180; frame for, 181; and its construction, 184; iron barrels, 184; three-wheeled trains, 185; dials, 216; illumination of, 222; regulating and setting the hands, 221; counterpoises for the hands, 221; hammers of, 188; mode of fixing them, 189; size and position of fly, 194; wire ropes for, 187; stops against overwinding, 189, 263; striking part, 198; lifting off hammers for ringing, 206; specification for, 243; Batch-wood clock, 183; larger ones, 193; smaller, 193.
Turret and church clock makers, 182.

UNIVERSAL joints, 225; new American, 226.
'Up and down' dial, 296.

VENTILATION of clock-rooms, 227.
Vertical watch escapement, 29; Gray's, 315.
Virgule escapement, 321.
Vulliamy, Mr., 91, 94, 181, 202; and the Westminster clock, 265, 266.

WAGNER'S continuous motion remontoire, 231.
Warners, bell founders, 269, 337, 351.
Watch cases, 333; dials, 333; escapements, 314—328.
Watch keys, best kind of, 293.
Watches, their early history, 290; general construction, 294, 295; balance-spring, 297; compensation of balances, 304; secondary, 307; dial wheels, 297; fusee, 293; mainspring, 290; regulation, 299; timing for position, 302; glass balance springs for, 302; keyless watches, 328; repeaters, 327; self-winding, 330; stop, 330; winding-stops, 295; American factories, 334.
Water clocks, 16, 155; used by the Greeks and Romans, 16; Lord Rosse's, 17; the one at Greenwich, 18.
Weathercock dials, 227.
Weights, stops for falling, 189; winding, 189, 263.
Weights and sizes of bells, 343, 346, 350.
Westminster bells, the, 339, 343, 361; their history, 269; their cost, 273.
Westminster clock, its history, 265; construction, 248; cost, 273; gravity escapement, 253; maintaining power, 254; regulation of the pendulum, 252; its rate, 263; striking parts, 257, 258; dials, 255; hands, 256; winding of,

261; curious suggestions for winding, 262; has no barometric error, 72.
Wheatstone's electrical clocks, 157.
Wheels for clocks, 95; train of, 20; bevelled, 283; skew-bevelled, 284; cast iron, 239; aluminium bronze for, 94.
Wheels for watches, numbers of teeth, 22, 133, 185, 198; for dials, 224, 297; in watches, 295; and see Teeth, 283.
Whitaker's almanack, sidereal time of noon in, 5.
White's gearing, 274.
Willis, Prof., his table of wheelwork, 277.
Winding keys for clocks, 137; should be long, 138; for watches, 293.
Winding stops for watches, 295; clocks, 189, 262.
Windows of bell-chambers, 387.
Wire ropes best for large clocks, 187.
Wooden beat pins, 131; pendulums, 59, 60.
Worcester cathedral bells, 348, 353; chimes, peculiar, 203; clock, 182, 206.

YEAR clocks, 138.
York Minster bells, 347, 363; great bell, a bad one, 338; quarter chimes, peculiar, 206.

ZINC in bell-metal, 364; zinc and steel pendulums, 54; better than mercurial, 70.

THE END.

PRINTED BY J. S. VIRTUE AND CO., LIMITED, CITY ROAD, LONDON.

Handbook of Specifications.

THE HANDBOOK OF SPECIFICATIONS; or, Practical Guide to the Architect, Engineer, Surveyor, and Builder, in drawing up Specifications and Contracts for Works and Constructions. Illustrated by Precedents of Buildings actually executed by eminent Architects and Engineers. By Professor THOMAS L. DONALDSON, M.I.B.A. New Edition, in One large volume, 8vo, with upwards of 1000 pages of text, and 33 Plates, cloth, 1*l.* 11*s.* 6*d.*

"In this work forty-four specifications of executed works are given. . . . Donaldson's Handbook of Specifications must be bought by all architects."—*Builder.*

Specifications for Practical Architecture.

SPECIFICATIONS FOR PRACTICAL ARCHITECTURE: A Guide to the Architect, Engineer, Surveyor, and Builder; with an Essay on the Structure and Science of Modern Buildings. By FREDERICK ROGERS, Architect. 8vo, 15*s.* cloth.

*** A volume of specifications of a practical character being greatly required, and the old standard work of Alfred Bartholomew being out of print, the author, on the basis of that work, has produced the above.—*Extract from Preface.*

Designing, Measuring, and Valuing.

THE STUDENT'S GUIDE to the PRACTICE of MEASURING and VALUING ARTIFICERS' WORKS; containing Directions for taking Dimensions, Abstracting the same, and bringing the Quantities into Bill, with Tables of Constants, and copious Memoranda for the Valuation of Labour and Materials in the respective Trades of Bricklayer and Slater, Carpenter and Joiner, Painter and Glazier, Paperhanger, &c. With 9 Plates and 47 Woodcuts. Originally edited by EDWARD DOBSON, Architect. New Edition, re-written, with Additions on Mensuration and Construction, and useful Tables for facilitating Calculations and Measurements. By E. WYNDHAM TARN, M.A., 8vo, 10*s.* 6*d.* cloth.

"Well fulfils the promise of its title-page. Mr. Tarn's additions and revisions have much increased the usefulness of the work."—*Engineering.*

Beaton's Pocket Estimator.

THE POCKET ESTIMATOR FOR THE BUILDING TRADES, being an easy method of estimating the various parts of a Building collectively, more especially applied to Carpenters' and Joiners' work. By A. C. BEATON. Second Edition. Waistcoat-pocket size. 1*s.* 6*d.*

Beaton's Builders' and Surveyors' Technical Guide.

THE POCKET TECHNICAL GUIDE AND MEASURER FOR BUILDERS AND SURVEYORS: containing a Complete Explanation of the Terms used in Building Construction, Memoranda for Reference, Technical Directions for Measuring Work in all the Building Trades, &c. By A. C. BEATON. 1*s.* 6*d.*

The House-Owner's Estimator.

THE HOUSE-OWNER'S ESTIMATOR; or, What will it Cost to Build, Alter, or Repair? A Price-Book for Unprofessional People, Architectural Surveyors, Builders, &c. By the late JAMES D. SIMON. Edited by F. T. W. MILLER, A.R.I.B.A. Third Edition, Revised. Crown 8vo, 3*s.* 6*d.*, cloth.

"In two years it will repay its cost a hundred times over."—*Field*

Horton's Measurer.

THE COMPLETE MEASURER; setting forth the Measurement of Boards, Glass, &c.; Unequal-sided, Square-sided, Octagonal-sided, Round Timber and Stone, and Standing Timber. Also a Table showing the solidity of hewn or eight-sided timber, or of any octagonal-sided column. By RICHARD HORTON. Fourth Edit. With Additions, 12mo, strongly bound in leather, 5*s.*

Horton's Underwood and Woodland Tables.

TABLES FOR PLANTING AND VALUING UNDERWOOD AND WOODLAND; also Lineal, Superficial, Cubical, and Decimal Tables, &c. By R. HORTON. 12mo, 2*s.* leather.

Nicholson's Carpenter's Guide.

THE CARPENTER'S NEW GUIDE; or, BOOK of LINES for CARPENTERS: comprising all the Elementary Principles essential for acquiring a knowledge of Carpentry. Founded on the late PETER NICHOLSON'S standard work. A new Edition, revised by ARTHUR ASHPITEL, F.S.A., together with Practical Rules on Drawing, by GEORGE PYNE. With 74 Plates, 4to, 1*l.* 1*s.* cloth.

Dowsing's Timber Merchant's Companion.

THE TIMBER MERCHANT'S AND BUILDER'S COMPANION; containing New and Copious Tables of the Reduced Weight and Measurement of Deals and Battens, of all sizes, from One to a Thousand Pieces, also the relative Price that each size bears per Lineal Foot to any given Price per Petersburgh Standard Hundred, &c., &c. Also a variety of other valuable information. By WILLIAM DOWSING, Timber Merchant. Third Edition, Revised. Crown 8vo, 3*s.* cloth.

"Everything is as concise and clear as it can possibly be made. There can be no doubt that every timber merchant and builder ought to possess it."—*Hull Advertiser.*

Practical Timber Merchant.

THE PRACTICAL TIMBER MERCHANT, being a Guide for the use of Building Contractors, Surveyors, Builders, &c., comprising useful Tables for all purposes connected with the Timber Trade, Essay on the Strength of Timber, Remarks on the Growth of Timber, &c. By W. RICHARDSON. Fcap. 8vo, 3*s.* 6*d.* cl.

Woodworking Machinery.

WOODWORKING MACHINERY; its Rise, Progress, and Construction. With Hints on the Management of Saw Mills and the Economical Conversion of Timber. Illustrated with Examples of Recent Designs by leading English, French, and American Engineers. By M. POWIS BALE, M.I.M.E. Crown 8vo, 12*s.* 6*d.* cl.

"Mr. Bale is evidently an expert on the subject, and he has collected so much information that his book is all-sufficient for builders and others engaged in the conversion of timber."—*Architect.*

"The most comprehensive compendium of wood-working machinery we have seen. The author is a thorough master of his subject."—*Building News.*

Saw Mills.

SAW MILLS, THEIR ARRANGEMENT AND MANAGEMENT, AND THE ECONOMICAL CONVERSION OF TIMBER. By M. POWIS BALE, M.I.M.E. With numerous Illustrations. Crown 8vo. [*In the Press.*

Spotted something that isn't quite right?

Let us know and we'll fix it for future reprints. And as a thank you for your diligence, we'll send you a free Sothis book of your choice.

publisher@sothis.press

This page intentionally left blank

Made in the USA
Monee, IL
08 November 2026

68061873R00246